CLIMATE DYNAMICS IN HORTICULTURAL SCIENCE

VOLUME 2

Impact, Adaptation, and Mitigation

CLIMATE DYNAMICS IN HORTICULTURAL SCIENCE

VOLUME 2

Impact, Adaptation, and Mitigation

Edited by

**M. L. Choudhary, PhD, V. B. Patel, PhD,
Mohammed Wasim Siddiqui, PhD,
and R. B. Verma, PhD**

AAP | APPLE
ACADEMIC
PRESS

Apple Academic Press Inc. | Apple Academic Press Inc.
3333 Mistwell Crescent | 9 Spinnaker Way
Oakville, ON L6L 0A2 | Waretown, NJ 08758
Canada | USA

©2015 by Apple Academic Press, Inc.

First issued in paperback 2021

Exclusive worldwide distribution by CRC Press, a member of Taylor & Francis Group
No claim to original U.S. Government works

ISBN 13: 978-1-77463-098-3 (pbk)
ISBN 13: 978-1-77188-070-1 (hbk)

Library of Congress Control Number: 2014919684

Library and Archives Canada Cataloguing in Publication

Climate dynamics in horticultural science.

Includes bibliographical references and index.
Contents: Volume 1. Principles and applications/edited by M.L. Choudhary, PhD, V.B. Patel, PhD, Mohammed Wasim Siddiqui, PhD, and Syed Sheraz Mahdi, PhD -- Volume 2. Impact, adaptation, and mitigation / edited by M.L. Choudhary, PhD, V.B. Patel, PhD, Mohammed Wasim Siddiqui, PhD, and R.B. Verma, PhD
ISBN 978-1-77188-031-2 (v. 1 : bound).--ISBN 978-1-77188-070-1 (v. 2 : bound).--
ISBN 978-1-77188-071-8 (set)
1. Horticulture. 2. Horticultural crops--Climatic factors. 3. Climatic changes. I. Choudhary, M. L., 1953-, editor

SB318.C42 2014   635   C2014-907285-6

Apple Academic Press also publishes its books in a variety of electronic formats. Some content that appears in print may not be available in electronic format. For information about Apple Academic Press products, visit our website at **www.appleacademicpress.com** and the CRC Press website at **www.crcpress.com**

ABOUT THE EDITORS

Dr. M. L. Choudhary

M. L. Choudhary, PhD, is currently Vice Chancellor of Bihar Agricultural University in Sabour, Bhagalpur, Bihar, India. He received a master's degree in horticulture from Banaras Hindu University, Varanasi, Utter Pradash, India, and his PhD in the USA. The Government of India deputed him for an advanced study on Hi-Tech Horticulture and Precision Farming in Israel and Chile. In a career spanning 30 years, he has held several executive positions, including Horticulture Commissioner, Ministry of Agriculture, India; Chairman, Coconut Development Board, Kochi (Kerala); Ministry of Agriculture, India; National Project Director, FAO; Visiting Scientist at Rutgers University, New Jersey, USA; Head of the Department of Ornamental Crops, IIHR, Bangalore; and Head of the Division in Floriculture at IARI, New Delhi. As Horticulture Commissioner of the Ministry of Agriculture, India, he has conceived, conceptualized, and implemented flagship programs such as the National Horticulture Mission and the National Bamboo Mission and Micro-irrigation.

Apart from his professional career, he was also Chairman and Member Secretary of various committees constituted by the Government of India as well as state governments. He has guided 18 PhD students and 21 MSc students in the field of horticulture and has published 124 book chapters, 19 books, and 109 research papers of national and international repute. Dr. Choudhary has represented India at various international forums in the capacity of Chairman and Member. He was also the Chairman of the Codex Committee of the Scientific Committee for Organic Standard of Ministry of Commerce. He has been conferred with 15 awards from various scientific and nonscientific organizations and government committees for his outstanding contribution in the field of horticulture/ floriculture. He has also been awarded the Fellowship of Horticultural Society of India. Dr. Choudhary in his 30 years of academic, research, and administrative career has visited more than 15 countries to participate in various professional meetings.

Dr. V. B. Patel

V. B. Patel, PhD, is presently working as University Professor and Chairman at Bihar Agricultural University, Sabour, Bhagalpur, Bihar. He has worked on developing leaf nutrient guides, nutrient management strategies, use of AMF for biohardening and stress tolerance, and a survey of indigenous germplasm of fruit crops. He developed the Leaf Sampling Technique and Standards as well as made fertilizer recommendations for several fruits. He has developed nutrient management through organic means for high-density planted mango.

Dr. Patel has organized eight national/ international seminars and workshops as convener, associate convener, or core team member, including the Indian Horticulture Congresses, the International Seminar on Precision Farming and Plasticulture, the National Seminar on Organic Farming, Seminar on Hitech Horticulture, and National Seminar on Climate Change and Indian Horticulture. He has guided three MSc and one PhD students. He earned his PhD from the Indian Agricultural Research Institute, New Delhi.

Dr. Patel has received a number of national awards and recognitions for his work in the field of horticultural research and development, including being named a Fellow of The Horticultural Society of India, New Delhi; Associate, National Academy of Agricultural Sciences (NAAS), New Delhi, India; Agricultural Leadership Award (2012) by Agricultural Today, Centre for Agricultural and Rural Development, New Delhi; Hari Om Ashram Trust Award (2007); Lal Bahadur Shastri Young Scientist Award, ICAR, New Delhi (2009); Young Scientist Award, NAAS (2005-06); Yuva Vigyanic Samman from Council of Science and Technology, Govt. Uttar Pradesh (2005); and AAAS Junior Award (2005) from the Indian Society for Plant Physiology, New Delhi. He has also received best research paper/ poster paper presentation awards from different organizations. He has published 41 research papers, 16 books and bulletins, and many popular articles, among other publications.

Dr. Mohammed Wasim Siddiqui

Mohammed Wasim Siddiqui, PhD, is an Assistant Professor/Scientist in the Department of Food Science and Technology, Bihar Agricultural University, India, and author or co-author of 21 peer-reviewed journal papers, 12 book chapters, and 16 conference papers. Recently, Dr. Siddiqui has initiated an international peer-reviewed journal, *Journal of Postharvest*

Technology. He has been serving as an editorial board member of several journals.

Dr. Siddiqui acquired a BSc (Agriculture) degree from Jawaharlal Nehru Krishi Vishwavidyalaya, India. He received MSc (Horticulture) and PhD (Horticulture) degrees from Bidhan Chandra Krishi Viswavidyalaya, Mohanpur, Nadia, India, with specialization in the Postharvest Technology. He was awarded the Maulana Azad National Fellowship Award from the University Grants Commission, New Delhi, India. He has received several grants from various funding agencies to carry out his research works during his academic career.

Dr. R. B. Verma

R. B. Verma acquired a BSc (Ag) from Banaras Hindu University, Varanasi; MSc (Ag) Hort (Vegetable Science); and PhD (Vegetable Science) from ND University of Agriculture and Technology, Kumarganj, Faizabad, Uttar Pradesh. Presently, he is serving as an Associate Professor-cum-Senior Scientist, Department of Horticulture (Vegetable and Floriculture) and Recruitment Officer, Bihar Agricultural University, Sabour, Bhagalpur, Bihar. Dr. Verma has been awarded with a Certificate of Honour FMASM in 2005 and Scientist of the Year Award in 2014. He has guided nine MSc and three PhD students. He has published three books and 45 research articles in the scientific journals of national and international repute.

CONTENTS

LIST OF CONTRIBUTORS

Tapan Adhikari
Division of Environmental Soil Science, Indian Institute of Soil Science, Madhya Pradesh, India

Aditya
Department of Extension Education, Bihar Agricultural University, Sabour, Bhagalpur, Bihar-813210, India

Kalyan Barman
Department of Horticulture (Fruit and Fruit Technology), Bihar Agricultural University, Sabour, Bhagalpur, Bihar-813210, India

E. Ulrich Berk
President, Deutsche Gesellschaftfur Homa Therapie, Haldenhof, Germany

B. P. Bhatt
ICAR Research Complex for Eastern Region, Bihar Veterinary College Campus, ICAR Parishar, Patna, Bihar, India

Dr. K. L. Chadha
No. 7281, B-X, Vasant Kunj, New Delhi, India

Vandna Chhabra
ICAR Research Complex for Eastern Region, Patna, Bihar, India

Jaipal Singh Choudhary
ICAR Research Complex for Eastern Region, Research Centre, Plandu, Ranchi-834010, Jharkhand, India

Bikash Das
ICAR Research Complex for Eastern Region, Research Centre, Plandu, Ranchi-834010, Jharkhand, India

N. A. Deshmukh
ICAR Research Complex for NEH Region, Umiam-793103, Meghalaya, India

Manoj Kumar Dwivedi
Department of Soil Science and Agricultural Chemistry, Bihar Agricultural University, Sabour-813210, Bihar, India

Abhijeet Ghatak
Department of Plant Pathology, Bihar Agricultural University, Sabour, Bhagalpur, Bihar-813210, India

A. Abdul Haris
ICAR Research Complex for Eastern Region, Bihar Veterinary College Campus, ICAR Parishar, Patna, Bihar, India

Md. Hedayetullah
School of Agriculture Science, Centurion University, Odisha-761211, India

A. K. Jha
ICAR Research Complex for NEH Region, Umiam-793103, Meghalaya, India

Anshuman Kohli
Department of Soil Science and Agricultural Chemistry, Bihar Agricultural University, Sabour-813210 (Bihar), India

B. Kumar
College of Agri-Business Management, GBUAT, Pantnagar, UK, India

Pramod Kumar
Regional Horticultural Research Station, Dr Y. S Parmar University of Horticulture and Forestry, Sharbo (ReckongPeo), District Kinnaur, Himachal Pradesh-172 107, India

Rajesh Kumar
National Research Centre for Litchi, Muzaffarpur-842 002 (Bihar), India

Ritesh Kumar
ICAR Research Complex for Eastern Region, Research Centre, Plandu, Ranchi-834010, Jharkhand, India

S. Kumar
ICAR Research Complex for Eastern Region, Bihar Veterinary College Campus, ICAR Parishar, Patna, Bihar, India

Shashi Bhushan Kumar
Department of Soil Science and Agricultural Chemistry, Bihar Agricultural University, Sabour-813210 Bihar, India

Vinod Kumar
Department of Food Science and Technology, Bihar Agricultural University, Sabour, Bhagalpur, Bihar, India

S. Kundu
Division of Environmental Soil Science, Indian Institute of Soil Science, Bhopal-462038, MP, India

Sudarshan Maurya
ICAR Research Complex for Eastern Region, Research Centre, Plandu, Ranchi-834010, Jharkhand, India

B. B. Mishra
Department of Soil Science and Agricultural Chemistry, Bihar Agricultural University, Sabour-813210 Bihar, India

Md. Abu Nayyer
Department of Horticulture (Fruit and Fruit Technology), Bihar Agricultural University, Sabour, Bhagalpur, Bihar-813210, India

S. V. Ngachan
ICAR Research Complex for NEH Region, Umiam-793103, Meghalaya, India

C. Nithya
Department of Entomology, Bihar Agricultural University, Sabour, Bhagalpur-813210, Bihar, India

R. K. Patel
NRC Litchi, Mushahari, Muzaffarpur, Bihar, India

V. B. Patel
Department of Horticulture, Bihar Agricultural University, Bhagalpur-813210, Bihar, India

R. K. Pathak
Homa Teacher, Five Fold Path Mission and ex-Director, Central Institute for Subtropical Horticulture, Lucknow, India

A. Subba Rao
Division of Environmental Soil Science, Indian Institute of Soil Science, Bhopal-462038, India

S. N. Ray
Department of Entomology, Bihar Agricultural University, Sabour, Bhagalpur-813210, Bihar, India

N. N. Reddy
Principal Scientist and Director, Central Research Institute for Dryland Agriculture, Santoshnagar, Hyderabad-500059, India

Richa Roy
Sam Higginbottom Institute of Agriculture, Technology and sciences-Deemed university, Naini, Allahabad, India

S. Dam Roy
Central Agricultural Research Institute, Port Blair, Andaman & Nicobar Islands-744101, India

Tamoghna Saha
Department of Entomology, Bihar Agricultural University, Sabour, Bhagalpur-813210, Bihar, India

Som Dev Sharma
Department of Fruit Science, Dr. Y. S.Parmar University of Horticulture and Forestry, Nauni, Solan, Himachal Pradesh, India

Mohammed Wasim Siddiqui
Department of Food Science and Technology, Bihar Agricultural University, Sabour, Bhagalpur, Bihar (813210) India

A. K. Singh
Division of Fruits and Horticultural Technology, Indian Agricultural Research Institute, New Delhi-110012, India

D. R. Singh
Central Agricultural Research Institute, Port Blair, Andaman & Nicobar Islands-744101, India

K. K. Singh
Scientist-F & Head, Agromet Services, Mausam Bhawan, IMD, Ministry of Earth Sciences, Lodi Road, New Delhi-110003

S. K. Singh
Division of Fruits and Horticultural Technology, Indian Agricultural Research Institute, New Delhi-110012, India

Shrawan Singh
Central Agricultural Research Institute, Port Blair, Andaman & Nicobar Islands-744101, India

Yanendra Kumar Singh
Department of Soil Science and Agricultural Chemistry, Bihar Agricultural University, Sabour-813210 (Bihar), India

A. K. Srivastava
National Research Centre for Citrus, Nagpur 440 010, Maharashtra, India

Lajja Vati
Department of Plant Pathology, G. B. Pant University of Agriculture and Technology, Pantnagar, U. S. Nagar, Uttarakhand-263145, India

B. Venkateswarlu
Central Research Institute for Dryland Agriculture, Santoshnagar, Hyderabad-500059, India

M. K. Verma
Division of Fruits and Horticultural Technology, Indian Agricultural Research Institute, New Delhi-110012, India

V. K. Verma
ICAR Research Complex for NEH Region, Umiam-793103, Meghalaya, India

Shailendra Kumar Yadav
Central Soil and Water Conservation Research and Training Institute (ICAR) Research Centre, Datia, MP-475661, India

Parveen Zaman
Bidhan Chandra Krishi Viswavidyalaya, Mohanpur, Nadia, WB-741252, India

LIST OF ABBREVIATIONS

AER	Agro-Eco-Regions
AERS	Agro-Eco-Subregions
AMF	Arbuscular Mycorrhizal Fungi
BC	Bacterial Count
BOB	Bay of Bengal
CCC	Cycocel
CEV	Centre of Excellence in Vegetables
CFCs	Chlorofluoro Carbons
COD	Chemical Oxygen Demand
DMI	Drip Method of Irrigation
DSS	Decision Support Systems
FAO	Food and Agricultural Organization
FC	Fungal Count
FMI	Flood method of irrigation
GHG	Green House Gases
HCFC	Hydrochlorofluorocarbon
HDP	High Density Planting
HFC	Hydrofluorocarbon
HOMA	Homa Organic Farming Technology
IARI	Indian Agricultural Research Institute
ICAR	Indian Council of Agricultural Research
IGBP	International Geosphere-Biosphere Programme
INM	Integrated Nutrient Management
IPCC	Intergovernmental Panel on Climate Change
ISM	Integrated Soil Management
MH	Maleic Hydrazide
MMI	Major and Medium Irrigation
NAA	Nephtalic Acetic Acid
NATP	National Agricultural Technology Project
nCu	Nano-Sized Copper
NICRA	National Initiative on Climate Resilient Agriculture
NMHC	Non-Methane Hydrocarbons

NP	Nano-Particles
OFT	On Farm Technique
PFC	Per Fluorocarbon
PPO	Polyphenol Oxidase
PRADAN	Professional Assistance for Development Action
PRO	Peroxidase
PZC	Point of Zero Charge
QUMP	Quantifying Uncertainties in Model Projections
ROS	Reactive Oxygen Species
SOC	Soil Organic Carbon
SOM	Soil Organic Matter
SPR	Surface Plasmon resonance
TSBF	Tropical Soil Biology and Fertility
TSS	Total Soluble Solids
UNFCCC	United Nations Framework Convention on Climate Change
VAM	Vesicular Arbuscular Mycorrhiza
VOC	Volatile Organic Compounds

PREFACE

The horticulture sector is very sensitive and is considerably impacted by climatic variability, which poses a threat to food security in the future. Productive and sustainable horticulture systems might help to reduce poverty in the context of climate change because they are a good source of income and nutrition. Therefore, to mitigate the ill effects of climate change and to increase and maintain food security, a holistic approach needs to be optimized. This includes understanding the climate change effect on crops, development of stress tolerant genotypes together with sustainable crop, and natural resources management, with implementation of all these efforts by sound policies. Meanwhile an increased temperature (2–4°C by 2100), rise in CO_2, droughts, and floods may become more frequent in the future; therefore, emphasis needs to be on climate smart horticulture with the aim of reduction of greenhouse gas emissions, enhanced resilience, and reduced wastes to help increase the productivity of small and large scale farmers. The approach of integrated crop management (ICM) might be a base for sustainable horticulture, which includes simulation modeling and remote sensing. Meanwhile multidisciplinary breeding with emphasis on warmer and drier environments needs to be utilized as a front line approach. This will ultimately bring sustainability in crop productivity and enhancement of resources use efficiencies. ICM can reduce the use of energy intensive inputs and can increase the efficiency of production. In general, the mitigation approaches, which have synergic relationship with crop productivity and climate change, need to be used to save our future.

This book deals with the prediction of the climate change effect on different horticultural crops. Inclusion of nanotechnology, soil fertility, biochar technology, and management strategies for soil and water as well as pollinators' biodiversity in vegetable crops in relation to climate change make this book very useful for the readers.

— **M. L. Choudhary, V. B. Patel,**
Mohammed Wasim Siddiqui,
and R. B. Verma

CHAPTER 1

GLOBAL CLIMATE CHANGE AND INDIAN HORTICULTURE

K. L. CHADHA

Ex-National Professor and DDG, Horticulture ICAR, New Delhi

CONTENTS

1.1   INTRODUCTION

Global horticulture has several challenges and opportunities. However, a new challenge, that is, climate change has drawn the attention of horticulturists worldwide. Climate change refers to a change in the state of climate that can be identified and persists for an extended period. United Nations Framework Convention on Climate Change (UNFCCC) defines climate change, as a change of climate, which is attributed directly or indirectly to human activity that alters the composition of the global atmosphere and which is in addition to natural climate variability observed over comparable periods. Change in Indian climate is consistent with global trends over the last 50 years and represents a serious threat to the health, environment, agriculture, and economy of the country and its residents.

1.2   GLOBAL CLIMATE CHANGE

1.2.1   GREEN HOUSE GASES

Increase in the concentration or amount of greenhouse gases in the atmosphere is one of the prime causes for climate change. Major greenhouse gases such as carbon dioxide, methane, nitrous oxide, sulpherdioxide, etc., are increasing in the atmosphere and rendering increase in global temperature. Agriculture contributes to global green house gas emission in the form of methane (CH_4), nitrous oxide (N_2O) and small quantity of carbon dioxide (CO_2). About 50% of total CH_4 emissions are from agricultural sector. Human activities such as burning fossil fuels, deforestation, industrial processes, and some agricultural practices have released large amounts of green house gases into the atmosphere. Ozone depletion is evident owing to manmade interventions like coolants in industrial and commercial operations. Production of energy-intensive agro-chemicals including fertilizers that add 0.6–1.2% of worlds' total CHG. After fossil fuel use, land use change and forestry, especially deforestation and degradation, are the next largest emitters of CO_2 (Baumert et al., 2009). CO2 is responsible for 77% of global warming over a 100-year period and hence the most important GHG (Climate Analysis Indicators Tool, 2011). Fast dwindling forest cover is not only adversely affecting biodiversity of horticultural crops but also creating an imbalance in atmospheric equilibrium (e.g., Western

Ghat). The CO_2 concentration has increased from a pre industrial value of about 280 ppm to 393 ppm in 2010. Similarly, the global atmospheric concentration of CH_4 and N_2O as well as other important GHGs, has also increased considerably.

1.2.2   TEMPERATURE

Average temperature on the earth surface has risen by 1.4°F over the past century, and is projected to rise another 2 to 11.5°F over the next hundred years. As per IPCC reports, temperature has shown an increase in 0.74°C between 1906 and 2005 (IPCC, 2007a). Global average sea level rose at an average rate of 1.8 mm per year over 1961 to 2003. There may be 0.5 to 1.2°C rise in temperature by 2020, 0.88 to 3.16°C by 2050 and 1.56 to 5.44°C by 2080 as projected by IPCC (IPCC, 2007b). Increase in global temperatures has been accompanied by the changes in weather and climate. Increase in the temperature might be more during winter than monsoon season towards the end of the century. Similarly, seasonal maximum temperature of Bihar, U.P. and Peninsular India likely to be increased. Whereas minimum temperature supposed to be higher Bihar, Maharashtra, Rajasthan, U.P. and Peninsular India. High temperature induces high evapo-transporation rate resulting in prolonged dry spell. Changes in temperature increase the frequency of climate extremes, for example, heat and cold waves.

1.2.3   WATER AVAILABILITY

It has been opined that the melting of ice glaciers, rise in sea water level, etc., will result in floods and cyclones, sea water intrusion into fresh ground water, acceleration of coastal erosion, submergence of coastal and island horticulture crops. On the other hand, a marginal increase, that is, 7–10% of annual rainfall in the subcontinent by 2080 has been projected. Climate change has had an effect on the monsoons too. India is heavily dependent on the monsoon to meet its agricultural and water needs, and for protecting and propagating its rich biodiversity. Scientists at IIT, Delhi, have already noted subtle changes in the monsoon rain patterns. They also warn that by the 2050s, India will experience a decline in its summer rainfall, which accounts for almost 70% of the total annual rainfall that is crucial to agriculture. The rainfall during summer (*Kharif*) season likely to be increased in eastern and central parts of India. Shift in surplus rainfall from west to east has also been reported. Primarily, increasing need, overuse of increasing population, covering more land under irrigation, and sand mining are the major causes resulting severe reduction in water resources for crops and regions.

Secondly, the continuous failure of rains and/or scarce raining is the reason for poor water recharge in wells that is not allowing the sufficient irrigation to the crops. Whereas, on other hand, IPCC goes on to estimate that even under its most conservative scenario, sea level in 2100 will be about 40 cm higher than today, which will cause an additional 80 million coastal residents in Asia alone to be flooded. The majority of those flooded will be in south Asia, particularly in Bangladesh and India. A one-meter sea level rise would result in nearly 6000 sq.kms in India being flooded, including parts of major cities such as Mumbai, Kolkata and Chennai. Sea level rise will affect the coastal zone in multiple ways, including the inundation and displacement of wetlands and lowlands, coastal erosion, as well as increased coastal storm flooding and salinization (IPCC, 2007).

All above factors are associated with the climate change and global warming and present the scenarios with extreme events like heat waves, cold spells, severe thunderstorms, tropical cyclones, storm surges, severe storms, drought, etc. Climate change has posed a threat due to rise in sea level and extreme weather events and threats to human health, water availability, and food security. Many countries even some states in India have been witnessing changes in rainfall, resulting in more floods, droughts, or intense rain, as well as more frequent and severe heat, cold waves, frost days, droughts, floods, etc., with immense impact on agricultural and horticultural crops.

1.3   INDIAN HORTICULTURE AND CLIMATE CHANGE

Horticultural crops like other agricultural crop are less vulnerable to the threat of climate change impacts. India has varying agro-climatic conditions that allow the cultivation of almost all types of horticultural crops thereby regarded as the best option for diversification and maximizing system productivity. Improved productivity and profitability compared to field crops have been observed in horticultural crops even under adverse climatic situations. Horticulture plays an important role in providing sustainable farm income, nutritional security, import revenues, and employment generation besides sustaining lives and livelihoods. Indian horticulture sector has also shown its impact in Indian economy. The contribution of horticulture sector to Indian GDP is nearly 30% from 15% area while of agriculture is 17%. The horticulture sector has been providing a stimulus for healthy growth trend in Indian agriculture by achieving growth in production by 7.6% per annum with an increase of about 3.8% in area per annum during the last decades ending 2011–12. India is presently producing 257.2 million tons of horticulture produce from an area of 23 million ha. The higher growth rate in horticulture has been brought about by improvement in productivity of horticulture crops, which increased by about 28% between 2001–2002 and 2011–2012. The supportive Govt. program for promotion of technology led horticulture development particularly use of high density plantations, protected cultivation, micro irrigation, quality planting material, rejuvenation of senile orchards and thrust on postharvest management have helped in increasing the productivity of horticultural crops.

India is now the second largest producer of fruits (11.8%) and vegetables (13.3%) in the world and has second largest area under floriculture after China. It is also the largest producer of banana, mango, sapota, papaya, acid lime, cauliflower, pea, okra, cucurbits, etc. It is considered to be the treasure house of medicinal and aromatic plants. It is a major player in coconut production in the world (1.90 million ha) with an annual production of nearly 13,000 million nuts (22.34% share) and is the largest producer, consumer and exporter of spices and spice products (4.02 million tons from an area of about 2.47 million ha). India is the leading producer of arecanut and accounts for 56% and 58% of the total area and production in the world. India is the largest producer (6.95 lakh MT), processor, consumer, and exporter of cashew.

The current demand for horticultural produce is increasing and is expected to accelerate with increase in population and per capita income besides health concerns. There is growing demand for high quality, high value produce throughout the country as well as the world. The competition within states and among countries is increasing. There are >20 fast growing cities with population above 4 million. The growth of fast developing multinational chains in food sector, horticultural production and retail chain, modernization of airports, road and highways has boosted the circulation of the horticultural produce. APEDA has taken initiatives to develop an end-to-end approach in production to exports of horticultural produce by establishing 24 AEZs for different horticultural crops. Similarly, the Ministry of Food Processing Industries has taken initiatives in establishing food parks, etc.

Horticultural crops also play an important role in ecologically fragile zones particularly hilly, coastal, rainfed and dry land areas. In order to provide sustainable food, nutrition and income security, sustaining lives, and livelihoods, the understanding on impacts of climate change and different adaptive strategies/ measures to minimize the adverse effect is important. In this paper, an effort has been made to compile the available information on response to observed impacts of climate related events on horticultural crops.

1.3.1 IMPACT OF CLIMATE CHANGE ON HORTICULTURAL CROP PRODUCTION

While most of the crops are likely to benefit from increase in CO_2, increase in temperature and water scarcity may affect agri-horticultural crop production thus challenging the food and nutritional security (Chadha and Kumar, 2012). The environmental changes projected due to climate change are likely to increase the pressures on Indian agriculture, in addition to the on-going stresses of yield stagnation, land-use, competition for land, water, and other resources, and globalization. Recent report of the IPCC and a few other global studies indicate a probability of 10–40 percent loss in crop production in India with increase in temperature by 2080–2100 (Rosenzweig et al., 1994; Fischer et al., 2002; Parry et al., 2004; IPCC, 2007b). Both lower and upper temperature is important for horticultural crop production. Lower temperature particularly cold waves have shown their impact on crop production in northern India recently. During winter season of 2002–03, it caused considerable losses crops like mango, guava, papaya, brinjal, tomato and potato (Sarnra and Singh, 2003). These studies assume that current management is continued to be followed and no or very low adaptation measures are taken up by the stakeholders. Some of the observed effects on fruit and vegetable productions are listed in Table 1.1.

TABLE 1.1 Effect of Climate Change on Fruit and Vegetables Crops

Crop	Changes in Climate	Effect on Crop
Apple	Sharp fall in temperature and rains during second fortnight of April (26°C for 17 days against 24°C)	Decrease in productivity. Poor fruit set. High temperature scenario advanced flowering by 16 days. Lack of chilling requirements leading to warmer climate
Banana	Dry spell during flower emergence & fruit sheds	Reduces crop duration
Citrus	Increase in 1–2°C beyond 25–30°C	Promotes vegetative instead of flowering flushes
	Untimely winter rain	Affects flower initiation and increases Psylla incidence
Mango	Low temperature (4–11°C)	Delayed panicle emergence
	High humidity (80%) and cloudy weather during January	
	Temperature regime of 27/13 over 21/4°C persisting for a long time	Higher number of perfect flowers
	Unprecedented cold wave	Reduction in yield by 100% and adverse effect on size and quality
Cauliflower	Increase in temperature	Delay in curd initiation
Onion	Temperature above 40°C	Reduction in bulb size
Tomato	High temperature after pollen release	Decreased fruit set and yield
Potato	High temperature	Yield loss (10–20%) and reduction in marketable grade of potato tuber
	Frost damage	Reduction in tuber yield (10–50%) depending upon intensity and coincidence with sensitive stages
Cumin	Frost during January (in Rajasthan)	Crop failure

1.3.2 CROPPING PATTERN

Irrigation water availability, which a main driver, determining cropping patterns that has reduced substantially due to climate change effects. Model projections have made it clear that irrigation water availability will potentially reduce due to changes in climate and irrigation extraction limits (Wang et al., 2011). These calls for changing the country's cropping pattern and intensity to enable farmers adapt and neutralize the adverse effects of climate change on agriculture and achieving the objectives of national food security. Saline-tolerant crops in the coastal belt and drought-tolerant crops in drought prone areas should be grown for achieving the food-security objectives. For example, sunflower could be grown in the coastal belt as a profitable saline tolerant crop known for its high yield. Likewise, carrot, beetroot, parsnip, and other root crops are relatively drought tolerant.

1.3.3 FLOWERING AND FRUIT SET

Flowering and fruit set are important events determining the yield of a crop. Climatic conditions at flowering have an effect on the fruit set and yield. Low temperatures and rainy springs may jeopardize pollination. Likewise, cold temperatures at flowering reduce the speed of pollen tube growth and shorten the effective pollination period (Sanzol and Herrero, 2001). However, relatively warm springs have often been assumed as a prelude of a good fruit set in several temperate fruit tree species. Fruit trees provide important adaptive values and tend to be more resilient to climate change due to their perennial nature. However, they too are affected by climate change in idiosyncratic ways. Climate change especially poses important difficulties for commercial production of fruit trees. Lots of research has been done to study the impacts of climate change on crops globally, but most of it is focused on studying major crops such as maize, wheat, and rice. Horticultural crops including fruits have mostly been ignored, partly due to lack of data and approaches for modeling.

Climate change will have both positive and negative impacts on fruits in tropical regions. In regions where the prevailing temperatures are already high, further increases in temperature will adversely affect the yield and quality of fruits. In regions where cold temperatures are one of the primary factors limiting crop production, temperature increases will be beneficial. The impact of temperature change can be clearly seen from the fact that the northern parts of India are warmer than the southern parts, with a general increase of 3–6°C over the base-period average (Lal et al., 1995; Lonergan et al., 1998). Studies carried out on perennial trees have to be contiguous and long range. Since experiments are carried out for short periods, these studies more often than not have become pointers rather than conclusive. The increase in temperature has been reported to affect the phenology of perennial trees. In certain regions where prevailing temperatures are already high, shifts in growing areas may take place. In the peninsular regions of India, it has been noticed that flowering was enhanced by a month in mango, thus affecting the fruit maturity and season of harvest. In the case of crops like guava, which has adapted very well to both tropical and subtropical climate environments, changes in temperature have contributed to shifting of fruiting season.

Under the influence of climate shift, both early and delayed flowering will be characteristic features in mango. Low temperature (4 to 11.5°C), high humidity (>80%) and cloudy weather in January delay panicle emergence whereas low temperatures during inflorescence development reduce number of perfect flowers. The perfect flowers were significantly higher in the temperature regime of around 27/13°C than in a temperature regime of 21/14°C in mango. If panicle development coincides with an unusual cold spell, mango production will face several problems. Owing to variations in temperature, unseasonal rains and higher humidity, fruit trees show altered flowering trends, with delays in panicle emergence and fruit set. Availability of hermaphrodite flowers for pollination and fruit set have an effect on yield due to pollen and stigmatic sterility. Temperature above 30°C induces maximum flower and fruit drop and high temperature after pollen release decreased fruit set and fruit yield in tomato. In papaya, higher temperatures have resulted in flower drop in female and hermaphrodite plants as well sex changes in hermaphrodite and male plants. Experimental findings indicate increased leaf production in banana with increase of 1–2°C temperature beyond 25–30°C, thereby reducing crop duration, and increasing production. Occurrence of frost during January in Rajasthan has been reported cumin resulting in total crop failure (Chadha and Kumar, 2012).

1.3.4   FRUIT QUALITY

Although, higher temperature regimes generally result in the best quality fruits, excessively high temperatures for extended periods of time are known to damage fruits result in delay of fruit maturation and reduction in fruit quality of grape. High temperatures also reduce color development. Higher daily temperatures were also related to a decrease in color hue values, that is, more red fruit. In the case of guava, it has been observed that red color development on the peel of guava requires cool nights during fruit maturation. Varieties like Apple Color guava, which have attractive apple skin color under subtropical conditions of North India, have red spots on the skin under tropical South Indian conditions. Observation in areas suitable for production of red color guava by Rajan (2008) revealed that the total soluble solids, fruit firmness, and percentage dry mass were negatively correlated with temperature during fruit growth. However, the relationship varied with the cultivar (Hoppula and Karhu, 2006). In subtropical and tropical fruit crops, there is a direct effect of the temperature the maturity and ripening of fruits. When there is sufficient moisture, TSS of fruit increases with increase in temperature. However, in fruits like passion fruit, increases in temperature do not increase TSS. Hence, the effect of different regimes of temperature can be different on different crops under subtropical and tropical environments.

1.3.5   INCIDENCE OF DISEASES AND PESTS

A number of factors affect the pest and disease scenario in fruits. High temperatures coupled with high rainfall and humidity help in building up ideal conditions for the growth of a number of disease pathogens. For example, the powdery mildew disease in mango caused by *Oidium mangiferae* Berthet is a sporadic but serious disease of mango inflorescence that can cause 80–90% losses of the crop in extreme cases. Optimal disease development occurs at 10–31°C and 60–90% RH. (Chhata et al., 2006) reported that high humidity (85–90%), moderate temperatures (maximum temperature of 25–26°C and minimum of 18–20°C) provided favorable condition for the initiation of disease. Correlations between weather parameters like maximum temperature regimes and sunshine hours had negative correlations with disease development, while minimum temperature, humidity, and wind speed had positive correlations. Diedhiou et al. (2007) found that fruits harvested during humid conditions were more heavily infested but a smaller number of fungal agents were involved; *Colletotrichum gloeosporioides* and *Phoma mangiferae* played the main role. In mango and guava, it has been observed that the incidence of fruit fly is much less at higher temperature regimes. However, a study conducted in India by Kumar et al. (2010) has shown that in mango cv. Chausa the rate of development of fruit fly increased with the increase in temperature from 20–35°C. The percent larval survival, adult emergence, and growth index also increased with an increase in temperature from 20 to 27°C and thereafter decreased up to 35°C. Thus, a temperature of 27°C was found to be ideal for survival and development of the immature stage and reproduction of *Bractocera dorsalis*. Post-harvest treatment of fruits at 48°C for 60–75 min has given good control of fruit flies (Verghese et al., 2002).

Downy mildew is found worldwide wherever grape is grown, occurring primarily where warm, humid conditions exist during the growing season. All common cultivated and wild species of grape as well as a few hosts outside the *Vitis* species are susceptible to this disease (Pearson and Goheen, 1988). Unseasonal rains coupled with higher temperatures during vegetative phase in grape also result in damage due to this disease.

1.3.6 SHIFTING OF PRODUCTION ZONES

Concurrently, climate change is shifting the habitat range of plants and animals (Pereira et al., 2010), including agricultural crops. For example, as average global temperatures increase, plant and animal populations may move to new latitudes with more favorable climates. The crops, which are productive in one area may no longer be so or the other way around.

Following issues will result in shifting of production zones due to changing climate effect:
- Changes in time to harvest for some crops and locations.
- Changes in the suitability and availability of cultivars for current and future production locations.
- Reduced availability and increased cost of irrigation water in most locations and in some seasons.
- Greater seasonal variability.

- Increased pest and disease incidence and 'new' pests, diseases and weeds.
- Damage from extreme events (rain, hail, wind and heat stress).
- Negative impacts on soils and crops due to extreme temperature and rainfall events (flooding).
- Changes in frost frequency.

1.4 ADAPTATION TO CLIMATE CHANGE

There have been several technologies, which are already available and can be useful for reducing the negative impacts of climate change. Development of adverse-climate tolerant varieties may take more time but agronomic adaptations, crop management, and input management can be readily used to reduce the climate related negative impacts on crop growth and production. Some such simple but effective adaptation strategies include change in sowing dates, management of plant architecture (in perennial crops), use of efficient technologies like drip irrigation, soil and moisture conservation measures, fertilizer management through fertigation, change of crop/alternate crop, increase in input efficiency, pre- and postharvest management of economic produce do not only minimize the losses but also increase the positive impacts of climate change. There is a lot of scope to improve the institutional support systems such as weather based agro-advisory, input delivery system, development of new land use patterns, community storage facilities for horticultural perishable produce, community based natural resource conservation, training farmers for adapting appropriate technology to reduce the climate related stress on crops, etc. All these measures can make the horticultural farmer more resilient to climate change.

1.4.1. OPTIMUM LAND USE

1.4.1.1 CHOICE OF CROPS

India is endowed with rich resources of water in various forms, which are underutilized or unexplored. Many areas are also flood affected or water logged. The risk of climate change also creates uneven distribution of rains resulting into both drought and flood conditions in Bihar. A systematic effort is required to harness more prosperity from ample available water in wetland, for example, North Bihar. A number of crops have been identified for adverse land situation such as Makhana, Singhara, Sludge mate and lotus, which can be harnessed for profitability from the wetland situation. Gorgon nut or Fox nut, commonly known as Makhana. It grows in water producing large floating leaves with a quilted texture, bright purple flowers and starchy white seeds. Ecologically makhana grows in shallow water bodies that have a certain amount of organic detritus accumulated at the bottom. The land situations and specific crop suggestion is given in Table 1.2.

TABLE 1.2   Land Situations and Suggested Crops

Condition	Suggested Crops
Water logged areas	*Makhana* and lotus.
Irrigated areas	Litchi, guava, cowpea, French bean, banana, lime, lemon and Betel vine.
Rainfed areas with good soil fertility	Mango-based cropping system (Mango, guava, cowpea, and French bean).
Rainfed areas with poor soil fertility	*Aonla*-based cropping system (*Aonla*, guava, turmeric).
High value and low water requiring crops	Seed spices (*Dhania, Methi, Kalajeera, Sounf*, etc.).
Flood affected areas	Japanese mint, vegetables, marigold.
Saline patches	*Bael, mahua, karonda, phalsa*, mint, citronella.
Ravinous areas	*Kair, khejri, karonda, etc.*

1.4.1.2   USE OF WASTELAND

At present, approximately 68.35 million hectare area of land is lying as wastelands in India. Out of these lands, Approximately 50% lands of area are under nonforest lands, which can be made fertile again if treated properly. It was unprotected nonforestlands, which suffered the maximum degradation mainly due to the tremendous biotic pressure on it. Selection of fruit crops for wastelands and development through fruit cultivation has been reported by Singh and Choudhary (2012). There are a number of fruit and vegetable crops, which can be cultivated under different wasteland categories (Table 1.3).

TABLE 1.3   Fruits and Vegetables for Wasteland Development

Wasteland Category	Fruits	Vegetables	Others
Sandy land	Guava, Manila tamarind, *khejri* Tamarind,	Chilli, cowpea, garlic, onion	*Aloevera, aswagandha, Vitex regundo.*
Salt affected land	*Aonla, bael, ber* drumstick,	Beet, carrot, chili, okra, *methi*, spinach	*Calotropis,* Fennel, henna (*Lawsonia inermis*).
Gullied and riverine land	Drumstick, *khejri, Phoenix* sp.	Chilli, cluster bean, cowpea, dolichos bean	*Ailanthus excelsa, Boswellia serrata, Bursera panicillata, Cassia fistula, C. siamea*, licorice, *kusum, sheesham.*
Undulating uplands	*Ber,* guava, olive, peach	Bell pepper (*Capsicum annum*), cluster bean, cucumber, French bean, watermelon, muskmelon, *parwal*	Henna, *Swertia chirata*, poplar
Mined and industrial wastelands	*Aonla, bael, ber*	Cucurbits	*Aloe sisalana*
Waterlogged areas	Lotus, *jamun,* tamarind, water chestnut	Leafy vegetables	*Khas khas*
Strip lands	*Karonda, jamun, ber*		Neem, S*imarouba glauca*

1.4.1.3   HIGH DENSITY PLANTING

Introduction of high-density planting is one of important technology to achieve high productivity per unit area both in short duration crops and perennial crops also to lower the impact of climate change. High-density planting in tree crops was first introduced in Europe in case of apple in sixties, with the development of Malling and Malling Merton Series of rootstocks. Subsequently, this was tried in other crops. Today, majority of apple orchards in Australia, Europe, and New Zealand use this intensive system of fruit production. Subsequently, this system was adopted in peach, plum, sweet cherry, and pear. The development of high density or intensive system has been made possible by the discovery and development of dwarfing roots like M9, M26, in case of apple, Quince A for pear, peach with almond hybrids (GF 677, 556) for peach and Cold in cherry. Other developments that contributed to high density planting are the use of growth regulators such as B-9, which dwarf the trees and encourage early bud production. In India, high density planting (HDP) has been standardized in several fruit crops to provide high yield, high net economic returns per unit area, more efficient use of natural resources (land, water, solar energy) and inputs resulting in higher yields. Successful harvests but selective adoption of HDP is already practice in mango, guava, banana, citrus, pineapple, pomegranate, papaya, cashew and coconut.

Technologies for Meadow orcharding through HDP and canopy management have been standardized in guava. Coconut-based high-density multispecies cropping systems have been developed for better stability of income. In citrus it was made possible by use of a dwarfing rootstock. In case of mango, development of Mallika and Amarpalli, which are medium vigor made high density planting possible. While in banana and pineapple intensive cultivation became possible by reducing distances between plants, rows and beds. In Dashehari mango at Pantnagar, high density planting has been achieved by closer spacing, canopy management and use of chemicals. Future high-density plantings have to be developed by development of dwarfing rootstock, interstocks, scion varieties as also a combination of canopy management and growth regulators. The success of high-density plantings also depends on optimum fertilizer and irrigation requirements. Efficient fertigation methods need to be standardized for such crops. Some examples of high-density plantations are given in Table 1.4.

TABLE 1.4   High Density Planting and Yield

Crop	Normal Spacing (m²)	Spacing (m²)	Planting Density (No. of plants/ha)	Yield (t/ ha)	National yield (t/ ha)
Pineapple	90×150	0.25×0.45×0.6	63,452	60	16
Banana	3 × 3	1.8 × 1.8	3086	114	37
Citrus	6 × 6	1.8 × 1.8	3000	20	9.3
Mango	10 ×10	2.5 × 2.5	1600	21	5.5
Papaya	3 × 3	1.25 × 1.25	6400	65	37.1
Guava	6 ×6	3 × 3	1111	18	10

1.4.1.4   VARIETIES FOR DIFFERENT TEMPERATURE REGIMES

For optimum utilization of land and land resources year round production of vegetables is advocated. Growing of vegetables throughout the year, promotes the household food and nutritional security of farm families through increased intake of homegrown vegetables. Such models are help-ful not only in meeting the nutritional needs but also to generate income of the farmers by producing vegetables during the off seasons. Varieties have been tuned in such a fashion to growth throughout the year by incor-porating the genome, which enables them to grow in different temperature regimes (Table 1.5).

TABLE 1.5   Varieties, Crop Cycle (Sowing Time, Maturity, Availability), and Yield of Some Vegetables

A. Cauliflower

Maturity (Aver-age Curd Initiation Temp.)	Varieties	Sowing Time	Availabil-ity	Yield (q/ha)
Early I (25–30°C)	Pusa Meghna	May end-Early June	Sept. to Oct.	95
	Pusa Kartik Sankar			115
Early II 20–25°C	Pusa Deepali	Mid June	Oct. to Nov.	110

Mid-Early 16–20°C	Pusa Sharad	July Aug.	Nov. to Dec.	250
	P. Hybrid-2			200
Mid-Late 12–16°C	Pusa Paushja	Aug. end	Dec. to Jan.	325
	Pusa Shukti			320
Late 10–16°C	P S B K-1	Sept.-Oct.	Jan. to Mar.	350
	PSB Kt-25			320

B. Radish

Variety	Sowing Time	Temperature	Availability	Yield (q/ha)
Pusa Desi	August to October	25–30°C	Sept. to Nov.	175
Pusa Mridula	Sept. to November	20–25°C	Oct. to Dec.	180
Jap. White	Oct. to December	15–20°C	Nov. to Jan.	250
Pusa Himani	Dec. to March	10–15°C	Jan. to Apr.	250
Pusa Chetki	April to August	20cha-35°C	Mar. to Sept.	170

C. Carrot

Variety	Sowing Time	Temperature	Availability	Yield (q/ha)
Pusa Vrishti	July to October	25–30°C	Oct. to Nov.	180–200
Pusa Rudhira Pusa Asita	Sept. to October	16–25°C	Nov. to Jan.	250–300
Pusa Yamdagini Pusa Nayanjyoti	Sept. to March	10–15°C	Nov. to Apr.	270–290
Pusa Vrishti				120
Pusa Yamdagini	March to April	25–35°C	May to Jun.	140
Pusa Nayanjyoti				160

1.4.1.5 PROTECTED CULTIVATION

India has varied climatic conditions in different regions, so the greenhouse and the supporting facilities are developed accordingly. About 40,000 ha of area are now estimated to be under protected cultivation of horticultural crops. Protectnet a comprehensive research program launched by the ICAR nearly four years ago to address some of the problems is at its fag end and the final results are awaited. Precision Farming Development centers in 19 institutions are involved in the development of technologies for protected cultivation and dissemination in different agroclimatic situations. However, nonavailability or inadequate power supply is a limiting factor in popularization and propagation of greenhouses. In Leh structures such as soil trenches and low cost polyhouses are quite useful. They are being used on a limited scale for raising early vegetables and flowers nursery production of early vegetable crops, extension of growing season, vegetable production during frozen winters, protection of valuable germplasm and cultivation of cucurbits, brinjal, capsicum and certain flowering annuals.

1.4.2 IMPROVING INPUTS USE EFFICIENCY

1.4.2.1 IRRIGATION MANAGEMENT

Water is a critical input for production of crops, and its efficient use is inevitable in context of achieving higher production from shrinking land resources. It is estimated that only 38% of the area is irrigated and with harnessing of all the resources of ground and surface water, with current efficiency of irrigation, not more than 60% of area could be irrigated. Micro-irrigation, popularly known as drip irrigation, is an efficient means of irrigation with high frequency application of water, where, water is delivered in and around root zone through a network of suitable emitting devices. Micro-irrigation economizes on use of water by 40–60% added with enhanced yield and quality of produce. For, short duration crops, surface installation as per layout and design are adopted while for perennial crops, subsurface installations are preferred. In India, microirrigation was practiced using indigenous methods such as bamboo pipe, perforated clay pipe, and pitchers but the modern system of drip irrigation is only a decade old. India has now emerged as one of the leading countries in using microirrigation technology. Among the crops, maximum adoption of drip system has been in coconut, followed by banana, grape, papaya, pomegranate, mango, and sapota. There is growing awareness for use of drip irrigation even for closely spaced crops especially tomato, capsicum, chilies and potato. Microirrigation not only saves the water but also is helpful in maintaining the hydration of root zone and efficient utilization of the water. The research in this area is focused for predicting the temporal and spatial variation of soil moisture, the minimal and optimal fraction of the soil volume to be wetted, and management system in different horticultural crops under varying weather and soil conditions, application of nutrients, critical stages of growth and development.

Fertigation means supplying both water and fertilizer to growing crop through drip irrigation. Through fertigation, many macro and micronutrients like nitrogen, phosphorus and potash as well as micronutrients, for example, Cu, Fe, Mg, Mn, Mo, Zn can be supplied directly to the active root zone. This system helps in improving productivity and quality of the produce and minimizes the losses of expensive nutrients. Fertigation though tested on a large number of horticultural crops is ideally suited for hi-tech horticulture production systems. Besides ensuring efficient delivery of two important inputs, it also exploits the synergism of their simultaneous availability to plants. When the drip system is adopted, the root system proliferates only in the moist zone below the emitter making this as a nutrient adsorption zone. One of the reasons for nonadoption of fertigation widely even where drip system has been installed is the nonavailability of soluble fertilizers at affordable cost. Other reasons include lack of scientific information on the frequency and time interval besides nonavailability of Govt. policies to promote fertigation. Some problems associated with this technology are that bad quality irrigation water rich in calcium and magnesium results in precipitation of phosphorus and adds to clogging of pores of the emitters. Because of the pH, carbonate, calcium and magnesium, the micronutrients have to be supplied in chelated form, which is costly. Clogging of emitters is another problem discouraging fertigation. The form of N is also important. Excessive use of NO_3–N results in iron deficiency. The fertilizers are also supplied as NPK complexes and freedom to choose required ratios is very limited. The grape, pomegranate, and banana growers in Maharashtra have also adopted fertigation to some extent. To exploit this technology fully, there is a need to remove the above constraints. Experience has proved beyond doubt that fertigation has potential for use in all horticulture crops, which, not only economizes on nutrient use, but future challenges of achieving higher productivity can be addressed through this technology. Therefore, there is a need to encourage fertigation through technological and policy support.

TABLE 1.6   Stress Tolerant Varieties

Crop	Variety	Stress
Lemons	Pramalini	Canker
	Sai Sarbati	Canker and Tristeza
	PKM-1	Canker and Tristeza
Carrot	Pusa Vrishti	Heat and humidity
	Pusa Meghali	Heat
Radish	Pusa Chetki	High temperature
	Pusa Himani	Low temperature
Turnip	Pusa Sweti	High temperature and humidity
Tomato	Pusa Sheetal	Low temperature (cold set)
	Pusa Sadabahar	Low and high temperature
	Pusa Hybrid 1	High temperature (hot set)
Cauliflower	Pusa Meghna	High temperature and humidity
	Pusa Kartik Sankar	High temperature and humidity
Cabbage	Pusa Ageti	High temperature

1.4.2.2   LOW INPUT FARMING

Low input farming is a holistic production management system, which promotes and enhances agro ecosystem health including bio-diversity, biological cycles and soil biological activities. The farming system emphasizes upon management practices wherein agronomic, biological and mechanical methods are used for sustainable production avoiding the use of synthetic materials. With increasing health consciousness and concern for environment, organic farming system has been drawing attention all over the world. As a result, there is widespread organic movement. Demand for organic products, especially in developed countries, has been increasing by leaps and bounds. Major components of organic farming are crop rotation, maintenance, enhancement of soil fertility through biological nitrogen fixation, addition of organic manures and use of soil microorganisms. Vermiculture has become a major component in biological farming, which is found to be effective in enhancing the soil fertility and producing large number of horticultural crops in a sustainable manner. Earthworks are also found useful for recycling the farm wastes and household wastes. In India, the old farming system largely was organic wherein crop rotation, choice of cultivars for the region, utilization of solar radiation for soil sterilization, etc. were used and soil fertility was maintained through organic manure. However, in quest for increasing the productivity, fertilizers and pesticides were used. Over use of fertilizers and chemicals have put forth a question mark on the sustainability of agri-horticulture in long run and it is calling attention for sustainable production to address social, ecological, and economical issues together. Northeastern Region, tribal areas, and hills have very low use of fertilizers and pesticides. This weakness of the regions can be converted into strength, especially for growing of organic produce. Organic farming would, not only provide hygienic horticultural produce but its impact on sustainability of agriculture would be rewarding. Therefore, it is essential that the organic movement be also given a focused attention, through integration of efforts made on scientific line to achieve sustainable production.

1.5 RESEARCH STRATEGIES

A lot of development to minimize the risk of climate change has taken place during the past decade. However, there are some researchable issues in relation to climate change for horticultural crops. Some of the issues need focused planning to generate information to adopt /mitigate the adverse effect of climate change.

1.5.1 GENERAL RECOMMENDATION

- Horticulture crops are vulnerable to climate variability and information on impact of climate is scarce and not documented systematically.
- Environmental changes projected due to climate change are likely to increase pressure on the Indian Horticulture therefore benchmarking of areas prone to climate change is needed urgently.
- Quantification of impact of major climate change parameters under field and environmentally controlled conditions is necessary.
- Future projections based on climate modeling on likely pattern of temperature and rainfall in the next 100 years.
- Prediction of changes to sustain horticulture crop production to achieve food and nutritional security in twenty-first century is needed.
- Development of proactive anticipatory research on commodities, crops, and varieties is needed.

1.6 RESEARCH PRIORITIES

1.6.1 OPTIMUM LAND USE

- Choice of crops in the relation to area and climate change pattern.
- Reclamation of wastelands by raising horticulture and agro-forestry plantation to sequester carbon.
- Promoting year round vegetable cultivation.
- Inter and mixed cropping.
- Integrated farming systems.

1.6.2 CROP IMPROVEMENT

- Identification of gene(s) for tolerance using genetic engineering.
- Development of varieties for changing climate, resistant/tolerant to abiotic and abiotic stress conditions.
- Promotion of low chilling varieties of pome and stone fruits.
- Study of horticulture crops and varieties for carbon sequestration potential.

1.6.3 PROTECTED CULTIVATION

- Research on protected cultivation should be taken up in collaborative mode with the SAUs.
- Centre of Excellence in Vegetables (CEV) should be established to develop package of practices for green houses cultivation of vegetables or flowers.
- Large-scale motivation and training to educated unemployed youths.
- Large-scale promotion of low cost structures with low pressure drip irrigation system with resource poor farmers.

1.6.4 MITIGATION STRATEGIES: WATER AND NUTRIENT USE EFFICIENCY

1.6.4.1 INTEGRATED NUTRIENT MANAGEMENT

- Soil and leaf testing laboratories be established in each district and leaf tissue analysis be used as a diagnostic tool for determining the sufficiency/deficiency levels of different nutrients.
- Biofertilizers/vermiculture should be used simultaneously reduce cost of inputs and improve the quality and production.
- Nutrient application should be based on application at right time and amount with proper placements and to improve fertilizer use efficiency and reduce emission of CH_4 and N_2O.
- Provide proper training to prepare vermi-compost and use of bio-fertilizers.
- Promote fertigation.

1.6.4.2 ALTERNATE FARMING SYSTEMS

- Strengthen research on different aspects of Organic farming.
- Crops with comparative advantage need to be identified.
- Farmers need training and financial support to prepare quality organic manure using gobar gas units, solar energy and bio-pesticides,

organic manures, green manure, vermi-compost and bio-fertilizers to encourage IPM/INM and nutrient management.

• Proper policy, incentives need to be put in place for carbon trading in case of farmers who adopt organic farming and conservation agriculture practices.

• Policy on crop residue and its effective utilization.

• Standardize suitable crop rotation schedules.

1.6.4.3   WATER MANAGEMENT

• Need for scientific water management to reduce costs and moderate CH_4 production.

• Emphasis on conservation of rainwater.

• Contingency plans for drought and flood situation.

• Improvement of water use efficiency for reduction of CHG through microirrigation.

• Promote in situ rainwater conservation and mulching.

1.6.4.4   IPM STRATEGIES

• Install automatic weather station appropriate locations in the state to establish a pest and disease forecasting advisory servers.

• Develop Ecofriendly disease and pest management strategies.

• Anticipate changes in pest and pathogen behavior to mitigate losses due to climate change.

1.6.4.5   TRANSFER OF TECHNOLOGY

• There is acute shortage of technology on climate change and manpower dealing with it.

• There is need to sensitize R and D workers on the climate change its likely effects and mitigation strategies.

• Introduce suitable and innovative extension mechanism for ensuring technology transfer and adoption in horticultural crops.

• Introduce suitable insurance models for damage to high culture crops from climate change.

KEYWORDS

- **Adaptation**
- **Climate Change**
- **Horticultural Crops**
- **Input Use Efficiency**
- **Mitigation Strategies**

REFERENCES

Baumert, K. A., Herzog, T., & Pershing, J. (2009). Navigating the Numbers, Greenhouse Gas Data and International Climate Policy. World Resources Institute, Washington, DC, USA.

Chadha, K. L., & Kumar, Naresh. (2011). Climate Change Impacts on Production of Horticultural Crops. In Dillon W. S. (eds).

Chhata, L. K., Jat, M. L., & Jeeva Ram Jain, L. K. (2006). Effect of Weather Variables on Outbreak and Spread of Leaf Spot Fruit Spot in Pomegranate, *Current Agriculture*, 30(1/2), 39–44.

Diedhiou, P. M., Mbaye, N., Drame, A., & Samb, P. I. (2007). Alteration of Post Harvest Diseases of Mango, Mangifera Indica through Production Practices and Climatic Factors, *African Journal of Biotechnology*, 6(9), 1087–1094.

Hoppula, K. B. & Karhu, S. T. (2006). Strawberry fruit quality responses to the production environment. *Journal of Food, Agriculture and Environment*, 4(1), 166–170.

IPCC, (2007). Climate Change 2007, Impacts, Adaptation and Vulnerability, Cambridge University Press, Cambridge, United Kingdom and New York, NY, USA.

Kumar, R., & Kumar, K. K. (2007). Managing Physiological Disorder in Litch,. *Indian Horticulture*, 52(1), 22–24.

Kumar, R., & Omkar Shukla, R. P. (2010). Effect of Temperature on Growth, Development and Reproduction of Fruit Fly Bractocera Dorsalis Hendel (Diptera Tephritidae) in mango, *Journal of Ecofriendly Agriculture*, 5(2), 150–53.

Lal, M., Cubasch, U., Voss, R., & Waszkewitz, J. (1995). Effect of Transient Increase in Greenhouse Gases and Sulphate Aerosols on Monsoon Climate, *Current Science*, 69(9), 752–763.

Lonergan, S. (1998). Climate warming and India. In: Dinar, A., Mendelsohn, R., Evenson, R., eds. Measuring the Impact of Climate Change on Indian Agriculture, World Bank Technical Paper No. 402, Washington, DC, Mall RK.

Pearson, R. C., & Goheen, A. C., eds. (1988). Compendium of Grape Diseases, APS Press, St. Paul, MN. 93.

Pereira, H. M., Leadley, P. W., Proença, V., Alkemade, R., Scharlemann, J. P. W., & Fernandez-Manjarrés, J. F. (2010). Scenarios for Global Biodiversity in the twenty-first century. *Science*, 330 1496–1501.

Rajan, S. (2008). Implications of Climate Change in Mango. Impact Assessment of Climate Change for Research Priority Planning in Horticultural Crops, Central Potato Research Institute, Shimla, 36–42.

Sanzol, J., & Herrero, M. (2001). The Effective Pollination Period in Fruit Trees. *Scientia Horticultureae,* 90, 1–17.

Singh, S. P. & Chaudhury, M. R. (2012). Production Technologies of Fruit Crops in Wasteland. Scientific Pub.

Verghese, A., Madhura, H. S., Kamala Jayanthi, P. D., Stonehouse John, M. (6–10 May 2002). Fruit Flies of Economic Significance in India, with Special Reference to Bactrocera dorsalis (Hendel). Proceedings of 6th International Fruit fly Symposium, Stellenbosch, South Africa. 317–324.

Wang, Y., Chen, Y., & Peng, S. (2011). A GIS Framework for Changing Cropping Pattern under Different Climate Conditions and Irrigation Availability Scenarios, *Water Resources Management,* 25(13), 3073–3090.

CHAPTER 2

CLIMATE CHANGE PREDICTION: UNCERTAINTIES AND ACCURACIES

K. K. SINGH

Scientist-F & Head, Agromet Services, Mausam Bhawan, IMD, Ministry of Earth Sciences Lodi Road, New Delhi – 110003; Email: kksingh2022@gmail.com

CONTENTS

2.1   INTRODUCTION

Global and regional models have been used for producing climate change scenarios with a special focus on the behavior (frequency and intensity) of extreme events like heat waves, cold spells, severe thunder storms, tropical cyclones, storm surges, severe storms, drought, etc. Extensive observational data over the past century and also the reconstructed data have been used in climatic change assessment. Better forecasting capability is central to an effective adaptation strategy, particularly in the Indian context where livelihoods are strongly related to the physical environment.

India is one of the world's most vulnerable countries to climate change. It is vulnerable to sea level rise and extreme weather events, and will increasingly face threats to human health, water availability, and food security. Additionally, about 12 percent (40 million hectares) of India is flood prone, while 16 percent (51 million hectares) is drought prone. Thus India is also vulnerable to potential climate change-induced shifts in precipitation patterns.

In India, climate/environment related observations are being taken, both on regular and campaign mode by various Central and State Government Departments, Universities, research institutions and some nongovernmental agencies. Efforts are required to bring all these sources of climate system related observations into a single national network for use by research community and climate service delivery.

2.2   OBSERVED TREND IN CLIMATIC EXTREMES

2.2.1   EXTREME RAINFALL EVENTS

Based on rainfall data during 1901–2005, annual normal rainy days varied from 10 over extreme western parts of Rajasthan to the high frequency of 130 days over northeastern parts of the country (IMD NCCRR, 3/2010). Both nonparametric test and linear trend analysis identified decreasing trends in the frequency of wet days in most parts of the country. Trend analysis of frequency of rainy days and heavy rainfall days shows significant decreasing trends over central and many parts of north India and increasing trends over peninsular India. Also the great desert areas of the country have experienced increase in number of wet days. One-day extreme rainfall intensity increased over Coastal Andhra Pradesh and adjoining areas, Saurashtra and Kutch, Orissa, West Bengal, parts of north-east India, east Rajasthan. Significant decrease both in intensity and frequency of extreme rainfall has been observed over Chhattisgarh, Jharkhand and some parts of north India. The flood risk also increased significantly over India.

2.2.2 TEMPERATURE TRENDS

Analysis of data for the period 1901–2011 suggests that annual mean temperature for the country as a whole has risen by 0.59°C over the period. It may be mentioned that annual mean temperature has been generally above normal (normal based on period, 1961–1990) since 1990. This warming is primarily due to rise in maximum temperature across the country, over larger parts of the dataset. However, since 1990, minimum temperature is steadily rising and rate of its rise is slightly more than that of maximum temperature. Spatial pattern of trends in the mean annual temperature (Fig. 2.1) shows significant positive (increasing) trend over most parts of the country except over parts of Rajasthan, Gujarat and Bihar, where significant negative (decreasing) trends were observed (Annual Climate Summary, 2009).

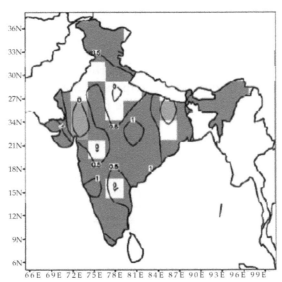

FIGURE 2.1   Annual mean temperature trends (°C/100 year) 1901–2011.

Season wise, maximum rise in mean temperature was observed during the Post-monsoon season (0.77°C) followed by winter season (0.70°C), Pre-monsoon season (0.64°C) and Monsoon season (0.33°C). During the winter season, since 1991, rise in minimum temperature is appreciably higher than that of maximum temperature over northern plains. This may be due to pollution leading to frequent occurrences of fog.

2.2.3 PRECIPITATION TRENDS

The all India monsoon rainfall and monthly rainfall for the monsoon months for the period 1901–2009 do not show any significant trend. Similarly rainfall for the country as whole for the same period for individual monsoon months also does not show any significant trend. During the season, three subdivisions viz. Jharkhand, Chhattisgarh, Kerala show significant decreasing trend and eight subdivisions viz. Gangetic West Bengal, West Uttar Pradesh, Jammu and Kashmir, Konkan and Goa, Madhya Maharashtra, Rayalaseema, Coastal Andhra Pradesh and North Interior Karnataka show significant increasing trends. The alternating sequence of three successive decade (30 years) having frequent droughts and never three decades having flood years are observed in the all India monsoon rainfall data. The decades 1961–1970, 1971–1980 and 1981–1990 were dry periods. The first decade (1991–2000) in the next 30 years period already experienced wet period.

However, during the winter season, rainfall is decreasing in almost all the subdivisions except for the subdivisions Himachal Pradesh, Jharkhand, Nagaland, Manipur, Mizoram and Tripura. Rainfall is decreasing over most parts of the central India during the premonsoon season. However, during the postmonsoon season, rainfall is increasing for almost all the subdivisions except for the nine subdivisions.

During the monsoon season, three subdivisions viz. Jharkhand (95%), Chhattisgarh (99%), Kerala (90%) show significant decreasing trends and eight subdivisions viz. Gangetic WB (90%), West UP (90%), Jammu and Kashmir (90%), Konkan and Goa (95%), Madhya Maharashtra (90%), Rayalseema (90%), Coastal AP (90%) and North Interior Karnataka (95%) show significant increasing trends. The trend analyzes of the time series of contribution of rainfall of each month towards the annual total rainfall in each year in percentages, suggest that contribution of June and August rainfall exhibited significant increasing trends, while contribution of July rainfall exhibited decreasing trend. However, no significant trends in the number of breaks and active days during the south-west monsoon season during the period 1951–2003 were observed.

2.2.4 CLOUD COVER OVER THE INDIAN SEAS

Both total and low cloud cover over Arabian Sea and the equatorial Indian Ocean are observed to decrease during the ENSO events. However, cloud cover over Bay of Bengal is not modulated by the ENSO events. On interdecadal scale, low cloud cover shifted from a "low regime" to a "high regime" after 1980, which may be associated with the corresponding interdecadal changes of sea surface temperatures over north Indian Ocean observed during the late 1970s (Rajeevan et al., 2000).

2.2.5 HEAT WAVE AND COLD WAVE

A significant increase was noticed in the frequency, persistency and spatial coverage of both of these high frequency temperature extreme events (heat and cold wave) during the decade (1991–2000) (Pai et al., 2004).

2.2.5.1 CYCLONES

1. A slight decreasing trend in the annual frequency of cyclones that formed over Bay of Bengal (BOB) during 1900–2009 is seen. But, there is a slight increasing trend in the annual frequency of severe cyclones that formed over the BOB during the same period. Also, there is an increasing trend in the intensification of cyclones to severe cyclones.
2. No trend is noticeable in the frequencies of cyclones and severe cyclones that formed over the Arabian Sea during the period 1900–2009.
3. Decreasing trend in frequency of cyclonic disturbances (depression and above) during 1891–2009 and increase in low-pressure areas during 1888–2009 have been observed.

2.2.5.2   DROUGHT

The results of mapping study (IMD 2010) of droughts are summarized as under:

1. In the North-west region of India, the probability of moderate drought varies from 12 to 30% and probability of severe droughts varies from 1 to 20% in most of the parts and about 20–30% in the extreme north-western parts.
2. In West Central India, the probability of moderate drought varies from 5 to 26% and that of severe drought varies from 1 to 8%.
3. In the Peninsular region, the probability of moderate drought varies from 3 to 27%, and that of severe drought varies from 1 to 9% in major parts.
4. In the Central North-east region, the probability of moderate drought varies from 6 to 37% and that of severe drought varies from 1 to 10%.
5. In the North-east region, the probability of moderate drought varies from 1 to 26% and that of severe drought varies from 1 to 3%.
6. In the hilly region, the probability of moderate drought varies from 9 to 31% and that of severe drought varies from 1 to 12% except in Leh and Lahul & Spiti.

In general, it can be concluded that in most parts of India, probabilities of moderate drought are in the range 11 to 20%. Major parts of India show probabilities of severe drought in the range 1 to 5%. In some West Central, Central North-east and North-east regions of India, no severe drought is experienced.

2.3   PROJECTED TREND IN CLIMATE EXTREMES

The projections indicate that above 25°N latitude, the maximum temperature may rise by 2–4°C during the 2050s and in the northern region the increase in maximum temperature may exceed 4°C. The minimum temperature in the 2050s is expected to rise by 4°C all over India, with a further rise in temperature in the southern peninsula. At an all-India level, little change in monsoon rainfall is projected up to the 2050s. There is an overall decrease in the number of rainy days over a major part of the country. This decrease is greater in the western and central parts by more than 15 days. The decreases in the number of rainy days over major parts of the country are also being observed in this study (IMD NCC 3/2010).

Simulations of future weather hardening were carried out using PRE-CIS model for three QUMP (Quantifying Uncertainties in Model Projections) for A1B scenario for the period 1961–1990 (baseline simulation) and for three time slices, 2020s (2011–2040), 2050s (2041–2070) and 2080s (2071–2098). Three PRECIS runs: Q0, Q1 and Q14 were carried out for the period 1961–2098 and were used to generate an ensemble of future climate change scenarios for the Indian region. It appears that there may not be significant decrease in the monsoon rainfall in the future except in some parts of the southern peninsula. Q0, Q1 and Q14 simulations project 16%, 15% and 9% rise, respectively, in the monsoon rainfall towards the end of the twenty-first century.

PRECIS simulations for 2020s, 2050s and 2080s indicate an all-round warming over the Indian subcontinent. Data indicates that Q14 simulations are warmer than the remaining two simulations. The annual mean surface air temperature rise by the end of the century ranges from 3.5°C to 4.3°C (NATCOM, 2012).

KEYWORDS

- **Climate Change**
- **Climatic Extremes**
- **Heat Wave and Cold Wave**
- **Precipitation**
- **Temperature**

REFERENCES

Attri, S. D., & Tgagi, A. (2010). "Climate Profile of India," IMD Met Monograph No. Environment Meteorology, 1/2010.

NATCOM, (2012). India's Second National Communication to the United Nations Frame Work Convention on Climate change, Ministry of Environment and Forests, New Delhi.

Pai, D. S., Thapliyal, V., & Kokate, P. D. (2004). "Decadal Variation in the Heat and Cold Waves over India during 1971–2000," Mausam, 55, 2, 281–292.

Rajeevan, M., Prasad, R. K., & De, U. S. (2000). "Cloud Climatology of the Indian Ocean based on ship observations," Mausam, 52, 527–540.

CHAPTER 3

CLIMATE CHANGE AND RAINFED HORTICULTURE

N. N. REDDY and B. VENKATESWARLU[1]

Principal Scientist and Director Central Research Institute for Dryland Agriculture Santoshnagar, Hyderabad 500 059;
E-mail: director@crida.in, [1]vbandi_1953@yahoo.com

CONTENTS

3.1. INTRODUCTION

Most countries of the world are prone to the effect of climate change, considering the indiscriminate population growth relying on agriculture and extreme stress on natural resources. Research on climate change revealed that increasing temperatures, incosistant rainfall pattern, and severe weather aberratation could drastically affect food production in the near future influencing agricultural GDP. The warming trend in India over the past 100 years was estimated to be 0.60°C. Recently, steady warming trend and unusual weather changes have been reported in India. Moreover, many regions such as Indo-Gangetic plains and coastal areas that are prone to frequent droughts and floods have been identified to be more vulnerable to climate change effects. Climate change impacts like drought, cold, heat, and floods have significant potential to affect the production of food crops like rice, wheat, and pulses and horticultural crops including fruits and vegetables, ultimately leading to shortages and higher prices. The Parliamentary Committee on Agriculture made a strong recommendation to strengthen research on climate change to ensure food and nutritional security in the country.

3.2   GREENHOUSE GAS MITIGATION

*Co-firing tree energy crop biomass at existing coal-fired power plants will achieve the **greatest reduction** of any renewable energy resource option, as:*

- Electricity produced from biomass fuel is carbon cycle neutral, just like the most advanced wind or solar energy technologies.
- Use of tree energy crop biomass also sequesters sizable amounts of carbon (e.g., a sustainable long-term storage through the trees' root system).
- Co-firing energy crop biomass fuel in base load power plants directly displaces %reduces coal use, which can achieve more than two times the Greenhouse gas reduction benefit of placing wind or solar power facilities on an integrated electricity power grid.

Biochar is an organic material produced via the pyrolysis of C-based feedstocks (biomass) and is best described as a 'soil conditioner.' This has 10 to 1000 times longer than residence times of most soil organic matter

(SOM). Biochar addition to soil can provide a potential sink for C. From a climate change mitigation perspective, biochar needs to be considered in combination with other mitigation strategies and cannot be seen as an alternative to reducing emissions of greenhouse gases. From a soil conservation perspective, biochar may be part of a wider package of established strategies.

Home gardening in the tropics imitate nature through integrating trees (fruit-bearing trees and fodder trees) and other perennials with a mixture of other crops. Home gardening is prevalent in many areas in India as mitigation and adaptation strategy to climate change.

3.3   RAINFED HORTICULTURE

In rainfed regions, horticulture helps in diversification, risk mitigation, value addition and enhancing farm income. With the introduction of National Horticulture Mission, there has been a sharp increase in area under horticulture in rainfed regions. Rainfed eco systems with 700 mm or more annual rainfall have relatively better opportunity for horticulture. The production and productivity of rainfed horticulture mainly depends on the quantum and distribution of rainfall. However, year-to-year productivity is also influenced by a number of climate factors like temperature, humidity and extreme weather events like cyclones, floods, hail storms, heat wave and cold wave. Rainfed horticulture also plays a stabilizing role in providing fodder for livestock when practiced as horti-pastoral system. In India, a number of horti-pastoral systems have been recommended which have demonstrated their usefulness both in terms of nutritional security and fodder productivity (Fig. 3.1). Mango, citrus, guava, custard apple, amla and tamarind are some of the important horticulture crops grown in semiarid and dry sub humid regions with supplemental irrigation. These crops are subjected to high climatic variability particularly the rainfall and temperature in a geographical location.

FIGURE 3.1   Horti-pastural system with fodder and sheep components in a mango orchard in Mahaboobnagar district, Andhra Pradesh.

3.4   IMPACT OF CLIMATE CHANGE AND CLIMATE VARIABILITY

The effect of climate is well known in fruit crops and each cultivar requires a specific climatic condition for their growth. Temperature is probably the most important component of the climate, which causes difference in growth and fruit characteristics. Among the climatic factors, temperature is an important parameter, which significantly affects the quality and production. Limitation of rainfall can be overcome by supplemental irrigation but the effect of temperature can be rarely managed. Understanding the influence of climate factors, particularly temperature on horticultural crops is essential for forecasting of pests and diseases (Sankaran et al., 2008).

   Climatic factors influence a number of physiological processes and the ultimate yield in fruit crops (Table 3.1). All fruit crops in general are subjected to weather extremes out of which drought and heat wave are important. Temperature is the most important factor that influences physiological processes and yields.

TABLE 3.1   Climate Change Impacts on Fruits in Rainfed Regions

	Physiological Effects	Yield
Drought	Drying of trees and vegetative growth is affected	Low yield
Flood	Anaerobic condition in root zone and death of plants	Low to no yield
Heat wave	Bud break and fruit growth is affected, e.g., Mango, Citrus	Low to medium yield

TABLE 3.1   *(Continued)*

	Physiological Effects	Yield
Cold wave	Mango malformation	Low yield
	Banana shooting and finger filling is affected	
Cyclone	Plantations are wiped out due to high winds and water logging. Pests and disease incidence increases	No yield
Hail storm	Physical damage to plants and fruit drop., for example, Banana & Mango	Low to no yield

3.5   IMPACTS ON DIFFERENT FRUIT CROPS

3.5.1   MANGO

Mango grows at temperatures as low as 0°C and as higher as 45°C, the ideal temperature range being 24–30°C during the growing season along with high humidity. At temperatures below 10°C and above 42°C, the growth is affected. Entrapment of back-radiated heat from the soil by the clouds is attributed for fruitlet and flower dropping in mango. Young trees and actively growing shoots are likely to be killed at 1°C. Flowers and small fruits are damaged if temperature drops below 4.5°C for a few hours. Higher temperatures during fruit development hasten maturity and tend to improve fruit quality. A cool and dry period, generally during winter, that slows or stops vegetative growth is essential for inducing flowering. Precipitation of 890 to 1015 mm in a year, if well distributed is ideal. Rains at flowering are harmful for mango causing total crop failure sometimes. Various workers have reported washout of pollen grains during flowering from different parts of the world. During rainy days, pollinating insects remain dull and effective pollination cannot occur. After continued moist weather, severe attack of mango hoppers and certain fungi cause heavy shedding of flowers and fruits and often result in crop failure. Pre harvest low light intensity due to cloudy weather reduce the contents of ascorbic acid and sugar in fruits.

Supplemental irrigation should be provided during prolonged droughts. Rainfall during the flowering period adversely affects fruit setting. Fog, cloudy weather at the time of flowering from November to February results in poor fruit setting and favors pest and disease incidence (Bose, et al. 1999). At Central Institute for Subtropical Horticulture, Lucknow, low temperature (4–11.5°C) in January, 2003 accompanied by high humidity (78%) and cloudy weather delayed panicle emergence. Winds may cause great damage to crop by way of fruit shedding, breakage of limbs or even uprooting of the entire plant from winds of high velocity. In places of frequent hail storms, mango cultivation is not feasible (Bhriguvanshi et al 2010). Maximum fruit setting is observed on panicles, which emerge first. Although some fruits set on panicle emerged subsequently but they are dropped due to sudden increase in temperature (9–33°C) and high wind velocity after Holi in March. The characteristic 'Jhumka,' that is, embryo abortion with aborted embryos is noticed in and around Lucknow, Malihabad and Kakori areas of UP causing around 30% losses.

Dashehari, a popular mango variety of Uttar Pradesh produces 40–50 days early crop when grown in Andhra Pradesh. This attribute is gainfully used for early marketing of this variety. Mangoes from coastal Maharashtra, Andhra Pradesh and Tamil Nadu are earliest to come in the market while late crop is produced in north-western India. Kanyakumari and some places in Madurai district of Tamil Nadu produce two crops of mango in a year. In the regions experiencing severe winter, mango malformation is a common problem (Sankaran et al., 2008). Alphonso, a popular variety from Konkan region is of export quality and it produces a physiological disorder called 'spongy tissue' when the soil temperature is high. In the coastal humid regions production of mango grafts through epicotyl grafting is highly successful because of high humidity.

3.5.2 CITRUS

The optimum temperature for citrus growth is 16–20°C. Citrus plants grow well in tropical and subtropical climates, with very high temperatures being detrimental. At least 700 mm of well-distributed rainfall is necessary. It has been observed that fruits grown in moist climate tend to have thinner peel and more juice than those grown in drier climate. Best quality fruits grow in semiarid, subtropical regions with supplemental irrigation. The fruit acquires a good color with warm days and cool nights even while growing on trees compared to humid tropics where even ripe fruits remain green. Soil moisture stress coinciding with maturation improves TSS and reduces acidity. Trees attain larger canopy in tropical climate than in subtropical climate. Mandarins produced in northeast India are flatter and smaller having no neck while fruits produced in central Maharashtra are larger with neck (Bose, et al., 1999). Cool weather in subtropics and moisture stress in tropics are known to condition a major part of the shoots to flower at one time during the year. More pronounced influence of temperature is seen on maturation of fruits.

3.5.3 GUAVA

In areas having distinct winter season, the yield increases and fruit quality improves. It grows best with an annual rainfall around 1000 mm restricted between June and September. However young plants are susceptible to drought and cold. Guava is widely distributed with highest productivity in Gujarat although best quality fruits are produced in Uttar Pradesh (Bose, et al., 1999). The fruit availability is markedly governed by agro climatic conditions (Sankaran et al., 2008). Among the tropical and subtropical fruits, Guava is more tolerant to environmental stresses. The optimum temperature lies between 23°C and 28°C.

3.5.4 CUSTARD APPLE

The temperatures above 40°C and below freezing point and low humidity are harmful for custard apple cultivation. No fruit set occurs in northern India due to high temperature and dry condition. For the growth and fruiting of custard apple, an annual rainfall of 800 mm is considered adequate. Although plants can tolerate extreme climatic conditions, yet for good fruiting, high humidity, occasional rains and warm temperatures are required (Bose et al., 1999).

3.5.5 AONLA

Aonla is a subtropical fruit. However it can grow under tropical conditions. Its cultivation is best suited in places with dry summer and frost-free winter. It is adaptable to a wide range of climatic conditions from seacoast to an altitude of 1800 m (Bose et al., 1999). It has been experienced that in the event of frost, aonla crop is severely affected and in young budded plants it has been observed that frost affect the scion portion and young grafts are dried. It has also been experienced that aonla matures before occurrence of frost and maximum produce is harvested. However fruits of late maturity varieties get affected by frost and low temperature (<2–3°C). The fruits become whitish in color, water starts oozing out of them and subsequently dry and turn black (More and Rakeshbhargava 2010).

3.5.6   TAMARIND

Tamarind grows well in tropical climate where in summers are hot and dry and winters are mild. It is not successfully grown in temperate and arid climate. It is a light demander and cannot be grown in shade. It is drought resistant but susceptible to frost. Ripening of fruits do not commence under cold climate (Singh, 1992).

3.6   ADAPTATION MEASURES FOR CLIMATE CHANGE IN RAINFED HORTICULTURE

3.6.1   MICRO SITE IMPROVEMENT

It is preferable to plant the fruit trees with the onset of monsoon in well prepared and well filled pits at suitable distances. Fruits like mango, tamarind, amla are given wider spacing. The pits should be of one cubic meter dimension filled with equal quantity of tank silt, well decomposed farm yard manure and good soil from the site, 2 kg. Single super phosphate, 2 kg neem cake or castor cake and 50 gm of folidol dust. Before filling the pit, dried leaves may be burnt in the pit to kill any pathogen inoculum in the site. Application of 10 kg bentonite at the bottom of the pit enhances availability of moisture to the root system. This helps in proper establishment of plants through better availability of moisture during the stress periods.

3.6.2 UTILIZATION OF TOLERANT ROOT STOCKS

Some fruit plants are susceptible to moisture stress but with the use of appropriate root stock their cultivation is possible. Root stock must possess deeper root system and have the capacity for water uptake even when little moisture is available. Root stocks play a major role in enhancing water use efficiency, biotic and abiotic stress tolerance and adaptation under climate change, scion compatibility, canopy architecture, fruit quality and absorption. If orchard soils are very heavy and hold too much moisture, a root stock with a shallow root system may be appropriate just as a deep rooted stock will be valuable in a sandy, drought prone soil. In future, root stocks having resistance to multiple abiotic stresses are needed for tropical and subtropical regions. This helps in stable productivity reducing input cost, combating stresses and successful utilization of problem soils.

3.6.3 USE OF FARM YARD MANURE AND POND SEDIMENTS

To improve the water holding capacity of sandy soils, application of FYM and pond sediments (tank silt) is beneficial for improving the water holding capacity.

3.6.4 IN SITU ORCHARD ESTABLISHMENT

It is advisable to sow the seeds or transplant polybag/polytube/root trainer/raised seedlings after pit preparation. After establishment of the plant, grafting/budding with scion shoots obtained from "elite clones" need to be carried out in the same or following year. This practice shall encourage better plant establishment.

3.6.5 MULCHING

Covering of plant basin with organic waste materials, black polythene strips or emulsions is termed as mulching. Mulching reduces the evaporation by cutting radiation falling on the soil surface and thus delays drying and reduces soil thermal regime during daytime. It also reduces the weed population. Continuous use of organic mulches shall be helpful in improving the organic content of the soil and thus the water holding capacity of soil shall also improve. In mango, citrus, aonla and guava, mulching of tree basin with FYM, paddy straw, groundnut husk and locally available materials have shown positive response in maintaining optimum moisture improving physical and chemical properties besides keeping the weeds under check.

3.6.6 IN SITU WATER HARVESTING

During rainy season the excess runoff can be diverted to tree basin in situ or in suitable structures ex situ, which further can be used for life saving irrigation. In sandy soils, in situ conservation while in heavier soils ex situ conservation should be popularized. The in situ harvested water remains stored deep into soil profile escape from evaporative losses and is available during critical period of crop. A number of catchment cropped area ratios and degree of slopes have been standardized for dryland horticulture crops. Percentage slope and catchment area have been advocated for fruits like guava, anola and custard apple. Trenches of the size 5×1×1 ft. dimension made on the upper side of the tree basin across the slope have also proved beneficial in custard apple at CRIDA for increasing the moisture availability and yields (CRIDA, 1988).

3.7 FUTURE RESEARCH STRATEGIES

- Breeding hybrids, varieties, root stocks tolerant to heat and drought stresses.
- Evolve cost effective water and soil management practices that conserve soil and rainwater.
- Studies on impact of climate change on dynamics of pests and diseases of fruit and vegetable crops.

KEYWORDS

- **Adaptation**
- **Climate Change**
- **Drought**
- **Dryland Horticulture**
- **Heat stress**
- **Mitigation**

REFERENCES

Bhriguvanshi, S. R. (2010). Impact of Climate Change on Mango and Tropical Fruits. In Challenges of Climate Change-Indian Horticulture, Singh, H. P., Singh, J. P., & Lal, S. S. (ed.). Westville Publishing House, New Delhi.

Bose, T. K., Mitra, S. K., Farooq, A. A., & Sadhu, M. K. (1999). Tropical Fruits. 1, Nayaprakash, 206, Bidhan Sarani, Calcutta, India.

CRIDA. (1988). Dryland Horticulture, Research Bulletin, 57.

More, T. A., & Bhargava, R. (2010). Impact of Climate Change on Productivity of Fruit Crops in Arid Regions. In: Challenges of Climate Change-Indian Horticulture (ed.) Singh, H. P., Singh, J. P., & Lal, S. S., Westville Publishing House, New Delhi.

Sankaran, M., Jaiprakash Singh, N. P., & Datta, M. (2008). Climate Change and Horticultural Crops. In Climate change and Food security (Datta, M., Singh, N. P., & Daschoudari, D., eds.) New India Publishing House, New Delhi, 243–265.

Singh, A. K., & Reddy, N. N. (2010). Natural Resource Management and the Ways to Overcome Abiotic Stresses in Fruit Crops. Paper Presented at National Seminar on Impact of Climate Change on fruit crops, PAU, Ludhiana. September 4–6, 2010.

Singh, S. P. (1992). Fruit Crops for Wastelands Scientific Publisher, Jodhpur, 215–221.

Srinivasata Rao, N. K., Laxman, R. H., & Bhat, R. M. (2010). Impact of Climate Change on Vegetable Crops. In Challenges of Climate Change-Indian Horticulture, Singh, H. P., Singh, J. P., & Lal, S. S. (Ed.). Westville Publishing House, New Delhi.

CHAPTER 4

CLIMATE RESILIENT HORTICULTURE BASED AGRARIAN LIVELIHOOD IN THE EASTERN REGION

BIKASH DAS[1,3], V. B. PATEL[2], S. KUMAR[1], and B. P. BHATT[1]

[1]ICAR Research Complex for Eastern Region, Bihar Veterinary College Campus, ICAR Parishar, Patna, Bihar, India;
[3]E-mail- bikash41271@yahoo.com
[2]Department of Horticulture, Bihar Agricultural University, Bhagalpur - 813210, Bihar, India.

CONTENTS

4.1   INTRODUCTION

The Eastern region of India occupies about 28% of the National geographical area and inhabited by about 35% of total population, spread over the states of Eastern UP, Bihar, Jharkhand, Orissa, Chhattisgarh, West Bengal, Assam, and plains of North Eastern States. The region is having about 1.24 times population density to that of national average.

The Eastern region can be divided into the following three distinct geographical entities:
1.   Plains of Eastern UP, Bihar, West Bengal, Assam,
2.   Hilly and Plateau regions of Eastern UP, Jharkhand, West Bengal, Orissa, Chhattisgarh, Assam.
3.   Coastal plains of West Bengal and Orissa.

The climate of the eastern region is tropical, hot and humid except in hilly areas with high rainfall. The eastern plains are endowed with rich basic natural resources viz., the most fertile land and abundance of water. The average annual rainfall in this region varies from 1008 mm to 3126 mm. The mean annual rainfall over the geographical domain of eastern UP, Bihar plateau, Bihar plains, Orissa, sub-Himalayan West Bengal, and Gangetic West Bengal is about 1008 mm, 1373 mm, 1203 mm, 1482 mm, 3126 mm and 1425 mm, respectively. Even though the region has rich rain, surface and ground water resources, they are grossly underutilized, with the result, large proportion of the cultivated area does not receive any irrigation water, and the farmers depend on the vagaries of the monsoon for crop production. As per available information only 43% of the net cultivated area in Bihar, 26% in Orissa, 22% in West Bengal and 9% in Jharkhand were irrigated as compared to 95% area irrigated in Punjab. Owing to poor utilization of water resources, the cropping intensity in the region is low, consequently, large tracts of cultivable land during rabi season, remain fallow in spite of the fact that crops in rabi season is relatively disease free and receive plenty of sunshine. Since a sizeable part of the cultivated area in eastern region do not have provision for assured irrigation, even short spell drought adversely affects the stability of agricultural production, thereby resulting in low productivity. Consequently, agriculture & horticulture development is much below its potential. As a result, the employment in agriculture sector is limited and a large proportion of the population still remains below the poverty line and suffers from malnutrition. The region is rich in rain, surface and ground water resources with plenty of sunshine and large tract of fertile land. However, owing to poor utilization of water resources, the cropping intensity in the region is low. Consequently large tracts of cultivable land remain fallow. Horticulture and horticulture based cropping patterns offer best land utilization option for these areas.

4.2 BIODIVERSITY

The region is endowed with an array of native flora and economically important horticultural species. The presence of a large number of ethnic groups and inaccessibility of the area following sustenance of agriculture has contributed to the preservation of horticultural crop diversity. The different agricultural universities and ICAR Institutes have taken up the task to preserve the diversity in their gene banks. The region has rich diversity of many fruit crops like *Citrus, Musa, Mangifera, Artocarpus, Emblica, Carica, Ficus, Grewia, Litchi, Buchanania, Morus, Syzygium, Tamarindus, Annona, Madhuca, Diospyrus, Ziziphus, Garcenia, Dellenia, Borasus, Averroha, Persia, Aegale, Avocado, Actinidia, etc.* The different Universities and ICAR institutes are maintaining the field gene bank of these fruit crops. Among the vegetables, the region has rich diversity of *Benincosa, Momordica, Cucumis, Trichosanthes, Cucurbita, Luffa, Coccinia, Solanum, Capsicum, Vigna, Lablab, Pisum, Colocasia, Amorphallus, Dioscoria, Abulmoschus, Amaranthus, Basella, Beta, Corriandum, Trachyspermum, Rumex, Nigella, Lepidum, Foeniculum, Euryale, Trapa, Bamboosa, Curcuma, Zingiver, Chenopodium.* The high vegetation area of Jharkhand, Chhattisgarh, West Bengal and Assam is endowed with a vast genetic diversity of different medicinal and aromatic plants such as *Solanum khasianum, Cassia angustalia, Diosocorea, Vinca, Bacopa, Withonia, Roulfia, Citrunella, Centela, Ociamum.* Out of the large gene pool few species has been tapped by the local Baidya and are being cultivated at larger scale in the region. These includes *Andrographis paniculate, Aloe vera, Asparagus racemosus, Bacopa monnieri, Chamomil, Nyctanthes arbortristis, Piper longum, Rauwolfia serpentina, Tinospora cordifolia and Withania somnifera.*

4.3   HORTICULTURE SCENARIO OF THE EASTERN INDIA

Horticultural scenario of the eastern region is undergoing sea change in recent years. Fruits and vegetable crops cover more than 70% of the total area under horticulture. The area under fruit crop is 1.20 million ha with an average productivity of 11.07 t/ha compared to national average of 11.3 t/ha. The total fruit production in eastern India is 13.39 million tons, which contribute 18.73% of total fruit production at national level. The last five years has witnessed 28.74% increase in total area under fruits and 39.63% increase total fruit production in eastern India. The increase in productivity in eastern India during the last five years was 10.6% as compared to 9.15% at national level. The higher rate of increase in productivity of fruit crops in eastern India can be attributed to higher rate of increase in the contribution of crops like banana (having high productivity level of 34.31 t/ha) to the total fruit production in the eastern states as compared to national level. At present, the area under vegetable production in eastern India is 3.97 million ha with a total production of 64.26 million tons. The average productivity of vegetable crops in eastern India is 16.17 t/ha as compared to national average of 16.74 t/ha. Eastern India contributes 48.06% of total vegetable production in the country. Last five years have witnessed 5.99% increase in area under vegetable crops and 13.78% increase in production of vegetable crops in eastern India. The productivity increase during the last five years in eastern India is 7.35% as compared to 6.84% at national level. This can be attributed to large-scale introduction of improved varieties of different vegetable crops along with high input based production technologies by the farmers. Share of potato to total vegetable crop production in eastern India is maximum. Hence, higher contribution of crops like potato (having high productivity level of 19.90 t/ha) to total vegetable production in eastern India can also be attributed to higher rate of increase in average productivity of vegetable crops in eastern India over that at national level. With respect to flower crops, the total area in eastern India is 0.0385 million ha with a total production of 0.1245 million tons of loose flowers and 3028.35 million numbers of cut flowers. Eastern India contributes 12.2% and 45.42% of total loose flowers and cut flowers produced in the country. The last five years have witnessed marked increase in the area and production of flowers both in eastern India and at National level. The area under floriculture crops increased by 59.14% in eastern India as compared to 41.31% at national level whereas the production of loose flower and cut flower increased by 129.72% and 175.10%, respectively in eastern India as compared to 53.11% and 117.03%, respectively, at national level. With respect to spices, eastern India contributes 12.60% of total spice production in the country with a total production of 0.48 million tons. The major spices produced in the region are ginger, chili, garlic, turmeric, coriander, fennel, fenugreek, cardamom (large). During the last four years, the spice production in eastern India increased by 17.36% and 5.45% increase in area. At national level the increase in spice production during the last five years was 8.4% with 4.1% increase in area under spices. Among the eastern states, the maximum spice production was recorded in Orissa. However, the increase in area and production during the last four years can mainly be attributed to significant increases in the state of West Bengal (21.35% increase in area and 64.82% increase in production). In case of medicinal plants, the total area and production in eastern India is 0.14 million ha and 0.08 million tons, respectively. Out of all the eastern states, Chhattisgarh contributes 87.07% of total production whereas Uttar Pradesh contributes towards 87.31% of total area under medicinal and aromatic plants. With respect to aromatic plants, the last five years have witnessed 97.8% increase in area and 554% increase in production in the eastern region as compared to 93.6% and 170.5% increase in area and production, respectively at national level. The large-scale increase in production of aromatic plants can be attributed to 605.04% increase in area and 808.94% increase in production in the state of Chhattisgarh. With respect to plantation crops, the total area in eastern region is 0.36 million ha with a total production of 0.73 million tons. Eastern India contributes only 6.12% of total production of plantation crops in India. The major plantation crops of the region are Cashew nut, Coconut and areca nut. Average productivity of eastern India is only 1.9927 t/ha as compared to 3.6538 t/ha at national level. Among the eastern states, the maximum area under plantation crops is in Odisha whereas West Bengal ranks first in terms of total production. Last five years have witnessed 2.1% reduction in production of plantation crops in the eastern region although 14.8% increase in area was recorded. At national level, there has been 2.0% and 5.9% increase in area and production of plantation crops during the last five years, respectively. Declining productivity in all the eastern states and 24.8% reduction in production in the state of Assam can be attributed to reduction in the production level of plantation crops in the eastern region. With respect to infrastructure for processing, a total of 620 number of units for processing of fruits and vegetables are present in eastern India as compared to a total of 5166 numbers at all India level. Among all the eastern states, the maximum number of processing units is present in the state of West Bengal (308) followed by Eastern Uttar Pradesh (170). A total of 1504 number of cold storages are available in eastern India compared to 5381 numbers at national level (27.95%). However, eastern India has 46.05% of total storage capacity at national level. Among the different eastern states, Eastern Uttar Pradesh has the highest number of cold storage units (556) whereas West Bengal has the maximum storage capacity (5,682,000 MT).

4.4 PROJECTED REQUIREMENT OF HORTICULTURAL PRODUCE IN THE EASTERN REGION

Considering 30% post harvest loss, the total availability of fruits in eastern region is 9.3752 million tons. Hence, the per capita availability of fruits in eastern region is about 65.96 g/day, which is much below than the dietary requirement of about 120 g/day (total population of 389.37 million). The present requirement of total fruit production to meet the minimum dietary requirement is 17.0542 million tons. Considering a population of 448.3210 million in eastern India by the year 2026, the total fruit requirement in eastern India will be 24.5455 million tons (with post harvest loss of 20% considering improved scenario of post harvest management of fruits and vegetables). To meet the gap, an additional area of 0.8621 million ha has to be brought under fruit cultivation considering increase in productivity to level of 12.0 t/ha. For this, about 344.84 million numbers of quality saplings will be needed.

At present the total vegetable availability in eastern India considering a post harvest loss of 30% is 44.9887 million tons. Hence, the per capita availability of vegetables in eastern India is 316.56 g, which is above the minimum dietary requirement of 300 g. At national level, there is a deficit of 38.67 million tons of vegetables to meet the minimum dietary requirement of entire population at present level of post harvest losses. The total vegetable requirement to meet the minimum dietary requirement of eastern India by 2026 will be 61.3639 million tons only (with post harvest loss of 20% considering improved scenario of post harvest management of fruits and vegetables). Hence, the region can play a major role to fulfill the vegetable demand at national level. For this, there will be urgent requirement to strengthen infrastructure on handling and transportation of vegetables from this region to other part of the country.

4.5   POSSIBLE IMPACTS OF CLIMATE CHANGE ON HORTICULTURAL CROPS

Horticultural crops play important role in ecologically sensitive hilly, coastal, rainfed and dry land area. Climate change has some impact on horticultural crops due to erratic rainfall, general warming and enhanced biotic and abiotic stresses. The IPCC has projected a temperature increase of 0.5 to 1.2°C rise in temperature by 2020, 0.88 to 3.16°C by 2050 and 1.56 to 5.44°C by 2080, depending on the scenario of future development in the Indian region. Apart from increase in temperature, climate change is projected to cause variations in rainfall, increase the frequency of extreme events such as heat, cold waves, frost days, droughts, floods, etc. Various plant processes like vegetative growth, flowering, fruiting and fruit quality are highly vulnerable to climate changes. The different fruit species with varying levels of adaptabilities to varying climatic conditions will also be influenced by effects of climate change. In case of citrus, under the situation of rise in temperature, it is expected that crops will develop more rapidly and mature earlier. In case of grapes, prolongation of the growing period and weak accumulation of sugars have been recorded due to total warming during the summer period. In mango, increases in temperature during December to March under north Indian conditions have been predicted to impair flower bud differentiation and fruit set. A comparison between rainfall pattern of current and predicted future climate (2030–2.050) indicated that in Dashehari growing areas, precipitation is likely to increase with major changes during May to October; especially rains during fruit bud differentiation are likely to adversely impact the process. Further, frequent rains occurring during May-June also may delay fruit maturity, reduce fruit quality and adversely impact postharvest life. In case of banana, it is expected that the areas favorable for the development of the Black Sigatoka disease will decrease under climate change scenario. In apple, significant decrease in average productivity in Kullu and Simla districts of Himachal Pradesh has been recorded which is attributed mainly to inadequate chilling required for fruit formation and growth. Reduction in cumulative chill units of coldest months might have caused shift of apple belt to higher elevations of Lahaul-Spitti and upper reaches of Kinnaur districts of Himachal Pradesh. With respect to effect of elevated atmospheric CO_2 concentration, photosynthetic down-regulation is expected in different citrus species under long-term CO_2 enrichment. However in grape, rise in CO_2 concentrations may strongly stimulate grapevine production without causing negative repercussions on quality of grape and wine. Similarly in mango and apple, substantial increases in tree biomass have been recorded under elevated CO_2 conditions.

4.6   CONSTRAINTS FOR HORTICULTURE DEVELOPMENT IN THE EASTERN REGION

Broadly the edaphic constraints in the region are listed below.
*   Low size of holding preventing introduction of improved production technologies like farm mechanization. Per capita availability of net cultivated land in the eastern region is lowest in the country. About 68% of the farm holdings are marginal to small ranging from 0.3 to 0.5 ha.
*   Soil acidity problem exist to a considerable extent in Chhattisgarh, Jharkhand and Assam.
*   Coastal soil salinity is also a problem in the states of West Bengal and Orissa.
*   Substantial area in the eastern region is also categorized as water congested or waterlogged where water remains stagnated for a longer durations.

For enumeration of specific constraints in different parts of eastern region, the entire eastern region can be geographically divided into: (1) Hilly, terai and plateau region, (2) Middle and lower gangetic plain and (3) Coastal region.

4.7   THE SPECIFIC CONSTRAINTS OF THE HILLY, TERAI AND PLATEAU REGION

4.7.1   EDAPHOLOGICAL CONSTRAINTS

*   The sloppy land in the region is the major contributing factor towards soil erosion and shallow depth of soil.

- Crusting of surface soil after rainfall.
- Low content of organic matter in the soil is one of the important factors for low productivity of different crops in the region.
- Lower availability of nitrogen due to leaching and lower content of available phosphorus due to fixation of phosphorus forming compounds like iron phosphate and aluminum phosphate.
- Toxicity of iron and aluminum and deficiency of Calcium, Magnesium, sulfur, boron, Zinc and molybdenum.

4.7.2 CLIMATOLOGICAL CONSTRAINTS

- High rainfall combined with high rate of leaching of nutrients has resulted in increase in soil acidity and higher accumulation of aluminum in the soil.
- Less number of sunny days, low temperature and persistent winter prevent successful cultivation of a number of horticultural crops like grapes, pomegranate, etc.
- Hail storm during spring and summer season result in severe damage to both fruit and vegetable crops.

4.7.2.1 BIOTIC CONSTRAINTS

- Hopper, shoot gall psylla, fruit fly, powdery mildew, anthracnose in mango, mite of litchi, wilt of guava, shoot and fruit borer in brinjal, powdery mildew in cucurbits, wilt of solanaceous vegetables, termite attack, rhizome rot in ginger.
- Low rate of success of plant multiplication due to prevailing low humidity and temperature in this zone.

4.7.2.2 DEVELOPMENTAL CONSTRAINTS

- Less than 21% of cultivated area is under irrigation.
- Economic backwardness of farmers.
- Low level of input uses 65.73 kg/ha $N+P_2O_5+K_2O$.
- Inadequate availability of seed and planting material of different horticultural crops.

4.7.2.3 TECHNOLOGICAL INADEQUACY

- Location specific soil and water conservation models.
- Suitable varieties of different fruit crops.
- Package of practice for cultivation of makhana under Chhattisgarh conditions.
- Suitable varieties of tuber crops such as Colacasia, Elephant foot yam, cassava for tribes.
- Diversification options in horticultural crops.
- Amelioration of multinutrient deficiency.
- Package of practices for horticulture based production system particularly high-density planting system under rainfed conditions.
- Package of practices for off-season vegetable cultivation.
- Package for in situ orchard establishment.
- Integrated nutrient management strategies for different horticultural crops.
- Protocol for organic farming in horticultural crops.
- Bio mulches for vegetable crops.
- Collection, characterization and evaluation of lesser-known wild crops.

4.8   THE SPECIFIC CONSTRAINTS OF THE PLAIN REGION OF EASTERN INDIA

4.8.1 EDAPHOLOGICAL CONSTRAINT

- Inundation for considerable period in the low-lying flood prone region.
- Low availability of phosphorus due to formation of calcium phosphate.
- Problem of soil sod city.
- Deficiency of micronutrients like iron, manganese, zinc, copper, boron.

4.8.2 CLIMATOLOGICAL CONSTRAINT

- Hot desiccating Westerly winds during summer season result in heavy damage to fruit crops like litchi.

4.8.3 BIOTIC CONSTRAINT

- Fruit drop in coconut, panama wilt in banana, powdery mildew and hopper in mango, fruit and shoot borer, mite and fruit cracking in litchi.
- Sudden dying of mango plants.

4.8.4 DEVELOPMENTAL CONSTRAINTS

- Small size of holding 77.6% of total number of holdings is marginal in size (<1.0 ha).
- Poor infrastructure development such as rural road and electricity supply.
- Inadequate post harvest handling facilities and poor market support, etc. does not enable growers to fetch remunerative price for their produce.

4.8.5 TECHNOLOGICAL INADEQUACY

- Production maximization through micro irrigation and fertigation.in horticultural crops.
- Use of bio regulator to manipulate the growth and development process in fruit crops to streamline the production.
- Development of INM strategies for different horticultural crops.
- Intercrop prescription for different fruit crops.
- Technology for protected cultivation of high value low volume crops.

4.9 THE SPECIFIC CONSTRAINTS OF THE COASTAL REGION OF EASTERN INDIA

4.9.1 EDAPHOLOGICAL CONSTRAINTS

- Soil salinity resulting in formation of salt encrustation.
- Deficiency of nitrogen and available phosphorus.
- Deficiency of micronutrients like iron, manganese, zinc, copper, boron.
- Impeded drainage.

4.9.2   CLIMATOLOGICAL CONSTRAINTS

- Natural calamities like cyclone and flood.

4.9.3   BIOLOGICAL CONSTRAINT

- Low efficiency of fertilizers.
- Powdery mildew and fruit fly in mango, bunchy top in banana wilt of solanaceous crops, fruit and shoot borer in brinjal, thripe in chilly.
- Germplasm erosion due to cyclone.

4.9.4   DEVELOPMENTAL CONSTRAINTS

- Low level of input use.
- Economic backwardness of farmers.

4.9.5   TECHNOLOGICAL INADEQUACY

- Unavailability of standardized varieties of fruit crops like mango, banana, etc. for local consumption as well as for export.
- Coconut based land use.
- Salt tolerant varieties of different horticultural crops.
- Utilization of water logged areas by introduction of water chestnut.

4.10   TECHNOLOGICAL OPTIONS AVAILABLE FOR CLIMATE RESILIENT HORTICULTURE PRODUCTION SYSTEM IN EASTERN INDIA

4.10.1   SELECTION OF APPROPRIATE CROP BASED ON CLIMATIC AND SOIL SUITABILITY

Selection of appropriate horticultural crops based on land, soil and climatic suitability for maximization of overall increase in production of horticultural crops in the country. To meet the annual growth rate in domestic demands for fruits and vegetables, a partial shift in agricultural policy and practices will be essential. The low and very low land productivity districts offer immense opportunity for production of hardy horticultural crops with minimum infrastructure support. For sustainable development and realization of high productivity levels, crop planning based on agro ecological considerations and efficient use of inputs has been attempted, only with partial success. Higher productivity levels from the eastern region (Bihar, eastern Uttar Pradesh, West Bengal and Assam) experiencing abundant water availability and good soil have not been harnessed fully so far. Under such circumstances, proper crop selection based on agro-climatic suitability and levels of consumption of inputs like fertilizers, pesticides and irrigation water needs careful consideration. Horticultural developmental activities through perennial fruit orcharding have already paid high dividends bringing stability in fragile ecosystems (e.g., apple in Himachal Pradesh, mango and cashew nut in the Western Ghats in Maharashtra and large cardamom in Sikkim). The niche potential of hill and plateau lands, if properly nurtured with scientific horticultural practices can bring fortunes and can convert the nonviable, subsistence farming to economically viable farming. The success story of the Konkan region in the Western Ghats in commercialization of mango, cashew, black pepper, etc. demonstrates the possibility of converting once barren hilly tracts into economically viable regions. Both Himachal apple and Konkan cashew nut have shown that productivity can be achieved on a sustainable basis without adverse effects on the environment. Both in the North Eastern Region and in coastal areas, diversification of cropped area with high-value horticultural crops has been successful. Coastal areas have already been used for plantation crops and spices; further growth is possible in the coastal ecoregions with transfer of relevant technologies.

4.10.2 OPTIMIZATION OF LAND USE IN HORTICULTURE PRODUCTION SYSTEM

Rapid increase in population of the country has necessitated parallel increase in productivity of horticultural crops as well as improving the land use efficiency. Techniques like fruit based multitier cropping systems, high-density orcharding in fruit crops, etc. have proven their effectiveness for improving the land use efficiency through better utilization of land, space and solar radiation. The multitier cropping system comprise of a combination of perennial and annual plant species as different components in the same piece of land arranged in a geometry that facilitates maximum utilization of space in four dimensions (length, width, height and depth) leading to maximum economic productivity of the system. Land use systems based on mixed tree crops have clear advantages over annual cropping systems for the maintenance of soil fertility in the humid tropics. The advantages include permanent soil protection; a more favorite environment for soil biological processes, which affect litter decomposition and soil structural improvement, and more efficient nutrient cycling. Therefore, the integration of deep-rooting trees into the production system increases the nutrient availability in the topsoil, if the quantity of nutrients taken up from below the crop-rooting zone is greater than the quantity stored in the tree biomass and undecomposed tree litter. Research conducted on different fruit based multitier cropping system have indicated a productivity level of 12 to 15 tons per ha against 8–1 0 tons per ha under traditional system of orcharding in the plateau regions of eastern India. Inclusion of different grass species in the horti-pastoral system is a boon for rainfed areas since animal contribute a significant proportion to the total productivity of rainfed farming system. Some of the grass species which can be grown under horti-silvi-pastoral system in rainfed conditions are *Sehima nervosum, Dicanthium annulatum, Heteropogon contortus, Chrysopogon fulvus, Iseilema laxum, Desmodium intertum, Desmodium uncinatum, Glycine wightii, Stylosanthes guinensis, Stylosanthes hamata, Stylosanthes humilis, Stylosanthes scabra, Lablab purpureus, Macroptilium ateropurpureum, Lasiurus sindicus, Cenchrus ciliaris, Cenchrus setigerus, Panicum antidotale, Atylosia* sp.

Tree components in fruit crop based multitier cropping system can be significant sink of atmospheric carbon due to their fast growth and high biomass production. Fruit tree based cropping system can arguably increase the amount of C stored in the lands devoted to agriculture while still allowing for growing of food crops. Study on carbon sequestration potential of an rainfed aonla-green gram based agri-horticulture system have indicated increase in soil organiccarbon from 8.35 t/ha to 13.42 t/ha in a period of 25 years. Similarly carbon sequestered by the system varied from 8.7 to 30.12 t C/ha whereas CO_2 equivalent carbon sequestered in the system varied from 31.89 to 110.47 t CO_2 equivalent C per ha during different years.

4.10.3   GREEN MANURING AND COVER CROPPING

Growing cover crops and green manure crops can be viewed as a type of crop rotation, where adding a nonrevenue generating crop between annual cash crops extends the growing season. Green manures, also referred to as fertility building crops, may be broadly defined as crops grown for the benefit of the soil. The terms cover crop and green manure are frequently used synonymously. They perform many similar functions and many of the same plant species are used as both cover crops and green manure crops. The main difference between the two is that the primary purpose of growing a cover crop is to protect the soil surface from raindrop impact, runoff, and erosion and the primary purpose of a green manure is as a soil-building crop to produce organic material for incorporation into the soil. Although green manuring crops have many roles, they are still often under utilized by today's organic farmers. However, recent emphasis on reducing the environmental impact of all farming systems has led to a growing interest from the conventional sector. Green manuring can bring a number of advantages to the grower like: (1) Adding organic matter to the soil, (2) Increasing biological activity, (3) Improving soil structure, (4) Reduction of erosion, (5) Increasing the supply of nutrients available to plants (particularly by adding nitrogen to the system by fixation), (6) Reducing leaching losses, (7) Suppressing weeds, (8) Reducing pest and disease problems, (9) Providing supplementary animal forage, (10) Drying and warming the soil. N (%), P_2O_5 (%), K_2O (%) composition of some green manuring crops are as follows: Sunhemp – 0.75, 0.12, 0.51; Mung – 0.72, 0.18, 0.53, Cowpeas – 0.71, 0.15, 0.58, Black gram – 0.85, 0.85, 0.53.

4.10.4   USE OF ORGANIC MATTER

The use of organic manures (farmyard manure, compost, green manure, etc.) is the oldest and most widely practiced means of nutrient replenishment in India. Prior to the 1950s, organic manures were almost the only sources of soil and plant nutrition. Owing to a high animal population, farmyard manure is the most common of the organic manures. Cattle account for 90 percent of total manure production. The proportion of cattle manure available for fertilizing purposes decreased from 70 percent in the early 1970s to 30% in the early 1990s. The use of farmyard manure is about 2 tons/ha, which is much below the desired rate of 10 tons/ha. At the present production level, the estimated annual production of crop residues is about 300 million tons. The production of urban compost has been fluctuating around 6–7 million tons and the area under green manuring is about 7 million/ha. Unlike fertilizers, the use of organic material has not increased much in the last two to three decades. The estimated annual available nutrient (NPK) contribution through organic sources is about 5 million tons, which could increase to 7.75 million tons by 2025. Thus, organic manures have a significant role to play in nutrient supply. In addition to improving soil physico-chemical properties, the supplementary and complementary use of organic manure also improves the efficiency of mineral fertilizer use. Combined application of FYM with organic concentrates has been found promising, which avoids bulky application of FYM. Composting manure is becoming popular. In comparison to manure, compost is a more stable product since almost the entire nutrient fractions are in inorganic form. Plants can uptake majority of nutrients in inorganic form.

4.10.5 CONSERVATION OF NATURAL RESOURCES IN HORTICULTURE PRODUCTION SYSTEM

Soil erosion removes topsoil, which is the richest layer of soil in both organic matter and nutrient value. Implementing soil and water conservation measures that restrict runoff and erosion minimizes nutrient losses and sustains soil productivity. Tillage practices and crop residue cover, along with soil topography, structure, and drainage, are major factors in soil erosion. Surface residue limits erosion by reducing detachment of soil particles by wind or raindrop impact and restricting water movement across the soil. Tillage practices manage the amount of crop residue left on the soil surface. Reduced tillage or no-till maximizes residue coverage. Water moves rapidly and is more erosive on steep slopes, so reducing tillage, maintaining surface residue, growing sod crops, and planting on the contour or in contour strips are recommended conservation practices. Using diverse rotations and growing cover crops also can reduce erosion. Soils with stable aggregates are less erodible than those with poor structure, and organic matter (including the activity of living soil organisms and fine roots) helps bind soil particles together into aggregates. Tillage breaks down soil aggregates and also increases soil aeration, which accelerates organic matter decomposition. Well-drained soils with rapid water infiltration are less subject to erosion, because water moves rapidly into and through them and does not build up to the point where it moves across the surface. Crop residues contain appreciable plant nutrients, which contribute to the maintenance of soil productivity when not removed. Conservation tillage may reduce crop yields, which may arise from intensive management, based on varying equipment, and long spectrum of weeds, insects, and disease problems combined with allopathic effects and decreased nutrients' availabilities. However, no-till and minimum tillage systems are more energy efficient than conventional tillage systems. Conservation tillage systems require less total energy to achieve approximately the same crop production levels as conventional tillage systems. No-till and minimum tillage reduce organic C losses from soil and reduce emission by using less fossil fuel.

4.10.6 RECYCLING OF CROP RESIDUE

Crop residues are parts of the plants left in the field after crop have been harvested and threshed or left after pastures are grazed. As two-thirds of all crop residues are used as animal feed, only one-third is available for direct recycling (compost making), which can add 2.5 million tons/year. In India, about 100 million tones of crop residues are available for recycling in agriculture annually. Crop residues are large reservoir of plant nutrient, improve physical and biological properties and protect soil from wind and water erosion. Most common crop residues are from wheat, maize, paddy, sugarcane trash, rice hulls, etc.

4.10.7 EFFICIENT WATER MANAGEMENT IN HORTICULTURE BASED CROPPING SYSTEM

The projected population of India is likely to range between 1.64 and 1.74 billion by 2050, when the world population is likely to reach 9 billion. Already, with a global share of 2.3% land, 4.2% of water, the per capita availability of resources in India is 4–6 times less than the world average. It is estimated that by 2050, about 22% of the geographical area and 17%

of the Indian population would face under water scarcity. Water supply for horticultural crops and livestock will face more intense competition among the multiple users of water, and therefore, investment in microirrigation for horticultural crops will need attention to mitigate the problem. Currently most of the water used to grow crops is derived from rainfed soil moisture, with nonirrigated agriculture accounting for about 60% of production in developing countries.

The strategies for increasing water use efficiency include appropriate integrated land and water management practices like: (1) appropriate scheduling of irrigation, (2) soil–water conservation measures through adequate land preparation for crop establishment, rainwater harvesting and crop residue incorporation, (3) efficient recycling of agricultural wastewater, (4) conservation tillage to increase water infiltration, reduce runoff and improve soil moisture storage, and (5) adequate soil fertility to remove nutrient constraints for maximizing crop production for every drop of water available through either rainfall or irrigation. In addition, novel irrigation technologies such as supplementary irrigation, deficit irrigation, drip irrigation and sprinkler irrigation can improve the water use efficiency of crops. Application of irrigation water at critical stages of the crop is one of the basic components of irrigation scheduling.

4.10.8 MICRO IRRIGATION FOR INCREASING WATER PRODUCTIVITY OF HORTICULTURAL CROPS

Drip irrigation practices are becoming popular now-a-days in horticultural crops due to their beneficial effects on water use efficiency and yield which has been successfully demonstrated in different parts of country. Since drip irrigation is a high capital-intensive technique, there are quite a few doubts among the farmers about its net profitability. No doubt, the initial capital investment for drip irrigation system is high. However, the fixed investment is not very high when compared with the benefits realized from the drip method of irrigation. The cost of the drip system varies depending upon the nature of the crops, its space, amount of water requirements, conditions of the terrains, discharge capacity of the emitters and, of course, distance from the source of water. It is clear from an estimate that the per hectare cost of drip irrigation is substantially high for the narrow spaced crops like vegetables and banana. For instance, per hectare cost comes to about Rs. 33,000 for banana, while the same varies only from about Rs. 9,000 to Rs. 16,000 for other crops like coconut, mango, grapes, pomegranate, etc. Though, the initial cost of drip irrigation system is higher, its relative cost is quite low when compared with the average capital cost in the surface system. For instance, during the Seventh Plan, the investment requirement to create one hectare of irrigation in Major and Medium Irrigation (MMI) was over Rs. 60,000 per hectare at current prices (1995). This when compared with drip irrigation does not amount to a losing proposition. Especially, one must take into account the benefit-cost ratio (B–C ratio) of any project for justification instead of comparing cost of investment alone. Studies conducted on benefit-cost ratio of drip irrigation show that the B–C ratio excluding water is up to 13 and it goes up to 32 when water saving is also accounted for the income side. This is substantially higher than the surface method of irrigation where B-C ratio varies from 1.8 to 3.9. Convinced with the advantage of drip irrigation, farmers have increased their area under drip irrigation significantly in the districts like Jalgaon and Nasik in Maharashtra in the recent years. Large numbers of farmers are willing to take drip irrigation for crops like banana and grapes. A comparative performance of Flood method of irrigation (FMI) and Drip Method of Irrigation (DMI) on different horticultural crops indicated that water saving in DMI over FMI can range between 0% (potato and water melon) to 84% in Okra. Although the saving in water is 0% in potato, drip irrigation has been found to increase yield by 46% and 179% over that of flood irrigation. Use of sub surface drip irrigation with placement of laterals at root zone below the ground level is a recent development and studies conducted on subsurface drip indicated that placement of lateral at a depth of 10 cm have resulted in increase yield of cucurbitaceous vegetables by 11%.

4.10.9 MULCHING FOR SOIL MOISTURE CONSERVATION

Mulching is an effective technique for minimization of evaporative losses. Use of plastic mulches has been successfully demonstrated to increase yield and quality of different horticultural crops. Extensive research with colored mulches has been conducted the experience so far clearly indicate scope of color mulches under Indian conditions. Studies on mulching at various centers of PFDC's all over India has indicated yield increases ranging from 9.2% (plum), 85.6% (tomato) by use of plastic mulches. The most serious problems associated with plasticulture relate to removal of used plastics from the field and its environment friendly disposal. Biodegradable plastics are made with starches from plants such as corn, wheat, cassava and potato. They are broken down by microbes. Biodegradable plastics currently on the market are more expensive than traditional plastics, but the lower price of traditional plastics does not reflect their true environmental cost. Field trials in India and abroad using biodegradable mulch on several crops have shown it performs just as well as regular plastic films and it can be safely plowed into the ground after harvest. In the absence of polythene, use of organic sources like weeds or straw have also been found beneficial in improving soil moisture status.

Application of externally produced organic mulches affects crop growth by influencing weed growth, soil conditions, soil nutrient status, and soil erosion: If weeds are likely to reduce crop yields, organic surface mulches might be able to suppress their growth by hindering and delaying germination (emergence) through shading. On the other hand, mulch may enhance soil conditions for plant growth in general so that also weeds are favored. Limited time for mulch decomposition is reason for the fact that mulch favors crop growth by enhancing soil physical conditions rather than by adding nutrients. Efficacy of organic sources as mulching material in terms of increase in soil water content, organic matter, yield and quality of produce have been successfully demonstrated in many horticultural crops like grape, apple, citrus, litchi, tomato, etc.

4.10.10   RAINWATER HARVESTING FOR LIFESAVING IRRIGATION OF HORTICULTURAL CROPS

In the areas with no alternate source of irrigation, technologies on rainwater harvesting can effectively be used for ensuring lifesaving irrigation particularly for newly planted fruit plants. Low cost rainwater harvesting structures like plastic lined *Dobha* has been developed and successfully demonstrated in farmers' fields for ensuring life saving irrigation to newly planted orchards. The technology has shown promise in farmers fields and hence required to be incorporated in different development schemes on horticulture development particularly under plateau and hilly regions with adequate rainfall.

4.10.11 INCREASING WATER PRODUCTIVITY BY MULTIPLE USE OF WATER

The Multiple uses of water, that is, using the available water sources for more than one uses percentage production system is inevitable to produce more with less water. Multiple use systems, operated for domestic use, crop production, aquaculture, agroforestry and livestock, can improve water productivity and reduce poverty. Under plateau conditions of eastern India, a model has been developed at Ranchi with plastic lined rainwater harvesting pond with litchi based multitier horticultural system planted in the command area (irrigated through gravity fed drip system from the pond) of the pond and production of fish in the pond and vegetables on the bunds measuring 3.0 m width around the ponds, supplementary irrigation to cereal production on a limited area of 50   25 m (0.125 ha) with surplus runoff storage during monsoon season. An analysis of climatic data indicated that after initial irrigation for plant establishment to 60 plants of litchi, 180 plants of guava up to end of June, enough water will be available for vegetable cultivation on about 1000 m^2 for two season (November–March and March–May). About 46.9% of water from 1.8 m deep pond will go as surface evaporation losses. Alternatively, the vegetables for the two seasons can be grown on 1500 m^2 without irrigation to any plant of fruit trees.

4.10.12 ADOPTION OF INTEGRATED FARMING SYSTEM APPROACH FOR IMPROVING SUSTAINABILITY

Traditionally, Indian farmers adopted integrated farming system approach for their livelihood. With industrialization, farmers were forced to become commodity farmers depending on their location, in the catchment of agribased industry like sugar factory, ginning mill, soya processing plant, rice mill, oil mill, dal mill, etc. Though, agro-climatic conditions are primarily responsible for the existence of particular crops and cropping pattern, industrialization, commercialization and mechanization have also played a major role in farmers decision making for growing particular crop or adopting a particular farming system. Integrated approach, however, had several distinct advantages as mentioned below:

  1. Security against complete failure of a system.

2.   Minimization of dependence for external inputs.
3.   Optimum Utilization of farm resource.
4.   Efficient use of natural resources sunlight, water and land, etc.

The major production systems in agriculture sector are as under:

1.   Arable farming system
2.   Horticultural production system
3.   Ago forestry production system
4.   Livestock based farming system
5.   Aqua production system (Fish production)
6.   Pastoral production system

Indian Council of Agricultural Research (ICAR) has delineated the country into 20 agro-ecoregions (AER) and 60 agro-eco subregions (AERS) using criteria of soils, physiography, bio-climate (climate, crops, vegetation) and length of growing period. ICAR under National Agricultural Technology Project (NATP) has identified five major agro-ecosystems, within each of the major ecosystem 2–4 different production systems, which require support in an interdisciplinary mode responsive to farmer's specific needs, were identified.

TABLE 4.1   Different Production Systems within the Five Major Agro-Ecosystems

Major Ecosystems	Production Systems
1. Arid Agro-Ecosystem	(i) Agri-silvi-horti-pastoral production system
	(ii) Livestock and fish production system
2. Coastal Agro-Ecosystem	(i) Fish and livestock production system
	(ii) Agri-horti production system
3. Hill and Mountain	(i) Agri-horti production system
	(ii) Livestock and fish production system
4. Irrigated Agro-Ecosystem	(i) Rice-wheat production system
	(ii) Cotton based production system
	(iii) Sugarcane based production system
	(iv) Dairying and fish production system
5. Rainfed Agro-Ecosystem	(i) Arable farming system
	(ii) Agroforestry production system
	(iii) Livestock based farming system

4.10.13 PROTECTED CULTIVATION

Protected cultivation of horticultural crops is also another approach for increasing land use along with resource use efficiency. The main purpose of protected cultivation is to create a favorable environment for the sustained growth of crop so as to realize its maximum potential even in adverse climatic conditions. Protected cultivation technology offers several advantages to produce vegetables, flowers, hybrid seeds of high quality with minimum risks due to uncertainty of weather and also ensuring efficient and other resources. This becomes relevant to farmers having small land holding who would be benefitted by a technology, which helps them to produce more crops each year from their land, particularly during off season when prices are higher. Thus, protected horticulture has great potential to enhance the income especially of small farmers if appropriate technological interventions are made.

Protected cultivation offers several advantages to produce horticultural crops and their planting material of high quality and yields, through efficient land and resource utilization. Fruits, vegetable and flower crops normally accrue 4 to 8 times higher profits than other crops. This margin of profit can increase manifolds if some of these high value crops are grown under protected conditions, like greenhouses, net houses, tunnels, etc. Protected cultivation has very high entrepreneurial value and profit maximization leading to local employment, social empowerment and respectability of the growers. Environmentally safe methodologies involving IPM tactics reduce the hazards lacing the high value products.

Protected Cultivation technology is a relatively new technology for our country. The total area covered under protected cultivation in our country is approx. 30,000 hectares. There has been a very good development in this area during the last five years. The leading states in the area of protected cultivation are Maharashtra, Karnataka, Himachal Pradesh, North-eastern states, Uttarakhand, Tamilnadu and Punjab. The major crops grown in the protected cultivation are tomato, capsicum, cucumber, melons, rose, gerbera, carnation and chrysanthemum. Nursery grown in the protected cultivation is becoming very popular venture for income and employment generation.

4.11   IMPORTANT PROTECTED CULTIVATION TECHNOLOGIES

- Plug Tray Nursery Raising Technology.
- Protected Cultivation Structures.
- Plastic Low Tunnel Technology for off-season Vegetable Cultivation.
- Protected Cultivation Technology of High Quality Tomato and Cucumber
- Insect Proof Net House Technology
- Walk in Tunnel Technology for Off-season Vegetable Cultivation.
- Low Pressure Drip Fertigation Technology.

A SUCCESS STORY OF HAND HOLDING BETWEEN RESEARCH INSTITUTION AND DEVELOPMENTAL ORGANIZATION FOR ENSURING LIVELIHOOD SECURITY OF FARMING COMMUNITY THROUGH TRANSFER OF HORTICULTURAL TECHNOLOGIES

The eastern plateau and hill agro-climatic zone has a total geographical area of 48.3 million ha spread over the states of Chhattisgarh, Orissa, Jharkhand, Madhya Pradesh Maharashtra and West Bengal in 66 districts. Constraints arising out of inefficient natural resource management coupled with poor soil fertility, low water holding capacity of soil with predominance of traditional varieties result in low productivity of different crops in the region. Deterioration of soil quality due to soil erosion is a major problem of the region. Different crops vary in their response to soil erosion. In general for every cm of top soil removed, the loss of crop yield in case of wheat, barley and mustard have been reported to be to the tune of 16, 14 and 9 kg/ha, respectively. In this context, growing of traditional crops like rainfed paddy under the large tract of uplands is resulting in unprofitability of production system. The area is largely dominated by tribal population. The agrarian scenario of the area is primarily rainfed paddy based. The vast upland generally remains unused or used for growing single crop like rainfed upland paddy or ragi. In the homestead level, some farmers with access to irrigation water, grow vegetables. In medium uplands with availability of residual moisture after harvest of paddy, pulse crop like black gram and pigeon pea are grown. This is resulting in gradual abandonment of farming by the tribal farmers, which is leading to increase in migration rate of tribal farmers.

PRADAN, THE DEVELOPMENTAL PARTNER IN THE PROGRAM

PRADAN (Professional Assistance for Development Action) a NGO is working among the tribal farmers of Eastern plateau and hill region. It was realized by PRADAN that establishment of fruit orchards can be a suitable option for effective utilization of uplands which either remain barren or being used for growing of single crop of rainfed upland paddy. Keeping this in view, it was conceptualized to introduce perennial tree based cropping system in the tribal farming system of eastern plateau and hill agroclimatic zone. The program on horticulture was initiated by PRADAN in the tribal areas of Purulia district of West Bengal and Gumla and West Singhbhum district of Jharkhand. But in some places, poor growth, establishment rate of plants was not encouraging in the initial years. Heavy mortality of young saplings, overcrowded plants and heavy incidence of disease and pest sometimes were the contributing factors for low level of motivation of staffs of PRADAN as well as farmers. Sufficient technical support through any ICAR institutions or SAUs for growing horticultural crops was not available either to the farmers or PRADAN. The sudden mortality, heavy pest infestation resulted in low confidence level of executives for implementation of different horticultural programs. At times, it appeared in the mind of PRADAN staffs that 'whether mango is really a possible options for marginal poor's or not?'

ICAR RCER, RC, RANCHI, THE PARTNER ON RESEARCH SUPPORT

The ICAR Research Complex for Easter Region, Research Centre, Ranchi has been working in the area of research and development in horticulture and agro-forestry for last 30 years. Out of its experience of working the eastern plateau and hill region since last 30 years it has been realized by ICAR RCER, RC, Ranchi that perennial crop based cropping system can provide a sustainable alternative for effective utilization of uplands of the region. Last three decades of research has resulted in development of large number of horticultural technologies for the region like identification of suitable varieties of different fruit crops, fruit based multitier cropping system for uplands, rain water harvesting for establishment of fruit orchards, nutrient management of horticultural crops, canopy management in fruit trees, integrated pest and disease management in fruit crops. Being the only center of ICAR to work on horticulture through natural resource management in the region, the ICAR RCER RC, Ranchi has constraints for reaching out to the large area in eastern plateau and hill region. It was felt by the center to join hands with developmental organizations for widening its outreach program.

COLLABORATION BETWEEN THE PARTNERS

Technical assistance on orchard establishment and management was sought by PRADAN from ICAR RCER, Research Centre, Ranchi during 2005–06. For up gradation and technical fine-tuning in the pomology sector, field visits were arranged by PRADAN for scientists of ICAR RCER to different areas in Gumla district of Jharkhand where horticulture program was being implemented. This helped the scientists to have a firsthand assessment of the progress made by PRADAN in promoting fruit cultivation in Jharkhand. The visits also helped the scientists of ICAR RCER, RC to assess the kind of technological support needed for strengthening the horticulture program of PRADAN. Preliminary study indicated four major concerns of the extension agency, PRADAN.
  1. Sudden mortality of 1 or 2 or 3 year old plants of mango in the field.
  2. Heavy incidence of pest inside the plant canopy.
  3. Plant mortality due to termite attack.
  4. Gummosis of plants.
  Upon discussion among other scientists working on mango, it was suggested by ICAR RCER, RC, R to use microbial antagonists in the soil like *Trichoderma*. It was also decided that *Trichoderma* should be used at the planting in all upcoming plantations. Sudden mortality of plants at the unset of rainy season was also observed. Hence it was also suggested by the scientists to avoid prolonged dryness at the root zone during the summer months. For this the staffs of PRADAN were given exposure to the technology of rainwater harvesting. Last four years of result indi-

cated reduction of plant mortality to the tune of less than 10% by use of *Trichoderma*. Large-scale demonstration of canopy management practices like center opening was made by scientists of ICAR RCER, RC, R in the farmers' fields. Although the farmers initially expressed hesitation in pruning of plants, due to effective group mobilization by PRADAN, the canopy management activity could be demonstrated in 56 farmers' fields of Gumla and Purulia in the presence of the scientist of ICAR RCER, RC. In the next year the farmers as well as executives of PRADAN were highly impressed by the performance of pruned plants. At present, canopy management has become a regular practice in young mango orchards planted by PRADAN. For management of termite, Bi fenthrin 2.5% could be successfully demonstrated for controlling the damage. Nutrient management in young plants, particularly use of micronutrients was stressed upon for minimization of gummosis. Based on the feedback of scientists, three workshops were organized by PRADAN at Purulia and Gumla in the presence of scientists from ICAR RCER, RC, R where the staffs of PRADAN as well as lead farmers were provided on-field training on techniques of canopy management, nutrient and water management, use of Trichoderma for establishment of young mango plants, winter protection of young plants, termite control, identification and management of mango pest. Looking at the positive impact of the on-field training programs, similar workshops as well as training were organized next year with different team members of PRADAN at Khunti, Lohardaga, Hazaribagh, Dumka districts of Jharkhand, Keonjhar district of Orissa and Dhamtari district of Chhattisgarh and monthly calendar of operation for management of the mango orchards were prepared with their participation. The members of different teams were provided exposure visit at ICAR RCER, Research Centre, Ranchi to appraise them of different technologies in horticulture and technology visioning. Regular monitoring was carried out on the part of PRADAN to ensure proper implementation of different recommended practices. For the mitigation of risk for life saving irrigation of mango, two events of capacity were organized at Purulia & Gumla to introduce the water-harvesting model "Jal Kunda" in the farmers field. Initially the model was tested for OFT (on farm technique) in Gumla & Jhalda block of Purulia. After seeing the success of this model, the model has been replicated in three locations of PRADAN (Gumla, Purulia & Dhamtari) with at least 150 farmers. After receiving response from farmers, now the fine-tuning is going on under the collaboration of ICAR, RCER, R and thematic unit of PRADAN on this water-harvesting model (Jal Kunda).

OUTCOME OF COLLABORATION

As a result of these overall activities, the farmers from Gumla district could sell mangoes worth Rs. 4.00 lakh during 2009–2010. In 2010–2011, the sale of mango from the region reached Rs. 14.00 lakh and during 2011–2012, the sale reached Rs. 24.00 lakh. At least five to six times income generation has taken place by the farmers from mango in the different location of PRADAN within 2–3 years. At present, this handholding has increased the confidence level of staffs of PRADAN for establishment of mango based production system in tribal region of Eastern plateau and hill region.

LEARNING FROM THE OVERALL PROCESS

Both of the agencies consider thoughts and realization of farmers as scope of development of own learning cluster. From the very beginning both agencies tried to document the farmers' response in each & every steps of intervention. Intercropping as an effective means of winter protection of young mango plants, use of polythene lined rain water harvesting structures for storage of water from streams to be used during summer season were some of the learning experience for ICAR RCER, RC, Ranchi out of this hand holding. Marked difference in the growth of plants where fruiting was taken up during the third year and during the fourth year was an interesting observation on the part of scientists. Unsuitability of highly degraded soil for growing mango plants was also a learning experience for the scientists of ICAR RCER.

Every observation and learning created a new scope of development. Unsuitability of mango in highly degraded soil now is creating a new scope of thinking of other fruit options for these specific regions. To make the product more marketable, now joint venture of ICAR, RCER, R and PRADAN is being undertaken on value chain exploration and technical up gradation in post harvest management of mango.

Hence, the program was a collaborative effort of ICAR RCER, Research Centre, Ranchi and NGO like PRADAN. The role played by ICAR

RCER was mainly on generation of technology and the implementing agency PRADAN has played the role of extension agent as well as conscious keeper of the dissemination process of technology in the farmers' field. Based on their expertise, the implementing agency (PRADAN) played a major role in mobilization of farmers' groups. This effective synergy resulted in dissemination of technology in to the tribal farming system with proper learning of the farmers. Since the tribal farmers have been trained about the technology based on experiential learning, the technology dissemination has become a sustainable one. The overall activity was a learning experience on the part of ICAR, RCER for successful transfer of technology by establishment of linkage with implementing agencies of developmental programs like NGOs through harnessing synergy between the two partners.

KEYWORDS

- **Biodiversity**
- **Biotic Constraints**
- **Climate Resilient Horticulture**
- **Edaphology**
- **Hilly**
- **Horticultural Crops**
- **Terai and Plateau Regions**

CHAPTER 5

CLIMATE RESILIENT HORTICULTURE FOR NORTH EASTERN INDIA

A. K. JHA[1,3], V. K. VERMA[2], N. A. DESHMUKH[1], R. K. PATEL[1], and S. V. NGACHAN[1]

[1]ICAR Research Complex for NEH Region, Umiam-793103, Meghalaya, India;
[3]E-mail: akjhaicar@yahoo.com

[2]Division of Fruits and Horticultural Technology, Indian Agricultural Research Institute, New Delhi-110012.

CONTENTS

ABSTRACT

North-Eastern states have 773.5 thousand hectare area under horticultural crops (fruits and vegetables), which is around 5.4% of the total area under fruits and vegetables in the country. From this area the region produces about 9908.2 thousand tones of fruits and vegetables with a productivity of 12.81 t/ha against the national productivity of 14.34 t/ha (Anon, 2011). The Region is known as the center of diversity for many horticultural crops like citrus, banana, minor fruits (*sohiong, sophie*, etc.), vegetables (Brinjal, king chili, Indian bean, kakrol and kartoli, chow-chow and dioscoria, etc.), spices (ginger and turmeric) and ornamentals (orchid anthurium and lily, etc.). North eastern region of India is facing the problem of heavy rains during the rainy season, which results into loss of fertile soil and flood in the valleys. Due to steep slope the drought during winter season becomes a major problem during November–March in last few years, the young orchard of *Khasi* mandarin has started declining in entire NEH Region. Erratic rainfall pattern, rise in temperature and change in relative humidity over last 15–20 years have led to emergence of new biotypes in different insect and pest species in most of the fruits and vegetables, besides affecting flowering, fruiting, production and productivity of these crops. Late blight of tomato and potato have become a serious problem due to early heavy rains and low temperature during April–May. Likewise, infestation of cabbage butterfly during March–April is becoming a serious problem in last few years in the region. Infestation of fruit fly in peach, guava, etc. is rising in last couple of years in Barapani conditions and because of this, there has been reduction in marketable yield of peach and guava up to 25–30%. These facts clearly indicate the change of the environment in the region. To mitigate the impact of climate change there is a need to focus on the conservation of genetic resources, development and production of climate resilient cultivars in different horticultural crops, demonstration and adaptation of technologies like, water harvesting, micro irrigation, protected cultivation, etc.

5.1   INTRODUCTION

The North-eastern region of India is one of the global biodiversity hotspots comprising of eight states, viz., Assam, Arunachal Pradesh, Meghalaya,

Manipur, Mizoram, Nagaland, Tripura and Sikkim which falls under the high rainfall zone and the climate ranges from subtropical to alpine. The area is extended from 88–97°E and 22–29, 30°N with a geographical total area of about 255,083 sq. km. It comprises of about 35% area under plain and the remaining 65% area under hills. The altitude varies from the low-lying plains of Bramhaputra (20 m) to around 6000 m in parts of Arunachal Pradesh. The region is characterized by difficult terrain, wide variations in slopes and altitudes, land tenure systems and indigenous cultivation practices. North-Eastern India has a total cropped area of 5.3 million hectares and a population of around 39 million. Out of 4.4 million hectare net sown area, roughly 1.4 million hectare lies in hilly sub region and around 1.3 million hectare area is affected by soil erosion problem. About 0.5 million-hectare area is under shifting cultivation in the North East region.

5.2 HORTICULTURAL SCENARIO

Horticulture is the backbone of agriculture sector of North East India and its potentiality is yet to be harnessed to its full extent. The total area under horticultural crops (fruits and vegetables) in the region is around 773.5 thousand hectare, which is around 5.4% of the total area under fruits and vegetables in the country. From this area the region produces about 9908.2 thousand tones of fruits and vegetables with a productivity of 12.81 t/ha against the national productivity of 14.34 t/ha. The area under various fruits and vegetable crops in the North Eastern region during 2009–2010 is about 367.7 and 405.8 thousand hectares with an annual production of 3404.4 and 6503.8 thousand tones having the productivity of 9.26 and 16.03 t/ha, respectively (Anon. 2011).

The region is considered to be the richest reservoir of genetic variability of large number of horticultural crops. The fruits grown in this region ranged from tropical and subtropical fruits (citrus, banana, papaya, pineapple, guava, and passion fruit) to temperate fruits (apple, pear, peach, plum, strawberry, kiwifruits, and nuts). Apart from these, there are certain underutilized fruit crops which are grown at a large scale in some or other part of the region are *Prunus nepalensis, Elaeagnus latifolia, Phyllanthus acidus, Marica*, etc. The enormous diversity makes the region a gene pool for the varietal improvement of many vegetable and spices crops. The important vegetable and spices crops grown in the regions are tomato, chili,

potato, brinjal, turmeric, ginger, cucurbitaceous crops, cole crops, leguminous and leafy vegetables, tuber and rhizomatous crops and lesser-known
vegetables like tree bean, tree tomato, drumstick, chow-chow, kartoli,
Flemingia vestita, etc. One of the areas in which there is tremendous potential for investment and development is food processing. There is ample
scope for setting up a small and medium scale fruit processing units in
North East India.

5.3   PROJECTED CLIMATE CHANGE AND ITS IMPACT ON HORTICULTURAL CROPS IN 2030S, NEH REGION (INCCA-2010)

5.3.1   CHANGES IN TEMPERATURE

The surface air temperature in this region is projected to rise by 25.8°C
to 26.8°C in the 2030s, with a standard deviation ranging from 0.8°C to
0.9°C. The rise in temperature with respect to the 1970s ranges from 1.8°C
to 2.1°C. Minimum temperatures are likely to rise from 1°C to 2.5°C and
maximum temperatures may rise by 1°C to 3.5°C. The Himalayan region
is likely to remain unaffected but changes in thermal stress are expected
in most parts of North-East region. In this region, the THI is likely to increase between April–October months with THI > 80.

5.3.2   CHANGES IN PRECIPITATION

The mean annual rainfall is projected to vary from a minimum of 940±149
mm to a maximum of 1330±174.5 mm. The increase in the 2030s, with
respect to the 1970s, is of the order of 0.3% to 3%. In the North-Eastern
region, the number of rainy days is likely to decrease by 1–10 days. The
intensity of rainfall in the region is likely to increase by 1–6 mm/day.
The trend in precipitation in the North-Eastern region exhibits considerable spatial variability in water yield in the 2030s but is in line with the
projected patterns of precipitation and evapotranspiration. As compared
to the 1970s, in the 2030s the northern parts of the region show a reduction in precipitation that varies from 3% in the north-western part of the
North-East to about 12% in the north-eastern part. The central portion of
the North-East shows an increase in precipitation varying from 0% to as

much as 25%. However, the majority of the North-Eastern region, except for Mizoram Tripura, Manipur and Assam, shows an increase in ET in the 2030s. As a result, the reduction in water yield for the Arunachal Pradesh is up to about 20%. There is an increase in the water yield to up to about 40% in Assam and Manipur.

5.3.3   CHANGES IN PRECIPITATION AND RAINFALL

The majority of the North-Eastern region, but for some parts of Mizoram, Tripura, Manipur and Assam, shows an increase in ET during the 2030s scenario. It is interesting to note that even those parts of Arunachal Pradesh that were showing a decrease in precipitation, show an increase in ET. This can only be explained by the occurrence of higher temperatures that enhance the evaporative force. However, the increase in ET ranges from a small fraction to about 20%. The reduction in ET in the southern portion is only marginal. The trend in water yield in the North-Eastern region is similar to the precipitation trend. The areas that have shown less increase in precipitation show a correspondingly low water yield. The reduction in water yield in Arunachal Pradesh is up to about 20%. An increase in water yield is seen in Assam and Manipur and the magnitude is up to about 40%.

The North-Eastern region also shows a considerable increase in sediment yield for the majority of the areas which are expected to see increase in precipitation. The increase in the sediment yield in the region is up to 25%. There are a few areas of Arunachal Pradesh that are expected to receive less rainfall and show a reduction in sediment yield of up to 25% under the 2030s scenario.

5.3.4   IMPACTS OF CLIMATE CHANGE ON HORTICULTURE

5.3.4.1   PRODUCTION SYSTEM

Little work has been done with respect to impact of climate change on Horticulture. The simulation analysis indicates that the potato yields are likely to be marginally benefited up to 5% in upper parts of NE region due to climate change influence, but in the central part, the yields are projected to reduce by about 4% while in the southern parts of NE region, the negative impacts will be much higher. Wurr et al. (1996) reported that by increasing the ambient temperature approximately 4°C above normal gave consistently earlier maturity of lettuce, delayed cauliflower curd initiation by up to 49 days and increased the final number of leaves in cauliflower by 36. The crop like lettuce, celery, cauliflower and kiwi grown under high temperatures matured earlier with lower harvest index than the same crops grown under low temperatures. The increased temperatures in Sambalpur, India, have delayed the onset of winter. As a consequence, cauliflower yields have dropped significantly (Pani, 2008). Where growers commonly harvested 1 kg heads, inflorescences are now smaller, weighing 0.25–0.30 kg each. Reductions in yield drive up production costs, an effect also observed in tomato, radish and other native Indian vegetable crops.

Rosenzweig et al. (1996) reported decline in yield of orange cultivar Valancia due to excessive heat during the winter which might be counteract by the rise in the CO_2 in the atmosphere, however fall in potato production with increased CO_2 and changes in planting date were estimated to have minimum compensating impacts on simulated potato yields. Moderately high temperatures do not appear to limit photosynthesis and dry matter production in potato (Midmore and Prange, 1992; Reynolds et al., 1990; Wolf et al., 1990b), but partitioning of photosynthate away from leaves and toward tubers is likely to be decreased under these conditions (Basu and Minhas, 1991; Krauss and Marschner, 1984; Wolf et al., 1990a). Nonetheless, the potato contains considerable genetic variability and plasticity, providing a basis for optimism with respect to adaptation to warmer climates (CIP, 1991; Levy, 1986 a, b; Manrique, 1989; Manrique et al., 1989; Reynolds and Ewing, 1989a, b).

Temperate horticultural crops in Himachal Pradesh were the worst hit during heat wave in summer, 2004. About 50% reduction in green tea leaves was noticed in April 2004 when compared to 2003 and 2005 in Himachal Pradesh due to increase in maximum temperature of the order of 2.1 to 7.9°C in March (Prasada and Rana, 2006). Flowering of apple is advanced by 15 days under high temperature scenario. A large-scale flower drops was seen in April due to acute moisture stress. Heavy rainfall during second forth night of April accomplished by sharp fall in temperature result in poor fruit set. The optimum temperature for fruit blossom and fruit set in apple is 24°C while the region experienced 26°C for 17 days. In contrast, some part of Jammu, Punjab, Harayana, Himachal Pradesh, Bihar, Uttar Pradesh and North Eastern States experienced the unprecedented cold wave during 2002–2003. The crop yield loss varied between 10 and

100% in the horticultural crops like mango, the fruit size and quality were adversely affected in many crops. However, temperate crops like apple, plum and cherry gave higher yield due to extended chilling. The damage was more in low laying area where cold air settled and remains for a longer time in ground (Sharma et al., 2004). Awasthi et al. (1986) have indicated that irregular bearing behavior of Starking Delicious apple is largely influenced by climatic conditions. The rains and hails during flowering adversely affect the fruit-set where as moderate temperature of 20°C with relatively low rains during flowering results in good fruit-set.

Field vegetable production systems contribute to climate change through emission of the greenhouse gases CO_2 and N_2O. Since field vegetables like all other plants fix atmospheric CO_2, the net emission of CO_2 from vegetable production systems will be insignificant, especially when high-yielding varieties are used, crop residues are not removed from the field, inorganic fertilizers are replaced by organic manures and reduced tillage is applied. N_2O emission can be reduced by increasing the efficiency of N use by the vegetables (Neeteshan et al., 2010).

5.3.4.2   QUALITY ASPECTS

Production and quality of fresh fruit and vegetable crops can be directly and indirectly affected by high temperatures and exposure to elevated levels of carbon dioxide and ozone. Chan et al. (1981) and Picton and Grierson (1988) observed that high temperature stresses inhibited ethylene production and cell wall softening in papaya and tomato fruits. On the other hand, cucumber fruits showed increased tolerance to high temperature stress (32.5°C) with no change in in vitro ACC oxidase activity (Chan and Linse, 1989). Tip burn in lettuce is a disorder normally associated with high temperatures in the field, which can cause soft rot development during postharvest. Black heart in potato occurs during excessively hot weather in saturated soil. The symptoms usually occur in the center of the tuber as dark-gray to black discoloration. The translucent fruit flesh in pineapple appears due to high temperature. Exposure of tomato fruits to temperatures above 30°C suppresses many of the parameters of normal fruit ripening including color development, softening, respiration rate and ethylene production (Buescher, 1979; Hicks et al., 1983).

Hogy and Fangmeier (2009) studied the effect of high CO_2 concentrations on the physical and chemical quality of potato tubers. They observed that increase in atmospheric CO_2 (50% higher) increased tuber malformation in approximately 63%, resulting in poor processing quality, and a trend towards lower tuber greening (around 12%). Higher CO_2 levels (550 μmol CO_2/mol) increased the occurrence of common scab by 134% but no significant changes in dry matter content, specific gravity and underwater weight were observed. Prolonged exposure to CO_2 concentrations could induce higher incidences of tuber malformation, increased levels of sugars in potato and diminished protein and mineral contents, leading to loss of nutritional and sensory quality.

Exposure of crops with ozone changes the carbon transport system in the underground storage organs (e.g., roots, tubers, bulbs). Normally carbon gets accumulated in the form of starch and sugars, both of which are important quality parameters in both fresh and processed crops. If carbon transport to these structures is restricted, there is great potential to lower quality in such important crops like potatoes, sweet potatoes, carrots, onions and garlic (Felzer et al., 2007). High concentrations of atmospheric ozone can potentially cause reduction in the photosynthetic process, growth and biomass accumulation. Ozone-enriched atmospheres increased vitamin C content and decreased emissions of volatile esters on strawberries. Tomatoes exposed to ozone concentrations ranging from 0.005 to 1.0 μmol/mol had a transient increase in β-carotene, lutein and lycopene contents (Moretti et al., 2010).

Due to heavy rain the loss of essential nutrient like Ca and Mg is a common problem with the toxicity of heavy metals. Recently the emphasis has been placed on N and Ca, the nutrients most closely associated with fruit quality. Possible responses include precision horticulture with more targeted nutrient management. This in turn will require improved understanding of application efficiencies and the timing and magnitude of nutrient demand in order to synchronize fertilization more closely with plant requirements (Neilson et al., 2010). Higher concentrations of atmospheric carbon dioxide (CO_2), for example, may actually benefit potatoes as increased CO_2 stimulates the development of underground biomass in potato plants, with tuber weight and number both increasing significantly. Higher levels of atmospheric ozone (O_3) also seem to benefit the crop, resulting in more of the antioxidant ascorbic acid in tubers.

5.3.4.3 PEST AND DISEASES

Higher rainfall and humidity is congenial for increasing disease intensity in potato, such as late blight (*Phytophthora infestans*), especially when combined with longer growing seasons. Bacterial wilt may also increase as the climate becomes warmer and wetter and potato pests, including disease-carrying aphids, will survive at higher altitudes. With the rise in temperature and humidity the new biotype of diseases will emerge which is tolerant to varied climatic condition. Soil born disease like bacterial wilt (*Ralstonia solanacarum*) will be more problematic which can grow up to 40°C (Masao Gato, 1992).

Jeong et al. (2010) in his investigation under future climate change condition treatments in growth chambers experiment on four major diseases of chili pepper including two fungal diseases, anthracnose (*Colletotrichum acutatum*) and Phytophthora blight (*Phytophthora capsici*), and two bacterial diseases, bacterial wilt (*Ralstonia solanacearum*) and bacterial spot (*Xanthomonas campestris* pv. *vesicatoria*). Treatments with elevated CO_2 and temperature were maintained at 720 ppm ± 20 ppm CO_2 and 30°C ± 0.5°C, whereas ambient conditions were maintained at 420 ppm ± 20 ppm CO_2 and 25°C ± 0.5°C. Pepper seedlings or fruits were infected with each pathogen, and then the disease progress was evaluated in the growth chambers. According to paired t-test analyzes, incidence of bacterial wilt and spot were increased on pepper by 24% and 25%, respectively. Intensity of Anthracnose got decreased and while intensity of Phytophthora blight slightly increased, but the fungal diseases were not statistically significant, suggesting that bacterial diseases on chili pepper will likely to be more serious in the future.

5.4 OTHER OBSERVATIONS

- The heavy rainfall from April to October causes heavy soil erosion and nutrient loss as well as results in manifestation of different types of insect-pest and diseases and as a result the crop productivity is reduced to a great extent.
- The long dry spell from November to March results in moisture stress in many of the horticultural crops and because of this the young orchard of *Khasi* mandarin has started declining in entire NEH Region.

- During last 3 to 4 years the incidence of fruit flies is becoming a serious concern in peach and guava. Control of fruit flies by appropriate means is a major challenge among the Entomologists.
- In underutilized crops powdery mildew and blight are new record in *Prunus nepalensis,* which was never noticed until 2–3 years back.
- Cabbage butterfly is another serious problem in cole crops during February–April, which was not recorded before.
- The late blight becomes a serious problem in tomato and potato during summer season (April–June) due low temp and heavy rains, which provide congenial environment for disease development.

5.5   MITIGATION MEASURES

5.5.1   WATER MANAGEMENT

The scarcity of water (during month of November to March) is one of the most limiting factors for the crop production in the region. During this period, to get better crops the farmers can adopt following measures.

5.5.2   DEVELOPMENT OF WATER RESOURCES

5.5.2.1   RAINWATER HARVESTING

Rainwater is the most limiting production factor for horticulture. Therefore, conservation and management of rainwater in the form of in situ soil moisture, ex situ rainwater harvesting as surface and subsurface storages and its efficient use assume highest importance for stabilizing horticulture production and enhancing. Water harvesting and integrated watershed management have shown promising response and offer great potential for promoting horticulture. The ex situ rainwater harvesting technologies include dug out pond, storage tank, nala bunding, gully control structure/check dams/bandharas (weirs), water harvesting dams, percolation tanks/ponds, subsurface dams/barriers, etc.

5.5.2.2   *CONSERVATION OF SOIL MOISTURE*

In order to minimize the losses of water due evapotranspiration, following practices along with cultural operations like hoeing and weeding can be adopted.

5.5.2.3   *IN SITU SOIL MOISTURE CONSERVATION*

It can be achieved by increasing infiltration with the profile modifications, mulching, keeping soil surface rough, contour trenching, interterrace land treatment, etc., different in situ rainwater conservation measures are trenching, catch pit, V-ditch, jalkund, micro catchment, pits with crescent bunds, etc.

5.5.2.4   *MULCHING*

Mulching reduces evaporation of soil moisture, weed growth and prevent drastic fluctuation in soil temperature and formation of soil crusts. Organic mulches like bark, cocoa shells, compost, leaves, newspaper, peat moss, straw, sawdust and wood chips can be incorporated into soil after every crop. Now 100 micron thick polythene mulch film is used for tomato and capsicum of these mulches, 5 cm diameter holes are made at desired spacing for planting crops. Mulching with drip further enhanced the crop yield to the tune of 10–20% and controlled weed up to 30–90% (Anon., 2001).

5.5.3   *EFFICIENT UTILIZATION OF WATER*

5.5.3.1   *DRIP IRRIGATION*

Drip irrigation is one of the most efficient irrigation methods for orchard and plantation crop where it saves 30–70% irrigation water and increase yield by 25–80%. Drip irrigation and mulching are helpful to combat water scarcity. Micro irrigation is the system that provides water in and around root zone of plant with help of emitters. A typical micro irrigation system has dripper, distribution line and control head system. By a micro irrigation, we can maintain optimum soil moisture in root zone of crop

(Bahadur and Singh 2001). The losses of water can be minimized by using of sprinkler and drip irrigation system. Under fan pad polyhouse condition the crop yield can be improved by 65.40% in tomato and 67.70% in capsicum through drip system of irrigation along with water saving of 43.23% and 42%, respectively (Singh et al., 2010).

5.5.3.2   FERTIGATION

The application of fertilizer through irrigation system (fertigation) has become a common practice in modern irrigated agriculture. Increase yield, improvement in quality of product, irrigation and fertilizer use efficiencies and protection of soil environment are some of the main characteristics of this method, which made it popular throughout the world. The reason why fertigation become the state of art in vegetable cultivation is that nutrient can be applied to plants in the correct dosage and at the time appropriate for the specific stage of plant growth (Clark et.al., 1990).

5.5.4   SOIL MANAGEMENT

Due to excess rains during the month of June–September the soil nutrient like, Ca, Mg, K and N getting lost from the soil by the process of leaching and developed into laterite soil with rich in metal ions like, Fe and Al, etc. by the process of cauterization. Most of the soils of northeastern are acidic in nature with pH range of 4.0–5.5 due to the above factors and process most of the crops face problems of deficiency of leached nutrient and toxicity of metal ions as well. In order to get better produce farmers can adopt following management practices as given below.

5.5.4.1   USE OF SOIL AMENDMENTS

To neutralize the effect of soil acidity and toxicity of metal ions on the crops, farmers can use the soil amendments like, lime.

5.5.4.2   APPLICATION OF MANURE AND FERTILIZERS

The application of fertilizer varied from crop to crop and soil pH. Hence, most of the soils of the region are acidic in nature so farmers need to avoid the application of acidic fertilizer. The application of FYM @ 15–20 tons per hectare is essential for better yield in vegetable crops.

5.5.5   CROP BASED MANAGEMENT STRATEGIES

5.5.5.1   TURMERIC

The institute has developed a turmeric variety Megha Turmeric-1 through clonal selection from the local genotype Lakadong, which is tolerant to leaf blotch and leaf spot. It mature in 300–315 days having an average yield of 27–30 t/ha. It contains 6.8% curcumin and dry matter content of 20–22%.

5.5.5.2   TOMATO

Megha Tomato-3 has been developed by the institute, which is tolerant to cold and bacterial wilt disease. The Fruits are round, smooth with uniform color development. The average yield of the variety is 45–55 t/ha with a TSS of 4.5–5.5 B and shelf life of 15–18 days at ambient condition.

5.5.6   OFF-SEASON PRODUCTION OF VEGETABLES

The experiment was undertaken to standardize a crop cycle for year round production of vegetables under low cost polyhouse condition. The experimental results revealed that crop cycle involving King Chilli (October–April) → Cucurbits + Okra (May–September) → Capsicum + Beans (October–January) → Lettuce + Chilli (February–May) → Cowpea + Palak + Tomato (June–September) is the best combination for off-season production of vegetables under polyhouse in Manipur. After second year, green manure crops should be grown for one season to enhance the soil fertility status. In this cycle, multiple crops have been selected to minimize the risk of crop loss. Solanaceous crops were not grown on the same area under polyhouse. If the night temperature in winter season goes below 4–5°C, charcoal may be burnt inside the polyhouse particularly at night period to maintain the inside temperature.

5.5.7 ROUND THE YEAR PRODUCTION OF VEGETABLES UNDER LOW-COST POLYHOUSE

Despite congenial temperature during rainy season, tomato cannot be grown under open condition due to heavy disease and pest infestation. However, tomato can be grown successfully under low cost polyhouse during May–September with proper crop management. Cucumber is identified as the best crop (immediately after tomato) during September–January. Capsicum followed by cucumber can be grown during January–May. Liming is essential for maintaining the optimum soil reaction (pH 5.5–6.0).

   A. Tomato: The high yielding and determinate type cultivars, Megha Tomato-1, Megha Tomato-2 and Megha Tomato-3 has been identified as most suitable cultivar for low cost polyhouse.
   B. Cucumber: Japanese Long Green and Kalyanpur Hara can be grown under low cost polyhouse.
   C. Capsicum: California Wonder, Bharat and Indira can be grown under low cost polyhouse.

5.5.8 CANOPY MANAGEMENT OF PEACH FOR EARLINESS

Among the low chilling varieties, Flordasun, Shan-e-Punjab and Partap (TA-170) found suitable for mid-hill conditions of north-east. Under mid-hills of Meghalaya, the ripening period coincide with early rains in the month of May. High infestation of fruit flies and brown rot during that time mars the eating quality, marketability and storage life of fruits. Ten years old grafted Peach cv. Partap (TA-170) was pruned on 30th October, 15th November and 30th November (normal pruning time) with 50% and 70% intensity. It was noticed that earliest fruits were harvested from 10 to 20 April (15 days earlier) in 30th October with 50% pruning severity as compared to normal pruning date where harvesting lasted up to 15 May. The yield, fruit size and quality were at par with normal pruning schedule.

Therefore, last week of October may be recommended for pruning of low chilling peach for early harvest of quality fruit.

5.5.9   OFF-SEASON PRODUCTION OF STRAWBERRY

Strawberry cv. Ofra can be produced 30–35 days earlier than normal period, when planted in low tunnels of 50% shade net (4.0 m × 0.90 m × 0.75 cm.) in the month of July or August and the period of fruit availability may be extended to 47 days from normal when planted in the month of November under UVS polythene tunnels (200 gauge). During study period, normal fruiting period under open condition was 18 Janunary to 15 March.

5.6   CONCLUSION

Climate change scenario becomes a serious concern for the global agriculture. North Eastern region of India is one of the major bio diversity hot spot and origin of many important species in which changes in weather parameters have profound impact on the livelihood of people and their ecosystem as whole. These problems can be mitigated by the selection of fruit crops and cultivar, like low chilling peach cultivar such as Shan-e-Punjab and Flordsun, Citrus species like Khasi mandarin, Assam lemon, passion fruit and strawberry, etc. in mid-hills and apple, pear and plum at higher altitude, and in plains like Guava, banana and papaya, etc. Apart from these, there are certain underutilized fruit crops which can be grown at a large scale in some or other part of the regions like, *Prunus nepalensis, Elaeagnus latifolia, Phyllanthus acidus, Marica*, etc. The important vegetable and spices crops which having potential, like tomato cold and drought tolerant cultivar Megha Tomato-3, Anthracnose tolerant chili cultivar Kasi Anmol, cucurbitaceous crops (Chow-chow, bottle gourd, ash gourd and pumpkin) potato (Kufri Megha) cole crops, leguminous (French bean, dolichos bean and cow pea) and leafy vegetables, tuber and rhizomatous crops and lesser-known vegetables like tree bean, tree tomato, drumstick, kartoli, *Flemingia vestita*, etc., and spices like turmeric and ginger to grow under varied weather condition. Protected cultivation of high value vegetable crops like, tomato, capsicum and cucumber provide means for round year production with the protection against low temperature and heavy rains. The drip system is most suitable system for efficient utilization of water during the period of moisture stress and especially for the horticultural crops. Fruit and vegetable crops also having potential in the sequestration of CO_2 from the atmosphere, which alone contributes about 60% of global warming. The appropriate of tactics to mitigate the impact of climate changes is necessary to go for conservation of our bio-diversity and provision of feed to our fast growing population. For this diversification towards horticulture crops and blending it with others like agronomic crops, animal husbandry, fish, pig, etc. together is necessary to get sustainable family income and conserve natural resources.

KEYWORDS

- **Crop Based Management Strategies**
- **Efficient Water Use**
- **Horticultural Scenario**
- **In-situ Soil Moisture Conservation**
- **Pest and Diseases**
- **Precipitation and Rainfall**
- **Quality Issues**

REFERENCES

Anonymous. (2001). Annual Progress Report, National Committee on Plasticulture Application in Horticulture, Department of Agri. and Co-operation, Min. of Ag., GOI, New Delhi, 98.

Anonymous. (2011). NHB Data Base, New Delhi.

Awasthi, R. P., Thakur, G. C., & Chauhan, P. S. (1986). Role of Meteorological Components on Flowering Time, Fruit-set and Yield of Starking Delicious Apple. Advances in Research on Temperate Fruits (Chadha, T. R., Bhutani, V. P., &. Kaul, J. L.), Dr. Parmar, Y. S., Uni. of Hort. and Forestry, Nauni-Solan, HP, 11–18.

Bahadur, A., & Singh, K. P. (2001). Micro-irrigation. A Vision Option for Water Economy, Souvenir, Indian Institute of Vegetable Research, Varanasi, 29–31.

Basu, P. S. & Minhas, J. S. (1991). Heat Tolerance and Assimilate Transport in Different Potato Genotypes, *J. Exp. Bot.,* 42, 861–866.

Buescher, R. W. (1979). Influence of High Temperature on Physiological and Compositional Characteristics Tomato Fruits. *LWT Food Sci. Technol.,* 12, 162–164.

Chan, H. T. & Linse, E. (1989). Conditioning Cucumbers to Increase Heat Resistance in the EFE System, *J. Food Sci.,* 54, 1375–1376.

Chan, H. T., Tam, S. Y. T., & Seo, S. T. (1981). Papaya Polygalacturonase and its role in Thermally Injured Ripening Fruit. *J. Food Sci.,* 46, 190–197.

Clark, G. A., Smajstria, A. G., Haman, D. Z., & Zazueta, F. S. (1990). Injection of Chemical into Irrigation System. Rates, Volumes and Injection Period, *Fla. Coop.Ext. Bull.* 250 12.

Felzer, B. S., Cronin, T., Reilly, J. M., Melillo, J. M., & Wang, X. (2007). Impacts of Ozone on Trees and Crops. *Compters Rendus Geoscience,* 339, 784–798.

Hicks, J. R., Manzano-Mendez, J., & Masters, J. F. (1983). Temperature Extremes and Tomato Ripening, *Proc. Fourth Tomato Quality Workshop,* 4, 38–51.

Hogy, P., & Fangmeier, A. (2009). Atmospheric CO_2 Enrichment Affects Potatoes, 2 Tuber Quality Traits, *Eur. J. Agron.,* 30, 85–94.

IPCC. (2007). Climate Change. In Solomon, S., Qin, D., Manning, M., Chen, Z., Marquis, M., Averyt, K. B., Tignor, M., Miller, H. L. (Eds.). *The Physical Science Basis. Contribution of Working Group I to the Fourth Assessment Report of The Intergovernmental Panel on Climate Change*: Cambridge University Press. 996.

Indian Network for Climate Change Assessment. (2010). Climate Change and India a Sectoral and Regional Analysis For 2030.

Jeong-Wook Shin, & Sung-Chul Yun. (2010). Elevated CO_2 and temperature effects on the incidence of four major chili pepper diseases. *Plant Pathol. J.,* 26(2), 178–184.

Keeling, C. D., Whorf, T. P., Wahlen, M., & Van der Plicht, J. (1995). Interannual Extremes in the Rate of Rise of Atmospheric Carbon Dioxide Since (1980). *Nature,* 375, 666–670.

Krauss, A., & Marschner, H. (1984). Growth Rate and Carbohydrate Metabolism of Potato Tubers Exposed to High Temperatures, *Potato Res.,* 27, 297–303.

Levy, D. (1986a). Tuber Yield and Tuber Quality of Several Potato Cultivars as Affected by Seasonal High Temperatures and by Water Deficit in a Semi-Arid Environment, *Potato Res.,* 29, 95–107.

Levy, D., Kastenbaum, E., & Itzhak, Y. (1991). Evaluation of Parents and Selection for Heat Tolerance in Early Generations of a Potato (*Solanum tuberosum* L.) Breeding Program. *Theor. Appl. Genet.,* 82, 130–136.

Levy, D. (1986b). Genotypic Variation in the Response of Potatoes (*Solanum tuberosum* L.) to High Ambient Temperatures and Water Deficit, *Field Crops Res.,* 15, 85–96.

Manrique, L. A. (1989). Analysis of Growth of Kennebec Potatoes Grown under Differing Environments in the Tropics, *Am. Potato J.,* 66, 277–291.

Midmore, D. J., & Prange, R. K. (1992). Growth Responses of Two Solanum Species to Contrasting Temperatures and Irradiance Levels: Relations to Photosynthesis Dark Respiration and Chlorophyll Fluorescence, *Ann. Bot.,* 69, 13–20.

Moretti, C. L., Mattos, L. M., Calbo, A. G., & Sargent, S. A. (2010). Climate Changes and Potential Impacts on Postharvest Quality of Fruit and Vegetable Crops, A review *Food Res. Int.,* 43, 1824–1832.

Morison, J. I. L., & Lawlor, D. W. (1999). Interaction Between increasing CO_2 Concentration and Temperature on Plant Growth. *Plant Cell Environ.,* 22, 659–682.

Pani, R. K. (2008). Climate change hits vegetable crop. http://www.expressbuzz.com

Picton, S., & Grierson, D. (1988). Inhibition of Expression of Tomato Ripening Genes at High Temperature, *Plant, Cell Environ.,* 11, 265–272.

Prasada, R., & Rana, R. (2006). A Study on Maximum Temperature During March (2004) and its Impact on Rabi Crops in Himachal Pradesh. *J. Agrometeorl.,* 8(1), 91–99.

Reynolds, M. P. & Ewing, E. E. (1989a). Heat Tolerance in Tuber Bearing Solanum Species, a Protocol for Screening, *Am. Potato J.* 66, 63–74.

Reynolds, M. P., & Ewing, E. E. (1989b). Effects of High Air and Soil Temperature Stress on Growth and Tuberization in *Solanum tuberosum*, *Ann. Bot.* 64, 241–247.

Sharma, J. S., Singh, G., & Ramakrishna, Y. S. (2004). Cold Wave During (2002–03) Over North India and its Effect on Crop, *The Hindu.* 6 January 10.

Shine, K. P., Derwent, R. G., Wuebbles, D. J., and Morcrette, J. J. (1990). Radiative forcing of climate, *Climate Change. The IPCC Scientific Assessment*, J. T. Houghton, G. J. Jenkins, and J. J. Ephraums, Eds., *Cambridge University Press*, 41–68.

Singh, A. K., Chandra, P., & Srivastav, R. (2010). Response of Micro-irrigation and Fertigation on High Value Vegetable Crops Under Controlled Conditions, *Indian J. Hort*, 67(3), 418–420.

Wigley, T. M. L., & Raper, S. C. B. (1992). Implications for Climate and Sea Level of Revised IPCC Emissions Scenarios, *Nature*, 357, 293–300.

Wolf, S., Olesinski, A. A., Rudich, J., & Marani, A. (1990b). Effect of High Temperature on Photosynthesis in Potatoes, *Ann. Bot.*, 65, 179–185.

Wolf, S., Marani, A., & Rudich, J. (1990a). Effects of Temperature and Photoperiod on Assimilate Partitioning in Potato Plants, *Ann. Bot.*, 66, 513–520.

Wurr, D. C. E., & Fellows, J. R. (1991). The Influence of Solar Radiation and Temperature on the Head Weight of Crisp lettuce. *J. Hortic. Sci.*, 66, 183–190.

Wurr, D. C. E., Fellows, J. R., Phelps, K., & Reader, R. J. (1993). Vernalization in Summer/ Autumn Cauliflower (*Brassica oleracea* var. *botrytis* L.). *J. Exp. Bot.*, 44, 1507–1514.

Wurr, D. C. E., Fellows, J. R., & Hambidge, A. J. (1995). The Potential Impact of Global Warming on Summer/Autumn Cauliflower Growth in the UK. *Agric. For. Meteorol*, 72, 181–193

Wurr, D. C. E., Fellows, J. R., & Phelps, K. (1996). Investigating Trends in Vegetable Crop Response to Increasing Temperature Associated with Climate Change. *Sci. Hortic.* 3–4, 255–263.

CHAPTER 6

CLIMATIC ISSUES AFFECTING SUSTAINABLE LITCHI (*LITCHI CHINENSIS* SONN.) PRODUCTION IN EASTERN INDIA

RAJESH KUMAR

Principal Scientist (Hort.) National Research Centre for Litchi, Muzaffar-pur-842 002 (Bihar), India; E-mail: rajeshkr_5@yahoo.com

CONTENTS

ABSTRACT

The study pertaining to climatic issues for sustainable litchi production indicated that the climate is changing and influencing the litchi enterprise in eastern India. The litchi production system is having high reliance on natural ecosystem services like soil fertility, pruning and training operation for canopy development, insect (honeybee) pollination for high fruit set, maintenance of soil moisture levels through rainfall and assured irrigation during its vegetative and reproductive phases. Therefore long-term changes to rainfall and its pattern, temperature (variation), soil moisture and air humidity regimes will greatly influence litchi production and will impact on fruit yield and quality. For sustainable litchi production, understanding of ideal climatic conditions is very much required for doing the relevance in the present perspective. Higher temperature, good sunshine and enough soil moisture in the rhizosphere improves fruit size and quality in this region. The aberrations in weather like prolonged cloudy weather and rains during the full bloom interfere with normal cross-pollination (as diminished activities of pollinators mainly honey bees) and fruit set in litchi, sometimes may cause total crop failure. Till date information on climatic issues concerning litchi production is entirely lacking, which needs to be generated urgently on the basis of earlier (but related) studies made, creating greater awareness, development of simulation models. Timely adoption of technologies needs to be given paramount importance for this fruit crop production. Hence, sustainable production is not only a matter of changing current management practices, changing varieties or changing cropping zones, but strategic adaptation and measures in the changed climatic perspective starts from creating awareness and building resilient systems. Current litchi production in this country and more particularly in eastern region needs to be both environmentally sustainable and economically viable for livelihood security and prosperity in the region.

6.1 INTRODUCTION

The litchi (*Litchi chinensis* Sonn.) is an important commercial, sub tropical fruit crop of India and belongs to the family Sapindaceae. It is known for its attractive, red color, good taste, excellent quality and high nutritive value, popularly called as "Queen" of fruits. Its original abode is China and till date this country is the largest producer of litchi in the world. History envisages that being very fastidious in its climatic requirement and a poorly researched crop, its spread from China to other parts of the world was slow till recent past. Now realizing the commercial viability, export potential and high nutritive and therapeutic values, much attention is being paid for its research and developmental aspects. India is second largest producer with relatively higher productivity and the best quality of litchi is produced in this country.

Bihar particularly is located in the eastern part of the country (between 83°-30' to 88°-00' longitude), and lies mid-way between the humid West Bengal in the east and the sub humid Uttar Pradesh in the west which provides it with a transitional position mainly in respect of climate. Though endowed with good soil, adequate rainfall and good ground water availability but yet to get realized its full production (agricultural/horticulture) potential. Its agricultural productivity is one of the lowest in the country, leading to rural poverty, low nutrition and migration of labor. This study is an overview of the state of knowledge applied to deal with the various climatic issues and concerns for sustainable litchi fruit production in Eastern India.

6.2   STATUS OF PRODUCTION AND PRODUCTIVITY

India and China account for 91% of litchi production in the world. Though, China is the largest producer of litchi, India enjoys the prominent position in the litchi production status map of the world in terms of production and productivity. Over the years, India has recorded significant growth in production and productivity of litchi, though trend is not consistent (Singh et al., 2012). Being specific in climatic and soil requirements, litchi has limited distribution. Cultivation intensity of litchi varies even within this country and its cultivation is restricted in the foothills of Himalayas from Tripura to Jammu and Kashmir and plains of Uttar Pradesh to Madhya Pradesh. Commercial cultivation of litchi in India is confined to north Bihar, submountainous region of Uttar Pradesh and Hoogly basins of West Bengal (Pandey and Sharma, 1989; Rai et al., 2001). The production area thus stretches from north to eastern region of the country with more concentrated locations lying in eastern region only. In this country, the fruit crop production statistics shows that litchi ranks 10th in area and 11th in production among fruit crops, but in commercial value terms, it rank sixth. Presently, it is grown in an area of about 74,000 ha, with productivity level of about 7.0–8.0 t/ha. Of the total production of litchi in India, up to 45.5 percent is contributed by Bihar (Singh et al., 2012). The second largest litchi producing state is West Bengal followed by Tripura and Assam An interesting feature of distribution of litchi in India is that maturity (of fruits) commences first in Tripura, followed by West Bengal then Bihar and lately up to June in northern states. The second week of May to first week of June is the real time for harvest of litchi fruits in the eastern and eastern region and more particularly litchi of Bihar starts maturing from 2nd to 4th week of May and continues up to first week of June. Though, the trend of production and productivity of litchi is constantly increasing in this country, but still there exists a wide gap in between the actual production and potential production. Litchi has ever-increasing demand in the national and International market but the present level of production is unable to cope up the demand. At present it is urgently require to deal with the issues concerned to intensifying the research needs and exploit the untapped potential.

6.2.1 CLIMATIC ISSUES AND CONSTRAINTS IN PRODUCTION

Cultivation of litchi is widely spread in eastern India, which provides livelihood opportunities to millions of people in this region. As litchis grow best in a subtropical climate with high summer temperatures, low and frost-free winter temperatures, the successful cultivation of litchi is influenced by climatic factors and any deviation from the normal drastically reduce the fruit yield and quality. The productivity and quality fruit production in litchi are strongly affected by the environmental parameters like temperature, rainfall, photoperiod/light intensity, and moisture content in the soil and atmospheric humidity. At present, litchi production requires to readdress the issues related to successful cultivation for economic viability, because the productivity continues to be low and a gap exists between potential and existing yield. The ratio in yield between the best-managed orchards and national productivity ranges between 2 to 4 times at different locations. The probable reasons for low yield seems to be adoption of traditional production systems, poor technological support and incidence of insect pests, coupled with poor postharvest management. The situation is still aggravated by threat of the ill effect of climate change and seasonal variation. The shortage of genuine planting material coupled with the long juvenile period of litchi is also the constraints. The low female/male flower ratio, premature fruit drop, and fruit cracking due to nonscientific water and nutrient management also add to low productivity and production of poor quality fruits. The litchi tree has luxuriant vegetative growth, which causes problems in harvesting and canopy management is lacking. Lack of scientific information on critical stages to balance vegetative and reproductive phase, that is, for flower bud differentiation, and requirements of water and nutrients also significantly reduces the yield. The shallow rooted crops due to only air-layered plantations. Proper pollination support to enhance fruit set and fruit yield (Menzel and Waite, 2005). The practices that can enhance postharvest life of fruits would be useful to achieve higher productivity but entirely influenced by the climatic conditions.

Its peculiar climatic requirements to complete the growth and production cycle are must and any deviation may certainly give erratic and undesirable behavior (outcome). Under the influence of climatic changed condition and occurring seasonal variation, effect of climatic factors on litchi production system should be carefully analyzed.

6.2.1.1 TEMPERATURE

The most imminent climate changes in recent times are the increase in temperatures, influencing the growth and productivity of litchi crop also. The effect of temperature on early and delayed panicle emergence flowering phases, low fruit set because of several abnormalities caused due to low night temperatures. Late flowering also reduces the fruit set, pollinators movement. Low winter temperatures are essential for bringing about the necessary physiological changes to stimulate flower initiation. Temperatures of below 2°C can cause damage on flowers and young shoots. The frost-free areas with a high summer, rainfall and humidity have been found most suitable. In addition, high temperatures during panicle development cause quick growth and reduce the number of days when hermaphrodite and male flower are available for effective pollination but no female flower phase at that particular juncture may lead to crop failure. Rising temperatures activity resulting into low fruit set (Bhriguvanshi,

2010). This crop suffers from a wide range of disorders due to sudden fluctuations in temperatures coupled with drought.

6.2.1.2   RAINFALL

Rainfall is one of the most important crop production parameters. Its uncertainty and erratic frequency distribution pattern will change the state of litchi production cycle, that are mainly reliant on rainfed production. In litchi production, high summer rainfall promotes maximum fruit growth and yield. Adequate water must be available from fruit set until the fruit can be harvested, that is from April to May. Sufficient soil moisture must be available during the flowering period and for 5 to 6 weeks after fruit set because this is the most important period of fruit set and the start of cell division in the young fruitlet, especially in the skin and the young embryo. The rainy season in this region generally commences from June July and lasts till the middle of August. The total annual rainfall varies significantly in this region and hampers the fruit production and quality harvest. As high humidity and heavy rain during aril development lead to excessive absorption of water by fruit (aril) and may aggravate many disorders like fruit cracking and fruit drop.

6.2.1.3   PHOTOPERIOD/LIGHT INTENSITY

The litchi crop is very much responsive to the light intensity and photoperiod for its enhanced quality fruit production. Light intensity in recent years have shown the distinct variation when compared to different orchards with the location and orchard plant density. The flowering and bearing behavior leading to fruit yield and quality have been correlated with the sunshine hours and light intensity (Menzel and Waite 2005). There are reports that light may limit flower development in dense orchard. Persistent cloud cover at flowering time can also be a problem. The characteristic demand of light by litchi is determined by season, in which these are adopted to be grown for obtaining good yields. The average seasonal variation, disturbing light and its intensity influence flowering and fruit set. Heavy shade for one-week increases fruit drop. Overall, the productivity is driven by the amount and distribution of light with the interaction of other factors. Solar radiation is essential for litchi and the seasonal variation cause adverse effect on overall quality production and productivity, as evidenced by the characteristic color pigmentation on the pericarp receiving different amount of shade inside the canopy. The process of photosynthesis in this evergreen plants depend on environmental conditions and the physiology of the leaves. The distribution of light and leaf nitrogen within the tree usually a good indication of potential photosynthesis. The seasonal variation due to day to day changes in weather conditions have been found to influence the rate of photosynthesis and its contribution in growth and fruit production.

6.2.1.4   WINDS

The initial establishment, proper growth and performance of litchi have been found affected by nature of winds. Litchi plantation at its initial stage experiencing hurricanes/high wind storm hamper establishment. The speedy wind, storm and squalls during the period of bloom and fruit development damage the crop severely, may cause complete crop loss. Hot winds during summer have an adverse effect both on fruit and foliage growth, which is more so in the areas adjacent to open areas that are not fully protected with windbreaks. Wind also lead to a loss of vigor and consequently, slower growth. Fruit drop at initial stages of its growth can be noticed due to high velocity of wind.

6.2.1.5   FOG AND FROST

Occurrence of fog a regular phenomenon in this region (Bihar) and fog definitely curtails light supply and reduces photosynthesis. Occasionally, spring frosts are particularly harmful to litchi plants. Frost may either kill the reproductive organs of the flower or completely destroys the blossoms, thereby influencing the fruit-set and ultimately the fruitfulness. Litchi young plants are extremely sensitive to cold and require frost protection. Frost has been regarded as one of the important factors responsible for causing drastic effect on bearing behavior and even leading to mortality of the plants/trees. The extent of damage by frost depends upon age of the tree, moisture content of the soil, condition of growth, actual timing of frost occurrence, severity and duration of the frost. It has been found devastating for litchi, growing in areas/places having mild tropical and subtropical climate.

High and frequent fluctuations in temperature, sunshine period, intense rainfall, severe storm or cyclone have the ability to devastate litchi crop, inundate, erode and leach orchard floor soil and ultimately damage the crop.

6.3   CLIMATIC RESILIENCE STRATEGIES AND SUSTAINABLE PRODUCTION

This fruit crop is very responsive to changes in different climate variables, seasonal variation, but farmers and Scientists involvement in generating short-term and midterm responses to cope with likely losses in yield and quality by mitigating the ill effects are also on. In order to better manage these processes and practices, impacts need to be properly assessed and improved adaptation strategies need to be tested, targeted and implemented. During only recent few decades, the growing concern of climate change and its impact has come to the forefront encompassing litchi production too. The general decline in potential production as well as erratic and inferior quality production in the recent past has warned to understand the nature and magnitude of change and litchi production system. Changes in seasonality may be one of the most difficult aspects of climate. These changes can have profound impacts on vegetative and crop success (Damour et al., 2007).

6.3.1   UNDERSTANDING OF IDEAL CLIMATIC CONDITION

The successful growing of litchi is an undertaking quite similar to some perennial fruit culture but it is exacting as to specific climatic requirement and thrive best and most productive in the subtropical to mild tropical type of climates. Litchi requires cool, dry, frost-free winter during growth and long summer with high relative humidity at the time of maturity. As, various studies have indicated that under field conditions, temperature, sunshine and soil moisture content are often correlated with the overall plant performance (Singh et al., 2012). Frost-free, cool, and dry winters, and humid summers, free from hot and dry winds characterize this growing zone. Integrated nutrient management based on soil and leaf nutrient standards for sustainable production has been worked out and timely essential practices have been focused on the regulation of the trees vegetative growth by physical and chemical means. Growth regulators are applied to discourage winter flushing in favor of flower initiation, high fruit set and fruit bearing leading to high yield and quality fruit production. Under proper conditions the litchi tree develops rapidly, has an extensive root system and will eventually reach a great age. On the other hand, if the soil is only 2–4 ft. deep, underlaid hard pan, or has a high water table, the trees will grow slowly, become stunted and will yield poor fruits. The key factors to consider when assessing the potential of different areas for litchi are temperatures in winter that affect flower initiation, temperatures and light levels in spring which affect fruit set and reliability of rainfall which affects fruit development. A deep, well-drained, nonsaline calcareous soils with proper texture, fertility and high organic matter have been found most suitable for this crop. After the fruits are harvested, the trees are regularly irrigated with water to maintain soil and require a dry period to phase change for panicle emergence. So the duration from end of rainy season to initiating of dry or stress condition, must stop supplying irrigation/water for the tree. After stages of fruit set and fruit development period, the tree should be steadily irrigated.

Normally temperatures below 20°C induce flowers, while flowering is irregular at higher temperatures. A short drought period in winter assist flowering, especially in the more tropical cultivars, but is not essential. Annual rainfall of 1200 to 1500 mm is probably required in the absence of irrigation. The other critical part of the crop cycle is fruit set that is reduced when temperatures fall below 20°C for extended periods during flowering. Occurrence of high temperature and water stress during fruiting period, especially in the lower elevation may give rise to many physiological disorders (many excessive drop and cracking of fruits), leading to total economic loss.

6.3.2 AWARENESS TO CLIMATE CHANGE

An awareness of climate change grows, so too will the demand for information and technologies that anticipate its impact. So will adoption level increase will enhance potential and production of quality produce by mitigating the ill effects of climate change and seasonal variability. There is need to promote growers/farmers to develop warning system through use of proper technologies, that too strategically in congruence with future climate change. Capacity building in surveillance, prevention, control and diagnosis of diseases in systematic manner is very much required. To develop database of litchi growers and litchi growing condition at region specific basis. There is also need to develop knowledge and understanding about the effect of climate change and seasonal variation on litchi production and productivity, as this fruit crop is very responsive to climate change conditions. It is urgently required to have farmers/grower and scientists involvement for this concern to manage its ill effects by suitable and possible processes and practices. To make policy support, strategic plan and implement specific measures to combat climate change in wider perspective, human resource development by proper training is required and also to undertake action research and extension activities to address issues such as poor soil, irrigation and limited resource base efficient utilization. The technological interventions through developing the demonstration plot in major growing areas to develop knowledge for better adoption of technologies.

6.3.3 CLIMATIC EVENTS FOREWARNING/FORECASTING

Making forecasts relevant to litchi production system, with focused investment and infrastructural development have supported the litchi production more pronounced and responsive. The complex interactions that govern the population dynamics and proliferation of invasive species, pests and disease make it extremely difficult for scientists to predict the extent of damage of litchi production under climate changed condition. As in case of warmer climate and changed rainfall pattern resulting frequent and undesirable flushing, litchi crop will become stressed and may be more susceptible to pest and disease outbreaks disturbing the economic viability. The recent developed climate ready technologies consisting of crops with specific traits such as drought, salt tolerance, management of agronomic practices, temperature variations, occurrence of rainfall, etc., can be suitably managed. It is required to properly exploit the developed forecast system as efficient and early warning system for successful litchi production.

6.3.4 IMPACT ASSESSMENT STUDIES

Impacts are dependent on both the important commercial crop and the environment conditions in the region. The relevant changes or deviation in desirable performance (quality fruit production) under increased temperature and changed rainfall conditions are important. The trend of research and development, technology developed and its updates, feasibility studies in the context of present scenario should be in the forms/way of recommendations, current use and adoption. This fruit crop is very responsive to changes in different climate variables, seasonal variation, but farmers and scientists involvement in generating short-term and midterm responses to cope with likely losses in yield and quality by mitigating the ill effects are also on. In order to better manage these processes and practices, impacts need to be properly assessed. The effects of climate on physiological aspects during particular period of growth phases and the evaluation of technologies with their adoption levels are to be analyzed properly for this fruit crop. The emphasis should be on water use efficiency to adapt to the hot and dry conditions, nutrient management to withstand adverse situation. Apart from these, measures to reduce the ill effects of the increase in temperature and CO_2 on litchi is required as because there is limited efforts made on these aspects. Hence the detailed studies are required to quantify the impact and assess the vulnerability of litchi crop growth and quality fruit production in the context of climatic issues and there is need to adopt several approaches and concerted efforts.

6.3.5 NEED OF SIMULATION MODELS

As among the perennial fruit crops, litchi is highly sensitive to changes in climate and climate variability. To assess the impacts of future climate on its production, simulation models are required, by integrating the all possible information regarding the ecology, growth and physiological development of crop, the local weather, the existing management practices, the soil characteristics. The proposed model should be developed for taking care of vegetative and reproductive growth, phonological aspects and other concerns for quality production and it should be specifically based on other models developed for perennial fruit crops.

6.3.6 TECHNOLOGY INTERVENTION AND BETTER ADOPTION

The timely adoption of technologies and application of proper input, also create suitable condition for normal vegetative and reproductive growth of the trees, which result in high fruit yield and better quality. The ideal climatic conditions can also be created by managing the orchard with intensive culture technique and proper infrastructural development, which will also be helpful in addressing the adverse impact of climate change (more particularly the variable climatic conditions) in litchi production and productivity. It should be mainly based on comparing the ideal climatic conditions projecting the potential differences with technological interventions in particular set of environmental conditions at region specific basis.

The adaptation strategies as planned under should include mainly to increase nutrient and water use efficiency, modify the nutrient application rates, nutrient availability by use of soil amendments and cultural practices, modifying the canopy architectural design, rejuvenation of old senile unproductive orchards, intercrops and need based intercultural operations, use of mulching and recommended type of mulching material, nutrient management with more emphasis on organic farming, water management strategically for enhancing water use efficiency, use of integrated pest management practices to control the pest complex preventing the significant loss due to their damage.

6.3.6.1 NEW PLANTATION

The efforts towards mitigating the ill effect of climate change is all about trying to minimize the amount of green house gases, we release into atmosphere. The first or rather foremost direct approach will be through activities such as to go for litchi plantation in suitable areas, to adopt practices and use inputs which promote the build up of organic matter (humus) in soils. The recommended spacing under normal and high density planting should be followed with utmost care and attention for better results.

6.3.6.2   AGRONOMIC PRACTICES

Proper leveling of orchard floor to increase use efficiency of inputs. The proven fact that fruit yields increase overtime, reflecting larger trees or possibly better agronomy. The orchard floor management for irrigation, nutrition and intercultural operations must be timely and skillfully done.

6.3.6.3   NUTRIENT MANAGEMENT

The increased use of more organic base input application or bio-fertilizers and less use of inorganic/chemical input/sources have resulted significant increase in the quality production. The substrate dynamics under integrated plant nutrient management (IPNM) should be taken care off in case of litchi production. The basis of soil and leaf nutrient analysis should be taken care of for modifying fertilizer application for solving the problems due to deficiencies and disorders (Kumar, 2008; Singh et al., 2011; Sinha et al., 1999). Improved fertilizers management can be translated into carbon credits to farmer will be an effort to climate resilience.

6.3.6.4   WATER MANAGEMENT

Efforts to protect rainfall catchments and existing water resources, expand water harvesting and improve water storage, timely irrigation and conservation practices will become increasingly important for many perennial fruit crops including litchi. Providing irrigation during critical stages of crop growth and conservation of moisture reserves are the most important interventions for bearing behavior and quality production of litchi.

6.3.6.5   POLLINATION BENEFITS

Litchi is a highly cross-pollinated crop and cross-pollination is rule for good fruit production. Honey-bees have been found significantly responsible for enhanced pollination and fruit set, leading to high quality fruit production. The added advantage by rearing and keeping honey bees of *Apis mellifera* (Italian honey bees) species in the litchi orchards can also provide litchi honey and other by-products of economic significance.

6.3.6.6   CANOPY MANAGEMENT AND REJUVENATION

Canopy management through suitable pruning and training make the proper balance between vegetative and reproductive growth to give annual crop of quality produce. The litchi tree in this region has found to give best performance with open umbrella or semicircular canopy shape. Canopy management is an important aspect to increase input use efficiency and high quality fruit production. Reiterative pruning and canopy rebuilding through rejuvenation technology of old senile unproductive orchards have been found to be transformed into young bearing orchard with enhanced quantum of quality production (Kumar, 2012).

6.3.6.7   USE OF GROWTH REGULATORS

Growth controlling chemicals should be used carefully to retard shoot and canopy volume. Growth retardants have been employed in litchi like paclobutrazol, cycocel (CCC) and maleic hydrazide (MH), etc., but requires further confirmations for wider practice. Use of planofix or nephtalic acetic acid (NAA) has been found to reduce fruit drop and enhance fruit yield (Sinha et al., 1999).

6.3.6.8   INTERCROPPING AND MULCHING

Mulches also increase soil organic matter, improve soil structure, increase water retention and help reduce fluctuations in root temperature. The crop management practices like mulching with crop residues and plastic mulches help in conserving soil moisture. It also controls the growth of weed. In some instances excessive moisture due to heavy rain and untimely rain becomes major problem and it could be overcome by growing intercrops, light intercultural operations and raised basin making.

6.3.6.7   USE OF IPM

Leaf curl or mite is a major problem, requires removal and burning of infected parts cuts down losses, and spraying of miticide. The bark eating caterpillar, which attack older trees and may be controlled by plugging the holes with petroleum, nuvan soaked cotton and sealing the holes with mud. Spraying of insecticides (metasystox, dizinon, amichlorapid, etc.) can take care of minor insect pests. Protection of fruits from birds and bats are again serious problem, can be protected by nylon net cover.

6.3.6.8   WINDBREAKS

Climate change mitigation measures by growing/developing windbreaks for litchi orchards are very common and essential measures for successful commercial litchi production.

6.4   CONCLUSION

Litchi production system in sustainable manner in the present day context clearly showed that the climatic issues are very much vital and influencing the potential production and plantation too under this region. Considering the importance of this fruit crop in the region, efforts are made to provide technological support through research and promoting sustainable production by proper scientific intervention and better technologies adoption pertaining to climate change perspective. The strategic measures and technologies suggested with respect to adaptations to climate change for sustainable production is not only a matter of changing current management practices, changing varieties or changing cropping zones, but adaptation to climate change, starts from building resilient systems under present conditions. Current litchi production in this country and more particularly in eastern region needs to be both environmentally and economically sustainable and provide the basis for enhanced quantum of quality production and livelihood security in the region wherever is grown.

KEYWORDS

- **Canopy Management**
- **Climatic Issues**
- **Climatic Resilience Strategies**
- **Forecasting**
- **Photoperiod**
- **Production Constraints**

REFERENCES

Bhriguvanshi, S. R. (2010). Impact of Climate Change on Mango and Tropical Fruits. In Singh, H. P., Singh, J. P., & Lal, S. S. (editors). Challenges of Climate Change-Indian Horticulture, Westville Publishing House, New Delhi, India, 224.

Damour, G., Vandame, M., & Urban, L. (2007). Longterm Drought Modifies the Fundamental Relationship between Light Exposure, Leaf Nitrogen Content and Photosynthesis Capacity, In Leaves of Lychee (Litchi chinensis). J. Plant Physiology. Doi. 10.1016.

Kumar, Rajesh, (2008). Managing Physiological Disorder in Litchi, Technical Bulletin-5, NRC for Litchi, Muzaffapur, Bihar.

Kumar, Rajesh, (2012). Litchi Mein Chatrak Prabandhan (Hindi). Technical Bulletin. Published by NRC on Litchi, Muzaffarpur, Bihar.

Menzel, C. M., & Simpson, D. R. (1995). Temperatures above 2°C Reduce Flowering in Lychee *(Litchi chinensis* Sonn.), *Journal of Horticultural Science,* 70, 981–987.

Menzel, C. M., & Waite, G. K. (2005). Litchi and Longan; Botany, Production and Uses (Eds.). CAB Publishing.

Pandey, R. M., & Sharma, H. C. (1989). The Litchi. Indian Council of Agricultural Research, New Delhi, 1–80.

Rai, Mathura, Nath Vishal, Dey, P., Kumar, S. & Das, Vikash. (2001). Litchi (ed.). CHES (ICAR), Ranchi, Jharkhand.

Singh, G., Nath Vishal, Pandey, S. D., Ray, P. K., & Singh, H. S. (2012). The Litchi. FAO of the United Nations, New Delhi, India.

Singh, H. S., Vishal Nath, Singh, A., & Pandey, S. D. (2011). Litchi-Preventive and Curative measures (Book). Published by SPSS, New Delhi.

Sinha, A. K., Singh, C., & Jain, B. P. (1999). Effect of Plant Growth Substances and Micronutrients on Fruit Set, Fruit Drop, Fruit Retention and Cracking of Litchi cv. Purbi, *Indian Journal of Horticulture*, 56, 309–311.

CHAPTER 7

CLIMATE CHANGE: RESILIENT ISLAND HORTICULTURE

D. R. SINGH[1], SHRAWAN SINGH, and S. DAM ROY

Central Agricultural Research Institute, Port Blair, Andaman & Nicobar Islands-744101;
[1]E-mail: drsingh1966@yahoo.com

CONTENTS

7.1 INTRODUCTION

Climate change is a crosscutting issue, affecting a multitude of sectors and activities like agriculture, water, health, tourism, land use, ecosystems management, disaster risk reduction, and gender, and requires cross-sectoral approaches for adaptation. To sustain livelihoods, people use a wide range of social, physical, natural, financial and human resources. To analyze livelihoods-climate linkages, those resources most important to livelihoods and short-term coping and longer-term adaptation must be identified, keeping in mind that different group will rely on different resources. It is widely recognized that there are differentiated impacts of climate change and that the poor will be hit hardest (Allaby, 1989).

The horticulture is well acknowledged segment of agriculture, which provide livelihood and contribute significantly in regional economies. Fruits and vegetables are the best resource for overcoming micronutrient deficiencies, enhancing livelihood, creating employment and market opportunity (Shon et al., 2003). Most of horticultural crops prefer cooler temperatures, thus productivity is lowest in the hot and humid tropical regions. These crops are generally sensitive to environmental extremes, and thus high temperatures and limited soil moisture are the major causes of low yields in the tropics and will be further magnified by climate change (Brklacich, et al., 1996).

7.2 STATUS OF HORTICULTURE IN ISLANDS

The geographical location of Islands (6°N to 14°N latitude on 92°E to 94°E longitudes) is bestowed with humid tropical climate. Climate of the Islands is that of warm humid tropics with temperature ranging from 23.2 to 30.7°C, relative humidity from 80 to 90% and average rainfall of about 3,000 mm, during both monsoons from May to December (Saldanha, 1989). Forest covers 86% while only 6.1% is under agriculture. Its climatic conditions favor cultivation of large number of tropical and sub tropical horticultural crops (Singh et al., 2005). The creditability of horticultural sector in Islands has been well established in improving factor productivity, employment opportunities, livelihoods and economic conditions of the farmers and also ensuring nutritional security to the Islanders. The major horticultural crops in islands are coconut, arecanut, vegetables, fruits, root and tuber crops, floriculture, mushroom, medical plants, etc., and the sector contributes nearly 24.5 percent of national share of agricultural GDP and is considered as third pillar of Island economy (Singh et al., 2011a).

Unfortunately, the productivity of horticultural crops in islands still remains near to stagnant over the years (Fig. 7.1). But, the demand has increased manifold due to increased population and tourist inflow. Resultantly, islands are facing scarcity of locally grown horticultural products and largely depend on shipment from mainland. This results in peculiar phenomena like ship to mouth, monotonous dietary pattern and huge transportation losses with high cost of perishable food items. This led Islands to a high cost economy and is proving a deterrent to tourist inflow as well as adding to the cost of development. Further, after the Supreme Court order for banning of logging of forests, production base and the employment opportunities have shrunken.

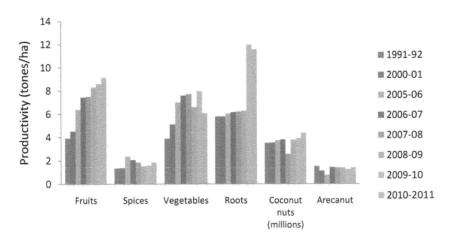

FIGURE 7.1   Trend in productivity of horticultural crops in islands.

A study on carrying capacity for Andaman and Nicobar Islands (2009) projections indicate that islands will be able to meet the local demand of horticultural crops by 2021 will continue to do so for increasing population and tourists (Table 7.1). But, over the years the area under horticultural remains nearly constant however, significant rise in area was observed in vegetables and flowers (Table 7.2) and larger proportions of the area (86%) is being used in dry season as Rice-vegetable cropping system.

Presently, around 81% of vegetable production in islands is harvested during dry season, that is, from December to April. Around 700 ha upland area is used for vegetable cultivation in rainy season. The most significant reasons are water stagnation in low lands and lack of suitable vegetable crops and varieties, which can tolerate water logging. The expected sea level rise may extend the water-logging situation from present period (May to December) to future (May to February). Even small reduction in growing period will have serious impact on selection of vegetables. Thus, sincere efforts were required to identify suitable crops, which are locally adaptive and can tolerate excess rains and stagnant water.

TABLE 7.1 Projected Demand and Supply of Food Items in Islands

Items	Projected Production and Supply Status (tones/year)		
	2011	**2021**	**2031**
Rice	21,864 (D-60%)	36,000 (D-33%)	40,000 (D-50%)
Pulses	1050 (D-344%)	1350 (D-332%)	1350 (D-440%)
Vegetables	29,687 (E)	36,171 (E)	41,577 (D-3%)
Dry season	24,803	26,296	28,014
Rainy season	4524	4875	5313
Protected cultivation	360	5000	8250
Root & tubers	9948 (D-134%)	29,616 (D-79%)	37,020 (D-72%)
Milk	16,531 (D-67%)	34,603 (D-15%)	43,254 (E)
Fruits	16,249 (S)	20,312 (S)	25,390 (S)
Meat & fish	28,048 (E)	29,630 (E)	31,373 (E)

D – Deficit; E – Excess; S – Sufficient.
(Carrying Capacity Group 2009 for Andaman and Nicobar Islands).

TABLE 7.2 Area and Production of Horticulture and Plantation Crops in A & N Islands

Crops	Area (in ha)								Production (in MT)							
	1991–1992	2001–2002	2005–2006	2006–2007	2007–2008	2008–2009	2009–2010	2010–2011	1991–1992	2001–2002	2005–2006	2006–2007	2007–2008	2008–2009	2009–2010	2010–2011
Fruits	3300	3700	2900	2800	3000	3005	3115.51	3160	12900	16700	18500	20800	22500	24949	26767.6	28772
Spices	–	1261	1812.6	1746	1700	1659	1662	1661	–	1726	4251.84	3,571	3120	2540.9	2631.56	3067.2
Vegetables	3400	3100	3668.9	4300	4000	4598.66	5201.4	5150	13200	15800	25682	32600	30800	30199	41500	31300
Roots	384	–	1701	1314	–	866.13	450	449	2216	–	10257.3	8077	–	5411.3	5390	5193
Coconut*	24115	25200	20927.02	21400	21500	21689.6	21760.22	21768	84400	89000	78000	81200	55210	81900	84970	95000
Arecanut	3100	4400	4046.44	4100	4100	4147.5	4152.5	4152	4700	–	3058	5800	5700	5720.5	5200	5800
Flowers	–	10	10	10	30	34.2	32.2	33.65	–	250	254	254	470	335.16	330.12	329.28

*Production as ('000 nuts).

7.3 TRIBAL COMMUNITIES

The archipelago of Andaman and Nicobar Islands is home of six primitive tribes namely Nicobarese, Great Andamanese, Jarawa, Shompen, Onges and Sentenales which are meeting up their food and medicinal needs from the flora and fauna of islands (Singh et al., 2011b). The Settlement plans also brought different communities including Bengali, Tamil, Telugu, Ranchi, Malayali, Odiya and others in islands from mainland India, which depict a mini-India phenomenon. About 38 of 572 islands inhabited having 94% landmass with about 3.5-lakh population. These communities are largely depends on local plant sources for meeting their food, feed and medicines (Rao et al., 2006). Some of the local plant foods like ferns, orchids and flowering plants are having specific climatic requirements for growth and developmental stages. The expected climate change associated factors like rising temperature, increasing humidity and frequent cyclones will be changing the crop physiology and adaptation. Resultantly, the crop structure and cropping sequences will be affected by the climate change phenomena. Thus, proper strategies need to be designed to ensure the traditional food resources for tribal communities in the expected phenomena of climate change. The underutilized fruits and vegetables are traditional sources of tribal diets need to be investigated with climate change associated factors, particularly for flowering and fruit setting parameters because, these two stages are most sensitive to climatic changes.

7.4 TOURIST DESTINATION AND REQUIREMENT OF QUALITY FOODS

Since the archipelago is emerging tourist destination and the demand of quality perishable food items like fruits and vegetables will be high in coming years. But, local production of commercial vegetables is confined to the dry spell of December to April. Around 22000 tones of vegetables and fruits are being transported from mainland every year. This is draining back the island economy to mainland and hindering the local development. But, changing climatic parameters like temperature, relative humidity, CO_2 level and frequencies of climatic adversaries will be further affecting the productivity of fruits and vegetables. Further, unscientific use of chemicals in locally grown fruits and vegetables deteriorate their quality, which are not liked by hotels and service class people. The use of chemicals and pesticides will be more in coming. Efforts need to be strengthened to develop suitable practices for production of quality fruits and vegetables.

7.5   INSECT-PESTS AND CLIMATE CHANGE

The realization of climate change phenomena will be manifested in emergence of new pathogens and insects, increase in incidence of diseases and pest attack and resultantly high use of chemical pesticides to control them will cause serious problem to ecosystem. The effectiveness of pesticides will be low at increased temperature which will results in speedy decomposition of organic matter in soil and affect soil fertility status. The expected climate change phenomena will affect the composition of species and numbers of 'beneficial insects,' which helps in management of harmful insect-pests. Thus, it is high time to start research programs for understanding the mechanism of emergence of new races/stains of pathogens and insects particularly in reference to the horticultural crops for island conditions. Further, the adaptive changes in beneficial insects and their numbers need to be investigated in reference to expected climate change phenomena for strengthening the ongoing efforts for declaration of islands as 'Organic farming zone.'

7.6   POST HARVEST AND PRODUCT QUALITY OF HORTICULTURAL CROPS

With expected climate change phenomena, the changes in agro-climatic parameters will affect the cell functioning thereby crop physiology. This will also affect the quality of the crop produce in terms of taste, appearance, flavor and content of micronutrients, phytochemicals and other dietary elements. This may due to increase in water content in plant cells, high incidence of photo-oxidation, synthesis of more reactive oxygen species (ROS) and secondary metabolites, which are having negative influence on product quality like tannins, phytate and polyphenols. This will cause astringency and rigidness in product quality and reduce post harvest life of the horticultural crops. The processing sector will also be affected due to changes in recovery of proximate components as well as health benefiting compounds.

7.7   CLIMATE CHANGE WATER MANAGEMENT IN HORTICULTURAL CROPS

Water availability and its management has now become core issue of research in view of its alarming rate of scarcity. The climate change causing erratic rainfall behavior and increase the frequency of dry spell. Thus, effective irrigation management practices need to be identified for island conditions. The initial investigations on drip irrigation and sprinkler irrigation methods showed significant influence of yield parameters of pineapple and okra, respectively in islands conditions. However, such kind of studies needs to be investigated in other horticultural crops for confirmations of findings so that these methods can be used for mitigating the expected changed climatic situations.

7.8   INTEGRATED FARMING APPROACH AND CLIMATE CHANGE

The scientific reports have amply proved that IFS approach is effective in mitigating climatic adversaries (Ravishankar, et al., 2007). Horticultural crops are being cultivated in 37,635 ha area (about 77%) of the total cultivated area in islands, predominated by coconut and arecanut. It depicts that the islands have horticultural based farming system. However, spices, fruits, vegetables, flowers (on upland) and rice (low land), fodder (on embankments) and fish component (in ponds) are found in this system. Intercropping of spices has given an additional return of Rs. 9299 and Rs. 29,428 from 0.05 ha of land under arecanut + black pepper and coconut + ginger systems respectively. An additional employment of 40 mandays was also generated under intercropping system compared to monocrops. The climate change phenomena will affect the selection of crops and their productivity in IFS system. Thus, efforts are required to see the possible threats to the existing IFS systems in islands from expected climate change. The research on fresh water and marine pond water based IFS models showed some hope that sea water as well as rain water flooded areas can be effectively used for generating livelihood in islands conditions. However, up scaling of the research efforts are required to substantiate the research findings.

7.9   HORTICULTURAL BIODIVERSITY AND CLIMATE CHANGE

Biodiversity is the degree of variation of life forms within a given ecosystem, biome or an entire planet. It is a measure of the health of ecosystems and is a function of climate. Long term changes in climate is known to have enormous impacts on plant diversity patterns in the past and are seen as having significant current impacts. It is predicted that climate change will remain one of the major drivers of biodiversity patterns in the future. Even though, the biodiversity will be an important contribution to both climate-change mitigation and adaptation. Consequently, conserving and sustainably managing biodiversity is critical to addressing climate change.

About 2500 angiospermous species including 132 species of orchids, 120 species of ferns, 300 species of medicinal plants and more than hundred species of fruits and vegetables are reported from different Islands. However, CARI has made significant attempts for tapping biodiversity of horticultural crops through generating information of nutritional value and as well as in developing varieties of crops like Green Orchid (var. Pretty Green Bay), Broad Dhaniya (var. CARI Broad Dhaniya), Sweet Potato (CARI SP-1, CARI-SP-2) and Greater yam (CARI Yamini) (Sankaran et al., 2012). But, the intense efforts are required to identify climate change resilient plant species from these genetic resources.

7.10   SOIL MICROBES AN DCLIMATE CHANGE

Islands have rich diversity in microbial population, which is helping in making equilibrium in nutrient cycling in islands ecosystem. Knowledge of microbial diversity and functionality of the soil, and entomopathogens are essential to understand soil health conditions and select potent microbes or pathogens to augment soil nutrition status or develop broad-spectrum bio-control agents. Some of the microbes were identified by CARI, which can tolerate higher level of salinity and temperature showing their potential in isolating genes for such traits. However, the research efforts need to be intensified for such identification of climate change resilient organisms from unexplored regions.

7.11   CLIMATE CHANGE AND NUTRITIONAL SECURITY IN ISLANDS

An increase in temperature may have significant effect on the quality of fruits, vegetables, aromatics and medicinal plants. The nutritional quality of cereals and pulses may also be moderately affected which in turn, will have consequences for our nutritional security. A combination of action on climate change adaptation and mitigation supported by research and technological development can reduce the threats to food and nutrition security. Climate change negatively affects food availability, conservation, access and utilization and exacerbates socioeconomic risks and vulnerabilities. Further, declining local production and probable disruptions caused by climatic hazards, the income generating opportunities and purchasing power will decrease for vulnerable populations, which ultimately affect the food and nutritional security. The CARI has made significant attempt in identifying nutrients rich underutilized fruit and vegetable crops, which can supplement the ongoing nutritional security schemes (Singh et al., 2005). However, efforts need to be intensified for finding high density nutrient sources and climate resilient alternative livelihood options for ensuring proper nutrition to the marginal sections (Singh et al., 2013).

7.12   CLIMATE CHANGE AND PLANTATION CROPS

The plantation crops are major source of income in islands and preliminary investigations indicate that climate change will affect the flowering and nut setting season. Further, the quality of nut water and copra content will also be affected. The islands have great diversity in coconut and arecanut, thus proper survey and vigilance should be ensured to see the influence of climate change and response of existing plantation diversity in islands. Further, efforts are required for identifying the suitable crop management practices, which can mitigate the influence of climate change and increase the productivity for plantations.

7.13 CLIMATE CHANGE AND MEDICINAL PLANTS

Tropical regions enjoy rich diversity of medicinal plants and their proper utilization depends on technological back up and scientific investigations. Most of island nations situated in this region are developing economies and rich in biodiversity but technologically poor which hinders tapping of potential bioresources for their economic benefits. CARI has initiated efforts in this direction by commercializing *Morinda citrifolia* having more than 200 phytochemical compounds (Singh et al., 2012). Similar efforts should be initiated at International level to exploit other native plants through International cooperation. Presently, the industry is harnessing the potential of medicinal plants *viz.* morinda, aloevira, kalmegh, brahami, tulsi, ashwagandha, opium, isbgol, aonla, etc. but their adaptation and recovery of phytochemicals will be affected by rising temperature. The climate change will certainly affect the composition and concentration of phytochemicals in plants but their change need to be investigated using different climate change modules (Singh et al., 2012).

7.14 CLIMATE CHANGE RESILIENT PERENNIAL HORTICULTURAL CROPS

Islands have fragile ecosystem and frequent soil manipulation along with climate change associated with intense rains will increase the soil erosion. This will cause serious threat to the island ecosystem. Thus, suitable perennial fruits and vegetable crops should be promoted in islands. These crops need less soil cultivation, bind soil properly, add more biomass, sequester carbon round the year and favor soil microbes. But, flowering and fruiting pattern of these crops mostly depends on climatic factors and slight change in temperature and relative humidity affect these stages. Thus, research efforts are required for identifying suitable interventions, which can ensure fruit setting and flowering in these crops, particularly perennial vegetables. The CARI has started efforts for bioprospecting and physiological aspects of underutilized perennial vegetables, which will be helpful in adding these crops in basket of climate resilient crops.

7.15   CLIMATE CHANGE RESILIENT GENES/ALLELES

Because of climate change, abiotic stress factors negatively impact the agricultural production systems world over. The islands are rich center of biodiversity and plant species are observed in different climatic situations, that is, dry season and wet season as well as in soil conditions like normal soil and seawater affected soils. These crops can be taken up as potential source of abiotic stress tolerance genes for transfer staple food sources. Further, the island biodiversity have some of the biotic stress tolerant plant species, which can be used as source of genes for transferring in commercial crops. The identification of CARI- Brinjal-1 as source of bacterial wilt resistance is a good example for commercial exploitation. These gene sources amply support the much-needed genetic enhancement for mitigating the climate change impact.

7.16   SEA AFFECTED LAND UTILIZATION

Around 8000 ha area is affected by seawater during *Tsunami* (2004), for its effective utilization a complete package of technology for rehabilitation was developed using Morinda as a representative crop. It included soil mounting, specific planting procedures, saline tolerant genotypes, grafting techniques, post planting practices, etc. The phytochemical analysis of Morinda was done for both normal soil grown plants as well as seawater engrossed lands. The observations of trainee farmers from different islands and scientific community were supportive for next stage adoption of the technology. Similar research efforts were required for other crops as well because seawater inundation will increase in coming years (Singh and Rai, 2007).

THE THRUST RESEARCH AREAS IN ISLAND HORTICULTURE TO MITIGATE CLIMATE CHANGE IMPACT

- Refinement of existing technologies for horticultural crops in islands for mitigating the impact of climate change on their productivity.
- Identification and utilization of climate change resilient crops and genes from existing biodiversity of horticultural crops.

- Strengthening laboratories with high-throughput techniques for enhancing the efficiency of ongoing research efforts in islands.
- The techniques identified for seawater affected lands need to be tested in different crops so that these can be used effectively.
- The disease and pest incidence will be high in changed scenario, so suitable IPM modules should be developed for minimizing the chemical load in ecosystem.
- Water management will be serious issue in coming years in island horticulture, so effective rainwater utilization strategies need to be developed.
- Successful use of Remote sensing technology in identification of rice cultivation need to be evaluated for coconut and arecanut plantations so that senile plantations can be converted into productive plantations.
- The potential of underutilized fruits and vegetables need to be harnessed in the context of climate change and food security.

KEYWORDS

- **Climate Change**
- **Integrated Farming Approach**
- **Nutritional Security**
- **Postharvest and Product Quality**
- **Tribal Communities**

REFERENCES

1. Allaby, M. (1989). Dictionary of the Environment. Third edition. New York University Press.
2. Brklacich, M. & McNabb, D. (1996). Estimated Impacts of Global Climate Change on Canadian Agriculture, Report Prepared for Agriculture and Agri-Food Canada.
3. D. R. Singh, Senani S. & Rai R. B. (2005). Nutritional aspects of underutilized fruits of Andaman and Nicobar Islands. CARI, Port Blair.
4. Sankaran, M., Singh, D. R., Singh, S., Damodaran, V., & Singh, L. B. (2012). High Yielding Varieties in Horticultural Crops Developed by CARI, CARI, Port Blair.

5. Rao, V. G., Sugunan, A. P., Murhekar, M. V., & Sehgal, S. C. (2006). Malnutrition and High Childhood Mortality among the Onge Tribe of the Andaman and Nicobar Islands, Public Health Nutrition, 9(1), 19–25.

6. Ravishankar, N., Pramanik, S. C., Jeyakumar, S., Singh, D. R., Nabisat Bibi, Shakila Nawaz, & Biswas, T. K. (2007). Study of Integrated Farming System (IFS) under different Resource Conditions of Island Ecosystem. Journal of Farming Systems Research & Development, 13(1), 1–9.

7. Saldanha, C. J. (1989). Andaman, Nicobar and Lakshadweep. An environmental impact Assessment. New Delhi, Oxford and IBH Publication Co. 144.

8. Shon, M. Y., Kim, T. H., & Sung, N. J. (2003). Antioxidants and Free Radical Scavenging Activity of *Phellinus baumii* (Phellinus of Hymenochaetaceae) extracts. Food Chemistry, 82, 593–597.

9. Singh, D. R., Rai, R. B. & Singh, B. (2005). The Great Morindaa Potential Underutilized Fruits in Bay Islands, the Daily Telegrams, Port Blair, April 24, 2.

10. Singh, D. R., Singh, S., Salim, K. M., & Srivastava, R. C. (2011). Estimation of Phytochemicals and Antioxidant Activity of Underutilized Fruits of Andaman Islands (India), International Journal of Food Science and Nutrition, 63(4), 446-52.

11. Singh, D. R., & Singh, S. (2012). Phytochemicals in Plant Parts of Noni (*Morinda citrifolia* L.) with Special Reference to Fatty Acid Profiles of Seeds. Proceedings of National Academy of Sciences-India Section B, Biological Sciences. Doi: 10.1007/s40011-013-0154-1.

12. Singh, S., Singh, D. R., Salim, K. M., Srivastava, A., Singh, L. B., & Srivastava, R.C. (2011). Estimation of Proximate Composition, Micronutrients and Phytochemical Compounds in Traditional Vegetables from Andaman and Nicobar Islands. International Journal of Food Science and Nutrition, 62(7), 765–773.

13. Singh, S., Singh, D. R., Shajeeda-Banu, V., & Salim, K. M. (2013). Determination of Bioactives and Antioxidant Activity in *Eryngium fetidum* L., A traditional culinary and Medicinal Herb. Proceedings of National Academy of Sciences-India Section B, Biological Sciences, 83(3), 453–460.

14. Singh, D. R., & Rai, R. B. (2007). *Morinda citrifolia*-An important fruit tree of Andaman and Nicobar Islands. Indian Journal of Natural Products and Resources, 6(1), 62–65.

15. Singh, R. S., & Singh, D. R. (2012). Noni Plant (*Morinda citrifolia*) Growth and Development Influenced by Ambient Temperature and Humidity under Sub-tropical Conditions of Varanasi, (India), Indian Forester, 138(4), 349–356.

CHAPTER 8

GLOBAL CLIMATE CHANGE: MYTH, REALITY AND MITIGATION

B. B. MISHRA[1] and RICHA ROY[2]

[1]Department of Soil Science and Agricultural Chemistry, BAU, Sabour, Bhagalpur – 813210; E-mail: bbmsoil@rediffmail.com

[2]Sam Higginbottom Institute of Agriculture, Technology and Sciences-Deemed University, Naini, Allahabad, India; E-mail: r.roy89@gmail.com

CONTENTS

ABSTRACT

The global climate change is an extraordinary complex phenomenon in earth's atmosphere. The challenges associated with its change cannot be tackled in isolation, but through integrated efforts, taking all six spectra together for developing the strategic planning. Logically framed six spectra are planetary physical, EM-nuclear, chemical, biological, pedogenic and anthropogenic spectra. Soil is rather the lowest boundary of the entire earth's atmosphere excluding the portions covered by oceans. The global climate change is either natural or man-made or evens both. The man-made challenges are manageable with participatory commitments in a planned way on site-specific basis through human intervention, pedogenic exploitation, etc. On soil surface, the role of conservation agriculture as well as horticultural community is of special interest to minimize greenhouse gas emission and for promoting the soil carbon stock. High-density orchards with proper soil management options may be a powerful mitigation option. Microorganisms may play vital role in synthesizing biofuels and C-sequestration. However, there are controversies in defining each spectrum in terms of mitigation.

8.1 INTRODUCTION

The combination of oceanic and atmospheric circulation virtually forms the global climate mainly by redistribution of heat and moisture. As a consequence, the areas surrounding the tropics are warm and relatively wet almost round the year, whereas in temperate regions, variation in solar input causes seasonal change. In the north hemisphere, such events may involve pronounced changes in temperature, whereas in the South hemisphere lands are located close to equator and majority of land surface is covered with water. The global climatic patterns are typically dynamic, where they are continually changing in response to solar radiation, atmospheric greenhouse gas concentrations and other related factors. The earth's elliptical orbit round the sun does shift under the influence of gravitational pull among the planets in solar system. The climate may be the average of weather condition, but for common understanding, it is the product of multicomponent and multidirectional interactions associated with interplanetary and extraterrestrial forces, which do form primarily the six distinctive spectra

(Mishra and Ghanshyam, 2009) viz. planetary physical, electromagnetic-nuclear, chemical, biological, anthropogenic and pedogenic spectra. There is often scientific controversy that carbon dioxide, methane (CH_4) or other greenhouse gases is hardly causing catastrophic heating of earth's atmosphere and as such, any disruption of earth's climate. Moreover, there is obvious scientific evidence that increase in atmospheric carbon dioxide may produce even beneficial impacts on plant and animal environments. Such reports need to be validated in order to overcome controversies.

Out of many such interaction events, human interferences as well as pedogenic exploitation do play vital role that may be manageable by integrating the activities inventoried in line with desired climate equilibrating with livelihood. Destruction of natural set up in any form following the emission of green house gases and other polluting materials in undesirable quantities would result in climate change on site specific basis. Any ray from a star reaching one's eyes in light years can tell only about the past events of the star. The technique to inventory the components associated is tedious to define and the strategic planning would be shaped logically in a balanced framework by integrating the defined activities so associated.

Soil is the lowest boundary of the entire earth's atmosphere, excluding the part covered with ocean, rock outcrop and surface construction, that undergoes interactions with incoming radiation including background nuclear counts as well as chemical, biological, physical and anthropogenic interferences, wherein soil science has a bridging role within the critical zone limits (Lin, 2006) in an open system to provide the food, water and environment and needs to be strengthened. Besides, soil has immense potential of storing carbon under a suitable pedogenic environment. The global climate change cannot be considered in isolation. It is an issue to be discussed on integrated basis taking different spectra into consideration. Pedogenic spectrum is one to be addressed precisely through anthropogenic intervention. Anthropogenic interactions in pedogenic environment need to be reviewed and updated systematically with further efforts enabling to understand how such complex system could be handled in management terms in order to rebuild a harmonizing global environment for sustaining the livelihood of the growing human population on long-term basis. The present paper is an attempt towards conceptual as well as logical explanation to the principles of strategic planning for combating the challenges of global as well as site-specific climate change.

8.2   LOGICALLY IDENTIFIED SPECTRUM ASSOCIATED WITH GLOBAL CLIMATE CHANGE

The earth's atmosphere is virtually an envelope surrounding the earth's lithosphere, wherein six spectra are proposed that play the major role in maintaining the earth's equilibrium. They are physical form and stability of the earth, that is, planetary physical spectrum, electromagnetic-nuclear spectrum besides chemical, biological, pedologic and anthropogenic spectra, wherein human interferences as well as pedogenic disturbances lead to impose challenges with equilibrium and stability of the earth's climate.

8.2.1   PLANETARY PHYSICAL SPECTRUM

The earth was not like what we see today. In reality, the continents were all part of single giant landmass. This fact is based on similarity between plant and animal fossils and rocks found in the eastern coastline of South America and western coastline of Africa. But, they are now widely separated by the Atlantic Ocean. The evidence of tropical plant fossils as coal deposits in Antarctica opens a conclusive discussion that the frozen landmass in the past might have been in close proximity to the equator, where climate was tropical with plenty of vegetation. When the landmass began to drift apart gradually in the long past, it might have caused an impact on climate by changing the physical features of the landmass and the position of water bodies even. Consequently, the separation of the landmasses changed the direction of wind flow including ocean's currents. The Himalayan range is still rising upward by about 1.0 mm every year, since the Indian landmass tends to move gradually and steadily towards the Asian land mass.

When a volcano erupts, it often throws out large volumes of sulfur dioxide, water vapor, dust and ash into the atmosphere influencing thereby the climatic patterns for years. These gases as well as dust particles do try to block the incoming radiation from sun and thus result into cooling. Besides, the SO_2 being combined with water forms the droplets of sulfuric acid. Mount Pinatoba in the Philippine islands underwent eruption in 1991 emitting huge amounts of gases into the atmosphere. Eruptions may often reduce the amount of solar radiation reaching the earth's surface and lowering the temperature even in the troposphere besides changing the pattern of atmospheric circulation.

The earth is tilted at an angle of 23.5° to the perpendicular plane of its orbital path, which is somewhat elliptical. In one half of the year during summer, the northern hemisphere remains tilted towards the sun, while in the other half during winter, the earth gets tilted away from the sun. As a rule, there is no seasonal variation, if there was no tilt.

The oceans are a major component of the climate system. The oceans do cover about 71% of the earth's surface and absorb almost twice as much of the sun's radiation as the atmosphere or the land surface and, thus, contribute to climate change appreciably.

The mass as well as morphology of the earth is gradually changing following the changes in gravitational and magnetic matrix in a way to maintain unique balance among magnetic field, gravitational matrix and incoming solar radiation. Our knowledge to this clue is by and large scanty. However, solar radiation is the sole regulator of major global climate change. But the amount of solar energy reaching the earth is not constant, but varies in several independent cycles. The overall physical scenario results into gradual modification in earth's atmosphere with lapses of time. With changing physical make-up of the earth, variation in incoming solar radiation and its impact on atmosphere is also changing to maintain an equilibrium, which seems to be at risk for human survival. Life is just a by-product of such equilibrium or simply a natural phenomenon as a result of unidentified interactions during course of global climate change. However, a group of physicists may approve and quantify the mode of equilibrium among magnetic field, gravitational matrix and incoming solar radiation that could be congenial to the existence of life on the earth. This truth has to be addressed precisely by establishing an index based on planetary physical components (gravitation, magnetism and their interactions with surrounding within exosphere) and its direct impact on existence of life. Earth's gravitation coupled with magnetism and incoming radiation may form the basis of integration to develop an indicator and changes thereof with corresponding changes in atmospheric climates. There is thus need for physicists, environmentalists, pedologists and life scientists to sit together to arrive at some concrete hypothesis for this endeavor. This will be a true beginning of a new science in a new era.

8.2.2   ELECTROMAGNETIC-NUCLEAR SPECTRUM

The electromagnetic spectrum represents the complete range of electromagnetic radiation. The region of the spectrum with a shorter wavelength than the color violet is referred as ultraviolet radiation, and the region of the spectrum with a longer wavelength than the color red is referred to as infrared radiation (Fig. 8.1).

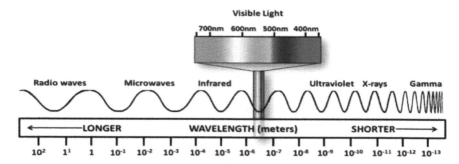

FIGURE 8.1   Electromagnetic solar spectrum (general view pasted from elsewhere).

The energy-energy interactions in the upper atmosphere seem to be of great importance in understanding the EM-nuclear spectra in relation to climate change. The transitional boundary of extraterrestrial and interterrestrial phases of the earth (exosphere) is subject to unique nuclear interactions in a given set of gravitational and magnetic matrix and this seems to be strange in our scientific wisdom. However, the changing physical spectrum seems to control the direction of the nuclear phenomenon and vice versa. Our knowledge to this is virtually scanty. The exosphere-ionosphere may be a huge future laboratory for our scientists, who could monitor how the climate change is becoming challenging to our survival and livelihood in days to come. Soil scientists may get exposed to the basic theme of such natural phenomenon and linkage thereof. The high altitude soils of the Himalayas are often subject to interaction with UV rays. Ozone depletion influenced on global biogeochemical cycles by increasing UV-B radiation at the earth's surface enhancing the global climate change has been documented over the past couple of years. Such elevated UV-B has significant effects on the terrestrial biosphere with important implications for the cycling of carbon, nitrogen and other elements (Zepp et al.,

2003). A colorless water through scattering looks white when flowing on mountainous landform as a stream, while a wet soil could allow light to pass through it facilitating phototrophic microorganisms to grow (Mishra, 1996). Hence, a systematic yardstick needs to be developed through integrated knowledge kits.

8.2.3 CHEMICAL SPECTRUM

As per report, there was perhaps no atmosphere with the origin of earth. However, the earth's primitive atmosphere was different in composition from that of today (Bohn et al., 1985). The trend of such change in atmospheric composition is obvious. Challenges in climate change are the consequences of atmospheric modification and it seems to be natural. Such changes may be or may not be congenial to the survival of lives on the earth. The magnitude and the rate of such change may be affected directly by planetary physical and nuclear spectra. The exosphere above ionosphere is a huge chemical laboratory due to energy-energy and energy-matter interactions. The maximum concentration of ozone is captured in stratosphere, which is sensitive to photochemical reactions. The troposphere or even surface soil is very sensitive to energy interaction. Chemical and biochemical changes are taking place in permutation and combination depending on associated environment. Green house gas emission for CO_2, N_2O and CH_4 are some other contributions of chemical spectrum. The industrial revolution in the 19th century, however, witnessed the large-scale use of fossil fuels. Changes in the atmospheric chemical spectrum are reflected by and large on pedogenic behavior considerably too.

The greenhouse gases include those listed in the Kyoto Protocol viz. CH_4, nitrous oxide (N_2O), hydrofluorocarbon (HFC), perfluorocarbon (PFC), sulfur hexafluoride (SF6) and those given in the Montreal Protocol include chlorofluorocarbons (CFCs), hydrochlorofluorocarbon (HCFC) and others. The tropospheric O_3 is synthesized and transformed by photochemistry in the atmosphere. The volatile organic compounds (VOC) like nonmethane hydrocarbons (NMHC) and oxygenated NMHC (alcohols, aldehydes and organic acids) affect the climate through production of organic aerosols and their involvement in photochemistry (formation of O_3 in presence of NOx and light. In the process of photosynthesis, for example, carbon dioxide reacts with water in presence of sunlight to form carbonic acid and oxygen. Besides, carbon dioxide dissolves in waters of the rivers, ponds, lakes and even oceans up to a critical point, where it attains an equilibrium with that in the atmosphere. The carbonic acid so formed is a powerful input in weathering processes of rocks and minerals. As a consequence, such weathering yields very specific bicarbonate ions, which often contributes to the formation of limestone beds mostly in the ocean floor. The process of weathering is promoted by warmer temperatures. More photosynthesis may cause additional acidity in the system. As explained by the US agency (National Oceanic and Atmospheric Administration), the chemistry of ocean acidification is well understood with the three concepts: (i) more CO_2 in the atmosphere means more CO_2 in the ocean (ii) atmospheric CO_2 is dissolved in the ocean, which becomes more acidic and (iii) resulting changes in the chemistry of the oceans disrupts the ability of plants and animals in the sea to make shells and skeletons of calcium carbonate, while dissolving shells already formed.

8.2.4   BIOLOGICAL SPECTRUM

The biological spectrum is the indicator of the existence of earth's atmosphere as well as life. The biodiversity has thus balancing trend with climate change. Extinction of many living races is indicative of the changing trend of earth's climate. New genes and species are emerging due to environmental factors both at macro and microlevels. There must be a reference point to begin with such complex biological system. Soil science has a big role to stimulate such biological management. Global soil biodiversity is now-a-days becoming priority to understand soil quality even. Voluminous reports are available for reference.

Many field trials across the globe indicate that organic manure compared to mineral fertilization is promoting the soil organic carbon and accordingly sequestering large amounts of carbon from the atmosphere to the soil. Lower emission of greenhouse gasses during crop production and enhanced carbon sequestration coupled with additional outcome of biodiversity makes the organic farming a boon towards combating the challenges of climate change.

8.2.5   ANTHROPOGENIC SPECTRUM

The men have immense potential to reorganize the natural set-up for self-contentment. However, the integrated efforts may lead to a concrete solution to discover such impacts. The present day speedy technological revolution is alarming. The ill-human treatments during past are the consequences of what we see today. Human interference is not confined to any closed terrestrial system. Major human activities result into major consequences as given below:

1. Deforestation, mining, erosion, landslide and bare lands devoid of vegetation.
2. Petroleum, coal and wood fuel burning with gas emission.
3. Shrinkage of natural landforms with nonfarming activities.
4. Loss of biodiversity through hunting, weeding and similar acts of eradication.
5. Nuclear contamination caused by human interferences.
6. Rapidly increasing human population and ill-techniques developed in selfishness.
7. Industrial products like CFC through technological up-gradation/ generation.
8. Decay of organic residues, garbage, etc. dumped interferences without recycling.
9. Electronic networking and war in hand damaging the natural stability.
10. Use of fertilizers, chemicals, insecticides in excess causing toxicity/pollution.
11. Large area under rice cultivation causing CH_4 emission.
12. Ill-human habits to contaminate/toxicate/pollute different foodstuff.

8.2.6 PEDOGENIC SPECTRUM

The land being the upper most surface of the earth's lithosphere is the lowest boundary of the earth's atmosphere. It is a sink for all types of incoming radiation, dumped organic residues, sequestering of almost all emitted green house gases, water, microbes and disposed materials. Human interferences are natural with soils. In particular, carbon may be sequestered in soil and carbon dioxide emission may be restricted through soil and soil water as under:

1. Roots by respiration and organic matter with decay result into CO_2 emission in soil that may get dissolved in soil water, if present in suitable proportion.
2. Organic matter during decomposition may get complexed and chelated with differing clay types restricting further decomposition.
3. Slowing down organic matter decomposition by using suitable mechanical and microbial techniques.
4. Humus materials either present or added to soils is resistant to further decomposition restricting CO_2 emission.
5. Special chemical arrangement to be made to divert emitted CO_2 during decomposition of organic materials in some nonemitting compounds in soils.
6. Low wavelength radiation (X-ray and gamma ray) often converts soil organic carbon to nitrogen and the mechanisms need to be quantified (Mishra et al. 2006a, 2006b, 2008).
7. EM radiation beyond 720 nm showing thermal properties may be reflected back to the atmosphere in order to suppress the soil organic matter decomposition.
8. Soil covered with vegetation round the year may buffer the diurnal temperature change and minimize the organic matter decay considerably (conservation agriculture).
9. Agro-forestry with high CO_2 demand on the river-banks may sequester CO_2 by its dissolution with river water preferentially.
10. Construction of highways and railway tracks on river-banks may facilitate green house gases to be sequestered under the influence of agro-forestry and river water.

Importantly, soil microorganisms are the major component of biogeochemical nutrient cycling and global fluxes like CO_2, CH_4 and N. The global soils are estimated to contain almost twice as much carbon as the atmosphere, making them one of the largest sinks for atmospheric CO_2 and organic carbon (Jenkinson et al., 1991).

8.3 STRATEGIC PLANNING TO COMBAT WITH CHALLENGES THROUGH INTEGRATED APPROACH

The global climate change is virtually an outcome of a complex phenomenon in an extraordinary open system that seems to be beyond ones control and can hardly be inventoried without system evaluation. The related issues, challenges and priorities are not so easy to enlist, since such exercise cannot be made possible in isolation. Moreover, its science is specifically specializing, but our approach to this effect has more or less been generalizing. Taking all six spectra (Fig. 8.1) into consideration, strategic planning may be formulated through integrated approach as below.

1. Mechanical as well as chemical efforts to minimize CO_2, CH_4 and N_2O emission.
2. Carbon sequestration by mechanical, chemical, biological and pedogenic manipulation.
3. Planting trees and agro-forestry of high CO_2-demand.
4. Green house gas emission from burning fossil, liquid, solid and gaseous fuels including other sources may be chemically used/sequestered in some other forms.
5. Minimizing load on petroleum and other fuel sources through different bio-fuel extraction.
6. Architectural manipulation providing greenery over buildings, lawn and similar constructions.
7. Adoption of conservation agriculture by keeping the land covered with vegetation round the year.
8. Restoration of biodiversity and forest plantation in all bare lands.
9. River-banks to be covered with agro-forestry with shrubs and trees of high CO_2 demand.
10. High-density orchards and garden plants following the management to keep the soil covered with shadow horticultural crops.
11. Impose legal ban against any construction or nonfarming activities in a productive land.
12. Construction of highways and railway tracks along the bank of river.

Top soil (edaphology) restoration is an issue to be addressed in the light of surrounding environment including slope gradient, parent materials, aspect, natural vegetation, effective soil depth, rainfall intensity and distribution, zeolitic materials, clay types and nature of land use and soil biodiversity. Soil developed on limestone is very susceptible to erosion as compared to basaltic soils. Conservation agriculture is a management system to keep a soil covered with vegetation round the year. But, our efforts must follow a system starting from evaluation of soil and its environment on total soil basis (pedology). With a comprehensive knowledge about the soil, let's move to its exploitation to minimize the emission of greenhouse gasses; for which technical options are available and readers may consult the relevant literature/text.

The conservation agriculture by keeping the land covered with vegetation round the year with least tillage may be an agenda to restore biodiversity and pedo-ecosystems. Selection of site-specific crop rotation based on quantitative land evaluation as well as cover crops between two main crops may need exhaustive experience with indigenous knowledge and this may enable agricultural scientists to work with farmers together in interactive environment. Interaction of light with soils is a new chapter (Photopedogenesis) to be tested as a soil based indicator of climate change (Mishra et al., 2006b; Mishra et al., 2008; Mishra and Ghanshyam, 2009). As Janseens et al. (2003) stated, the soil is one of the important sources as well as sinks of greenhouse gasses (GHGs) causing global warming and climate change. It contributes about 20% to the total emission of carbon dioxide through soil respiration and root respiration, 12% of CH4 and 60% of anthropogenic nitrous oxide emissions (IPCC, 2007). It is presumed that the global warming may influence global carbon cycle besides distorting the structure and function of ecosystem. The GHGs virtually trap the outgoing IR radiation from the earth's surface and raise the temperature of the atmosphere (Pathak et al., 2007, 2009, 2012).

The challenges of global climate change are very alarming that integrates earth's physical spectrum with pedogenic spectrum (earth's mass in changing gravitational-magnetic equilibrium) as affected by chemical, EM-nuclear, biological and anthropogenic spectrums in order to sustain the equilibrium of dynamic global climates (Mishra and Ghanshyam, 2009). Distorted equilibrium as caused by human interferences (man-made distortion and technologic imbalance) may lead to many catastrophic events. Agricultural interventions may mitigate the expected earth's disturbances by adopting the conservation agriculture, banning against land shrinkage/ land sealing. The orchards are being destroyed on way to sale the land for construction purposes in some parts of India. There must be a legal ban against such free practices (Fig. 8.2).

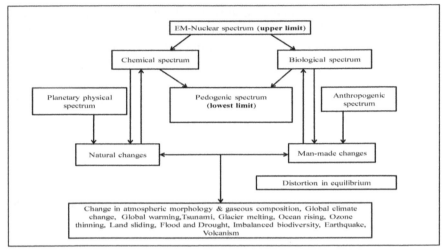

FIGURE 8.2 Principles of strategic planning to combat with challenges of climate change using integrated management inputs.

India has a network of rivers that cover large area under agriculture. Flood may often be disastrous. In order to sustain agriculture as well as maintenance of perennial rivers like Ganga, Kosi, Brahamputra, Gandak, Ghaghra and Mahananda in Bihar, a pilot trial may be undertaken to make ridges on both sides of the river banks followed by planting of agro-forestry (with high photosynthetic demand) and construction of highways and railways. The maintenance of railways, highways and agro-forestry would facilitate the maintenance of river-bank embankment and flood management in a big way.

8.4 CONCLUSION

The global climate change is a challenging task to be tackled not in isolation, but through integrated efforts, in which a high level global scientific team may be involved from all specialized sectors covering all spectra of interest. The challenges with global climate change are both natural as well as man-made. Natural consequences may be handled by gathering scientific know-how, whereas man-made impacts are manageable with participatory commitments in a planned way on site-specific basis. The past may entail the issues, while present will scrutinize the challenges and based on critically documented issues and challenges, our efforts should aim at recognizing the priorities for framing the action plan on integrated basis to look for a future healthy environment congenial to our survival and livelihood on sustainable foundation of the earth. In such endeavor, the pedologic interventions do deserve special attention in integrating all the rest spectra of relevance, since soil is the lowest boundary of the earth's atmosphere.

ACKNOWLEDGMENTS

Thanks for technical support to those whose contributions are used in some ways in order to enrich the theme of this chapter.

KEYWORDS

- **Controversies**
- **Global Climate**
- **Integrated Approach**
- **Mitigation**
- **Principles**
- **Strategic Planning**

REFERENCES

Bohn, H. L., McNeal, B. L., & O'Connor, G. A. (1985). Soil Chemistry, 2nd Ed. John Wiley & Sons, New York, 15–17.

IPCC. (2007). Climate Change 2007, Climate Change Impacts, Adaptation and Vulnerability, Summary for Policy Makers, Inter-Governmental Panel on Climate Change.

Janseens, I. A., Frelbauer, A., Ciais, P. et al. (2003). Europe's Biosphere Absorbs 7–12% of Anthropogenic Carbon Emissions. Science, 300, 1538–1542.

Jenkinson, D. S., Adams, D. E., & Wild, A. (1991). Model Estimates of CO_2 Emissions from Soil in response to Global Warming. *Nature* 351, 304–306.

Lin, H. (2006) Clarifying Misperceptions and Sharpening Contribution, In the Future of Soil Science, IUSS, The Netherlands, 80–83.

Mishra, B. B. (1996). Theory of Photopedology, J. Indian Soc. Soil Sci. 44, 541–543.

Mishra, B. B., Heluf, G., & Sheleme, B. (2006a). Photopedogenesis New Chapter in Soil Science. Theatre Presentation (New Frontier), 85/457b, 18th WCSS, 9–15 July, 2006, Philadelphia, USA.

Mishra, B. B., Heluf, G., & Sheleme, B. (2006b). Photopedogenesis Concept and Application, J. Food Agric. Environ. 4, 12–14.

Mishra, B. B., Heluf, G., & Sheleme, B. (2008). Photopedogenesis New Frontier in Soil Fertility Evaluation, Indian Fertilizer Scene Annual 2008, 97–106.

Mishra, B.B. and Ghanshyam. (2009). Are Challenges of Global Climate Change Manageable? International Conference on combating challenges of global climate change, New Delhi, 7–8 Feb., 2009, 42–44.

Pathak, H., & Wassmann, R. (2007). Introducing Green House Gas Mitigation as a Development Objectivein Rice Based Agriculture, (1) Generation of Technical Coefficients. Agric. Syst. 94, 807–825.

Pathak, H., Aggarwal, P. K. & Singh, S. D. (Eds). (2012). Climate Change Impact, Adaptation and Mitigation in Agriculture, Methodology for Assessment and Application, IARI, New Delhi.

Pathak, H., Jain, N. & Bhatia, A. (2009). Global warming mitigation potential of biogas plants in India. Environ. Monit. Assess. 157, 407–418.

Zepp, R. G., Callaghan, T. V., & Erickson, D. J. (2003). Interactive Effects of Ozone Depletion and Climate Change on Biogeochemical Cycles, *Photochem Photobiol Sci.* 2(1), 51–61.

... ...

CHAPTER 9

NANOTECHNOLOGY, PLANT NUTRITION, AND CLIMATE CHANGE

S. KUNDU, TAPAN ADHIKARI[1] and A. SUBBA RAO

Division of Environmental Soil Science Indian Institute of Soil Science, Bhopal-462038, India; [1]E-mail: tapan_12000@yahoo.co.uk

CONTENTS

ABSTRACT

Nanotechnology is emerging out as the sixth revolutionary technology in the current era after the Industrial Revolution of mid 1700s, Nuclear Energy Revolution of the 1940s, The Green Revolution of 1960s, Information Technology Revolution of 1980s and Biotechnology Revolution of the 1990s. It is a new frontier with challenges and opportunities that will extend our reach and enrich our lives. The word *Nanotechnology* has originated from a Greek word which means "dwarf" and nanometer is 1 billionth of a meter (1 nm = 10^{-9} meter). Nanotechnology is defined as the understanding and control of matter at dimensions of roughly 1–100 nm, where unique physical properties make novel applications possible (EPA, 2007). Utilizing basic principles of nano-science and nano-technology, we can contribute significantly towards development of new agro-production and protection technologies. It is worthy to consider the following relationship between particle size and surface area what we have known for a long-time about how the properties of particles change as they decrease in size to the micron to submicron size range. For spherical particles, the ratio of surface area 'A' ($A=4\pi r^2$) to volume 'V' ($V= 4/3\pi r^3$) is inversely proportional to the particle radius 'r,' i. e. $A/V = 3/r$. This relationship tells us that as a particle becomes smaller, its surface area becomes an increasingly larger component of its overall form. Thus, reduction of particle sizes of the naturally occurring minerals is an important means to increase their reactivity in soil vis-à-vis availability to the growing plants. Vast deposits of minerals, which are not suitable for industrial use, can be made useful as sources of plant nutrients for the crop production. Our investigation clearly indicated that low grade rock phosphates, can easily be made as a source of P to the plant when they are converted to nano-size (<100 nm). Similarly, vast deposits of gluconite/waste mica, dolomite and magnetite (as source of Ca & Mg), pyrite (as source of Fe and S), can be made useful for agriculture use with a reduced cost and without impairing damage to environment. A protocol was developed to fortify the Urea granules with a consortium of nano-particles of Zn, Cu, Fe, and Si using oleoresin. This protocol can successfully be used to deliver nano-particles of micronutrients along with urea. The nano-particles coated urea, thus produced, contained 43.84% N, 2.20 mg Zn/g Urea, 1.10 mg Fe/g Urea, 0.66 mg Cu/g Urea and 1.06 mg Si/g Urea. Application of such urea @ 200 Kg/ha will supply, 440 g Zn, 220 g Fe, 132 g Cu and 212 g Si along with 87.68 kg N/ha to the crops. A protocol has also been developed to coat the seeds of maize, soybean, pigeon pea and ladies finger with nano scale (<100 nm) ZnO powder. The most important advantage of seed coating with nano ZnO is that it does not exert any osmotic potential at the time of germination of the seed, thus, the total requirement of Zn of the crop can be loaded with the seed. This protocol of seed coating with ZnO can be used by the seed producing agencies to produce customized seed for Zn deficient areas of the country.

9.1   INTRODUCTION

Currently, the effects of engineered NPs on human health and on entire ecosystem is not much explored ever, though naturally occurring NPs (e.g., nanoclays, iron oxides and volcanic ash) are ubiquitous in all ecosystems. It has also been demonstrated that naturally occurring NPs have important local, regional, and even global consequences. For examples, it has been shown that airborne, NPs (and large dust particles) link the deserts of Africa with the productivity of the open ocean and help sustain the nutrient demands of terrestrial ecosystems worldwide (Chadwick et al., 1999) we now know that naturally occurring NPs are even present in interplanetary and interstellar space (Hochella, 2008). They have also been abundant on earth since its formation, were part of its formation (Becker, 2006) and life from the beginning has evolved in their presence. Emerging research is suggesting that many organisms synthesize NPs. As analytical tools for the detection of NPs improve, we may find that biogenic NPs are ubiquitous and bio-geochemically vital across the living planet.

9.2   DEFINITION OF NANOTECHNOLOGY

Nanotechnology is defined as the understanding and control of matter at dimensions of roughly 1–100 nm, where unique physical properties make novel applications possible (EPA, 2007). The British Standard Institution (BSI, 2005) and American Society for Testing and Materials (ASTM, 1985) defined nanotechnology as "Design, characterization, production and application of structure, devices and systems controlling shape, size and composition at the nanoscale." Banfield and Zhang (2001) suggested that *nanoparticles* might be defined based on the size at which fundamental properties differ from those of the corresponding bulk material. The commonly used definition of "dissolved" is in most cases operationally defined by all compounds passing through a filter, in many cases with a cutoff at 0.45 μm (450 nm). The colloidal fraction is defined as having a size between 1 nm and 1 μm, (Buffle, 2006) therefore overlapping with the nano-particles (NP). A wide variety of materials can be used to make such nanoparticles, such as metal oxide ceramics and silicates, magnetic materials, lyposomes, dendrimers, emulsions, [etc.] The size dimension of some naturally occurring materials are given in Table 9.1.

TABLE 9.1 Size Dimensions of Different Natural Materials

Name	Size
Molecules	< 1 nm
Colloids	1–1000 nm
Nanoparticles	1–100 nm
Virus	10–100 nm
Red Blood Cell	2000–5000 nm
Bacteria	250 nm-1000 nm
Tissue Cell	10,000 nm
Width of Human Hair	80,000 nm
Intracellular Spaces in Seed Coat Parenchyma	< 10,000 nm

Nano particles (NPs) are generally defined as materials that are <100 nm size (0.1 μm) in at least one dimension. This means that nano materials can be three-dimensional particles (Spherical, Cuboids etc), or two-dimensional particles (Ultra thin film) or one-dimensional (fine rods). Their chemical (reactivity, solubility, etc.), mechanical (elasticity, hardness, etc.), electronic (conductivity, redox behavior) and nuclear (magnetic) properties often change as a function of size. These changes can be, and often are dramatic. This is precisely what leads to their exceptional scientific and commercial value and their anticipated enigmatic behavior in already extraordinarily complex earth environment.

9.3 UNIQUE FEATURES OF NANO-PARTICLES

Because of the small size, the physical, chemical, electronic properties of nano-structures changes as a function of size and are very different from that of their bulk counterparts. Due to the small size of nano-particles, there are more atoms on the surface compared to the interior of the particles, which leads to a large surface to volume ratio which in turn leads to higher reactivity of nano-particles. For example, if a cube of 1 mm size is broken into cubes of 1 nm size, the total volume remains the same, but the surface area is increased by 10^6 times. One of the principle ways in which a nano-particle differs from a larger or bulk material is that a high proportion of the atoms that are associated with a NP occur at the surface. As surface area increases in comparison to the volume, the behavior of the atoms on the surface of the particle becomes more potent as compared to those atoms that are inside the particle. Once particles become small enough they exhibit quantum mechanical behavior. Because their size is smaller than the order of wavelength, nanoparticles do not obstruct light.

The large surface to volume ratio also results in more interaction between atoms in intermixed materials in nanoparticles, which may lead to increased strength, increased heat resistance, etc. Melting points of nano-material decreases for clusters smaller than a few hundred angstroms. For example the melting temperature of gold decreases by approximately a factor of two when the cluster size is reduced from 10 nm to 2 nm. The melting point of gold in bulk is 1337°K whereas melting point of gold nano-particle (~2 nm) is 650°K. Magnetic Properties of nano-clusters are also very different from that of the corresponding bulk material. For example nanoclusters of certain materials like Pd, Na, K and Rh are ferromagnetic, where as in bulk form, these elements are paramagnetic. Super-paramagnetism is a phenomenon that arises from the small size of nanoclusters. Discretezation of energy level is an important property of nano-particles. In bulk, the overlapping of the molecular orbitals of a large number of atoms results in a continuum of energy levels or energy band. But in nano-particles, due to fewer atoms the overlapping of their orbitals is not much and thereby exhibit discrete energy levels. Thus discretization of the electronic energy levels takes place in nano-particles along with an increase in the electronic band gape energy, which in turns, results in interesting optical and electronic properties in nano-particles. Most of these changes are related to the appearance of quantum effects as the size decreases, and are the origin of phenomena such as the super-paramagnetism, Coulomb blockade, surface plasmon resonance, [etc.] Surface Plasmon resonance (SPR) is a collective excitation of the electrons in the conduction band near the surface of the nano-particles. The surface of the nano-particles is like plasma having free electrons in the conduction band with positively charged nuclei. The position of the specific Plasmon absorbance band indicates the presence of specific size of nano-particles. Specific Plasmon absorbance at 412 nm (λ_{max}) indicates the presence of silver nano-particles. Similarly, a single absorbance band at 529 nm confirms the presence of gold nano-particles (7–20 nm). The wavelength of the absorption peak depends on various factors like the particle size, dielectric constant of the surrounding medium and inter particle distance. For example, with the increase in diameter of gold nano-particles from 22 nm to 100 nm, the λ_{max} value increases from 526.5 nm to 540 nm. As mentioned earlier, the origin of novel properties in metal nano-particle is attributed to an increase in surface to bulk atoms and appearance of surface plasmon resonance. The dependence of properties on the shape of the particle originated due to different reasons: the surface energy, electronic structure and the stability of different facets. The spherical particles of gold absorb electromagnetic spectrum in the visible region whereas nanotriangles of gold absorb in the near infrared region. On the contrary, gold nanorods show two plasmonic structures: longitudinal plasmon (owing to electronic oscillation parallel to the longitudinal axis) and transverse plasmon (owing to electronic oscillation parallel to the transverse axis).

9.4 STABILITY OF NANO-PARTICLES

The stabilization of ultra fine or nano-particles in suspension is very important for both controlling the particle size and for developing process based application of these suspensions to achieve a desired result. The formation, stabilization and sedimentation of nano-particles depend upon the discreet steps of nucleation, condensation and coagulation into larger particles. Therefore, stabilization requires the optimization of these competing factors. When the size of the particle is reduced to the nano level, the ratio of surface to the bulk atoms increases thereby increasing energy of the system as a whole, leading to a decrease in the system stability. A number of laboratory studies indicate that nano-particles will aggregate to some extent following their release to water. The reported size distribution of selected ENMs in water is presented in Table 9.2. It can be seen that the hydrodynamic Particle size is much greater than the individual particle size in the dry phase, indicating that the aggregation is a common process for ENMs in water.

TABLE 9.2　Reported Size Distribution of Selected Engineered Nanomaterials (ENMs) In Water

ENMs	Individual Particle Size (nm)	Hydrodynamic Particle Size (nm)	References
Ag	26.6 ± 8.8	216	Griffitt et al. (2008)
Cu	26.7 ± 7.1	94.5–447.1	Griffitt et al. (2008)
Al	41.7 ± 8.1	4442	Griffitt et al. (2008)
Co	10.5 ± 2.3	224–742	Griffitt et al. (2008)
Ni	6.1 ± 1.4	44.9–446.1	Griffitt et al. (2008)
TiO_2	20.5 ± 6.7	220.8–687.5	Griffitt et al. (2008)
ZnO	50–70	320 ± 20	Zhang et al. (2008)[10]
SiO_2	10	740 ± 40	Zhang et al. (2008)
Fe_2O_3	5–25	200 ± 10	Zhang et al. (2008)
Fe_2O_3	9.2	46.2	Baalousha et al. (2008)
Fe_3O_4	<10	120	IIIes and Tombacz (2006)
CeO_2	8	323–2610	Xia et al. (2008)
Al_2O_3	60	763	H.H. Wang et al. (2009)
CdSe/ZnS	2.1	˜12.5	Slaveykova et al. (2009)

The particle size distribution of commercially available nano-particles (Sigma-Aldrich) in aqueous system, measured by Laser Diffraction Particle Size Analyzer (Model–LS 13320) at 23°fixed detector angle clearly indicated (Table 9.3) that all the particles had dimensions more than the size claimed in dry phase (Tapan et al., 2009). This might be due to the fact that the negative charged surface of the particles had strong affinity towards water and the hydrated radii of the particles were much higher than their un-hydrated counterparts. Therefore it is very important to get stable suspension of nano-particles of desired size for harnessing the benefit of nano-particles.

The zeta potential of the nano-particles is another important parameter that has been extensively investigated for stability of nano-particles. High zeta potential (negative or positive) will impart stability to the nano-particles suspension, whereas nano-particles with low zeta potentials tend to coagulate or flocculate (Table 9.4) pH is a major factor determining the zeta potential of nano-particles. When pH is at point of zero charge (*pzc*) or isoelectric point, the nano-particles exhibit minimum stability (i.e., exhibit maximum coagulation/flocculation). When the pH is lower than the *pzc* value, the nano-particle surface is positively charged and the zata potential will increase with decreasing pH below the *pzc*. Conversely, at pH above *pzc*, the surface is negatively charged and the zeta potential will be more negative with increasing pH. The values of *pzc* of selected nano-particles are presented in Table 9.5. The ionic strength of the surrounding aqueous medium is also an important factor affecting the stability of the nano-particles. When ionic strength is increased and/or the zeta potential is reduced (both effects usually result from an increased salt concentration), attractive force between colloids will outweigh the repulsion, and the particles can then adhere each time they collide. The surface charge is also another important property that can dominance the migration of ENMs in porous media (Darlington et al., 2009). The soil particles are generally negatively charged. Thus, positively charge ENMs will be readily electro-statically attracted to the soil surface. Nano-particles with higher negative charges are believed more mobile in soil matrix because of the longer electrostatic repulsion between the nano-particles and soil particles and between nano-particles themselves as well. Therefore, various methods have been applied to modify ENM surface properties to control (enhance or restrict) the transport of ENMs in porous media/soil among which surface functionalization with hydrophilic functional groups (e.g., -OH and –COOH) and surface physical modification using polymers or surfactants are two commonly adopted methods.

TABLE 9.3 Size of Nano-Particles in Aqueous System

Nano particles	Distribution in different fraction		
	10%	40%	50%
Zinc Oxide (<100 nm)	<109.10 nm	109.10–188.00 nm	188.00–325.40 nm
Cu- oxide (<50 nm)	<126.70 nm	126.70–185.90 nm	185.90–275.80 nm
Iron Oxide <100 nm	<97.80 nm	97.80–176.20 nm	176.20–310.10 nm
H- Apatite (<200 nm)	75.40 nm	75.40–145.60 nm	145.60–283.50 nm
Cu, Fe, Zn Oxide (<100 nm)	116.90 nm	116.90–195.20 nm	195.20–360.80 nm
TCP (<200 nm)))	92.30 nm	92.30–232.20 nm	232.20–616.10 nm

TABLE 9.4 Effect of Zeta Potential on Colloidal Stability (Adapted from American Society for Testing and Materials, 1985)

Zeta Potential (mV)	Stability Behavior of Colloids
From 0 to ± 5	Rapid coagulation or flocculation
From ± 10 to ± 30	Incipient instability
From ± 30 to ± 40	Moderate stability
From ± 40 to ± 60	Good stability
More than ± 61	Excellent stability

Table 9.5. Reported Point of Zero Charges (pzc) for Several Engineered Nanomaterials

Nanoparticles	Particle size (nm)	pH_{pzc}	References
TiO_2	15	5.2	Zhang et al. (2008)
Fe_2O_3	9.2	9.1	Baalousha et al. (2008)
Fe_2O_3	<10	8	IIIes and Tombacz (2006)
Al_2O_3	60	7.9	Ghosh et al. (2008)
ZnO	50–70	9.2	Zhang et al. (2008)
SiO_2	10	1.8	Zhang et al. (2008)
NiO	10–20	9.1	Zhang et al. (2008)
MnO_2	50	2.4	Feitosa-Felizzola et al. (2009)
CeO2	157.4	>7.0	Necula et al. (2007)

9.5   BEHAVIOR OF NPS IN ATMOSPHERE

The atmospheric scientists describe NPs as ultrafine particles, a category consisting of particles smaller than 10 nm (referred to as nucleation mode) and those from 10 to 100 nm (referred to as Aitken mode). Particles in this size range originate mainly from combustion and photochemical reactions. Natural and incidental NPs dominate the number distribution of particles in the atmosphere in polluted urban, rural, remote continental and marine atmosphere, typical concentration ranges are 10^5 to 4 10^6, 2 10^3 to 10^4, 50 to 10^4 and 100 to 400/cm³, respectively (Pandis, 2006) NPs in the atmosphere are subject to coagulation, surface coating through condensation of semi volatile compounds and heterogenous reaction with gaseous pollutants. For NPs, coagulation is dominated by Brownian (thermal) motion that leads particles to collide with each other, grow in size, and shrink in number. Coagulation rates are low when particles are of same size but increases by orders of magnitude as the differences in size grows. The time scale for coagulation is dependent on the particle size of interest and the background particle size distribution. For example, in mono disperse population of 20 nm particles, the time scale for coagulation in urban and rural setting are around 10 min and 10 h (Tiwari, 2010). Particles are removed from atmosphere mainly by Brownian diffusion, gravitational settling (dry deposition) and wet deposition in precipitation to the earth's surface. Time scale for wet deposition is highly dependent on the frequency of precipitation events. During a precipitation event, atmospheric NPs may be removed via wet deposition mechanism in a matter of minutes to hours (Tiwari, 2010). The time scale for removal by dry deposition in an urban area in 20 h for 40–50 nm size particles (Berkowicz, 2004). The most common types of naturally occurring particles in the atmosphere consists of black carbon or soot, a combustion by product which is thought to very harmful because it contains fullerenic compounds and carbon nanostructures (Vander et al., 2000).

9.6   BEHAVIOUR OF NPS IN AQUATIC ENVIRONMENT

Engineered and natural NPs, after entering the aqueous environment, will interact with the ubiquitous natural aquatic colloids which affect the stability and subsequent environmental behavior of both NPs and aquatic colloid. Stability of NPs in aqueous environments is a key factor controlling their transport and ultimate fate in aqueous environments. Large aggregate of NPs will quickly precipitate out and their transport and bioavailability will be greatly restricted. However, well-dispersed NPs will be widely transported and have higher chances to interact with and cause potential harm to organisms. System pH is a major factor determining the zeta potential of colloids. When pH is at point of zero charge (pzc) or isoelectric point, the colloidal system exhibits minimum stability (i.e., exhibits maximum coagulation/flocculation). When the pH is lower than the pzc value, the colloid surface is positively charged and the zeta potential will increase with decreasing pH below the pzc. Conversely, at pH above pzc, the surface is negatively charged and the zeta potential will be more negative with increasing pH. High zeta potential (negative and positive) will impart stability to the NP suspension, whereas NPs with low zeta potential tend to coagulate or flocculate.

9.7   BEHAVIOR OF NPS IN TERRESTRIAL ENVIRONMENT

After entering into soil environment, NPs may be retained by soil matrix or break through soil matrix and reach ground water, which is also determined by the properties of NPs and soil. There are significant physical and chemical similarities between the most widely manufactured ENPs and naturally occurring nanoparticles, although in a number of cases the exact size, shape, and coatings/surface functional groups may be quite different from ENPs. Also while the term nanoparticle may not yet be widely used in ecology, earth scientists have been studying at least some major classes of natural nanoparticles for many decades. For example, it was realized years ago that the charged surfaces of clays in soil-an example of natural nanoparticles-were known to form electrostatic bonds with ions (e.g., NH_4^+, Ca^{2+}, K^+, and Mg^{2+}) that contribute substantially to soil fertility by preventing the loss of these vital nutrients to groundwater (Eriksson, 1952; Gieseking, 1939). Now modern nanoscience has become an integral part of soil science that goes far beyond the study of clay minerals (Hochella, 2008; Yuan, 2008). This and other aspects of nanogeoscience, as it is now called, have extensively developed relatively recently, particularly in the last decade (Banfield, 2001; Hochella, 2002; Hochella, 2008). This is because an exceptionally wide variety of nanoparticle exist on earth, and are in fact ubiquitous in both the biotic and abiotic compartments of earth (Banfield, 2005; Blango, 2009). Emerging research is suggesting that many organisms synthesize nanomaterials. Bacteria in sediments may synthesize electrically conductive pilli, called nanowires, for sensing neighbors or for transferring electrons and energy (Bargar et al., 2008; Fredrickson et al., 2006). Bacterial reduction of urantyl, U^{6+} (aq), to U (IV) oxide (uraninite) is an important bioremediation strategy (Bose et al., 2009; Manceau et al., 2008) found that wetland plants, or their symbionts, synthesize copper (Cu) nanoparticles in their rooting zone when grown in contaminated soils, thereby reducing Cu uptake. Dissimilatory metal-reducing bacteria even respire on iron oxide nanoparticles in anaerobic environments (Moore, 2006).

9.8   ENTRY OF NANO-PARTICLES INTO PLANTS

Not much research information is available on interaction of NP with plants. The uptake of many types of NPs in the bacterial cell (prokaryotes) is very much limited as they do not have mechanisms for transport of NPs across the cell wall but in eukaryotes, cellular internalization of NPs occurs through the process of endocytosis and phagocytosis (Obidzinska, 1998). Because seed coats have pores that exibit selective permeability, the interaction between particulate constituents and the plant may be limited until the radicles emerge and come into direct contact with the growth medium (Wouterlood et al., 2003). However, intracellular spaces (<10 μm) in seed coat parenchyma may be filled with aqueous media facilitating the transport of soluble nutrients as well as small particles to the embryo (Neill et al., 1999). It is unknown whether intracellular uptake is a requirement for causing phytotoxicity or beneficial effects.

Plants are an important component of the ecological system and may serve as a potential pathway for NPs transport and a route for bioaccumulation into the food chain. Plant cell wall acts as a barrier for easy entry of any external agent including nano-particles into plants cells. The sieving properties are determined by pore diameter of cell wall ranging from 5 to 20 nm (Sigg et al., 2008). Hence, only nano-particles or nanoparticle aggregates with diameter less than the pore diameter of the cell wall could easily pass through and reach the plasma membrane (Obidzinska, 1998; Xing, 2008). Certain NPs may increase the permeability of plant cell walls under stress and then permeate into the cells (Uzu). There is also a chance for enlargement of pores or induction of new cell wall pores upon interaction with engineered nano-particles, which in turn enhance nano-particles uptake. Further Internationalization occurs during endo-cytosis with the help of a cavity like structure that form around the nano-particles by plasma membrane. They may also cross the membrane using embedded transport carrier proteins or through ion channels. When nano-particles are applied on leaf surface, they enter through the stomatal openings or through the bases of trichomes and then translocated to various tissues (Goldbach; Tilney, 1991) After entering the cells, NPs may be able to transport between cells via plasmodesmata, which are microscopic channels of plants traversing the cell walls and enabling transport and communication between cells. Plasmodesmata or intercellular bridges were reported to be cylindrical channels about 40 nm in diameter (Jia et al., 2005). Thus, NPs with diameter less than 40 nm may enter and transport in the plant cells through the plasmodesmata once they are in the plant cells. In the cytoplasm, the nano-particles may bind with different cytoplasmic organelles and interfere with the metabolic processes at that site (Lee et al., 2008). Plant uptake can be a critical transport and exposure pathway of NPs in the environment. Interactions between plants and NPs, such as the mechanism of uptake and translocation and the interactions between the NPs and plant tissues at the molecular and cellular level, merit further investigations.

A recent study showed significant uptake of nano-sized copper (nCu) by *Phaseolus radiates* (mung bean) and *Triticum aestivum* (wheat), respectively (Xing, 2008). Transmission electron microscopy analysis showed that Copper nano-particle was absorbed and agglomerated into the cytoplasm of the root cells and the extent of absorption depended on the concentration of the copper nano-particle deposited on the roots' surface. Another study found individual ZnO nano-particles in endodermal and vascular cells of rygrass (Jin et al., 2008). Significant uptake, translocation and accumulation of Fe_3O_4 nano-particle in the roots and leaves of *Cucurbita maxima* (pumpkin) has also been reported without any effect on growth and development of the test species (Wilson et al., 2002). Therefore, some uptake of nanoparticles by plants is very possible. However, little is known about the maximum nano-particle size amenable for plant uptake, and how uptake kinetics and toxicity are affected by plant type and rhizospheric chemistry. Recent research highlights the importance of transition metals that adsorb to nano-particles and promote oxidative stress (Owen et al., 2008) whereas natural organic matter in soil or pore water can sorb, coat or stabilize nano-particle suspensions and affect their mobility, bioavailability, reactivity and toxicity (Mahendra et al., 2008; Olszyk et al., 2008). This illustrates the daunting challenge of quantifying and predicting the nano-particle properties and bioavailable concentration to which plant roots may be exposed in nature.

Limited studies have been conducted to examine the phytotoxicity of ENMs in plants. Carbon nano-tubes were observed to be adsorbed onto the root surface of several crop species, but no visible uptake of CNTs was shown However, Lin et al. (2009) observed the uptake, translocation and transmission of carbon nano-materials in rice (Oryza sativa L.) plants. Copper nanoparticles were found in root cell of mung bean (*Phaseolus radistus*) and wheat (*Triticum aestivum*) in the forms of individual nanoparticles and aggregates, and the bioaccumulation increased with increasing nanoparticles dose (Lin and Xing, 2008) also observed the adsorption of ZnO nanoparticles onto ryegrass (*Lolium Perenne*) root surface and confirmed the presence of the nanoparticles in the endodermis and xylem cells, indicating that the nanoparticles were taken up by the plant. But it seems that few (if any) ZnO nanoparticles were detected to transport from root to shoot in that study. However, a species of pumpkin was found to absorb, translocate, and accumulate of magnetite (Fe_3O_4) nanoparticles from a hydroponic solution throughout the plant tissues (root, stem, and leaves). (Owen et al., 2008) the plant uptake of magnetite nanoparticles significantly depended on plant variety and environmental condition. (Zhu et al., 2008) showed that lima bean (*Phaseolus limensis*) plants could not absorb magnetite nano-particles from the hydroponic solution, and the uptake and accumulation of the nano-particles were significantly reduced in the pumpkin (*Curcubita maxima*) plants grown in sand compared with those grown in the hydroponic solution and were not detected in the pumpkin plants grown in soil probably due to the sorption of nano-particles by soil and/or particle aggregation.

Clearly, interactions between plants and ENMs, such as uptake potential of different plant varieties, the effect of plant growth media condition, mechanisms of uptake and translocation, and the interactions between the particles and plant tissues at the cellular and molecular level, require further in depth investigation. Such studies will help us understand the plant uptake of ENMs as a potential transport and exposure route and its role in bioaccumulation through the food chain. There are many unknowns regarding the environmental fate, transport, exposure, ecotoxicity, and life-cycle of NPs and therefore, we need strong research support for the development and implementation of environmentally benign nanotechnology.

9.9   OLEORESIN COATED UREA FORTIFIED WITH NANO-PARTICLES

A protocol was developed to fortify the Urea granules with a consortium of nano-particles of Zn, Cu, Fe, Si using oleoresin. For easy dispersion of nano particle in water, ethyl alcohol was used instead of hexane to dissolve oleoresin for coating urea and then fortified with nano particles. This protocol can successfully be used to deliver nano-particles of micronutrients along with urea. About 12 g crude oleoresin was dissolved in 100 mL ethyl alcohol in a wide mouth bottle and to it 220 g urea was mixed and shaken for 5 min. After that whole content was transferred to plastic tray fitted snugly on a horizontal shaker and shaking operation was continued with maximum speed to evaporate the alcohol. When 80% of the alcohol evaporated, a mixture of nano-particles was uniformly sprayed over the tray using a 53 μm sieve. The mixture of nano-particles contained 1 g Zn as ZnO (<100 nm), 0.5 g Fe as Fe_3O_4(50 nm) 0.3 g Cu as CuO (50 nm) and 1 g Si as SiO_2 (<20 nm). Before spraying these nano-particles on the pine oleoresin coated urea, they were mechanically mixed well. After the addition of nano-particles, the shaking operation continued for 10 min and thereafter kept in oven at 50°C for hardening to get free flowing urea fortified with the consortium of nano-particles. Out of the total amount of nano-particles added, 48.42% of the nano-particles got adsorbed on the surface of oleoresin coated urea. The nano-particles coated urea, thus produced, contained 43.84% N, 2.20 mg Zn/g Urea, 1.10 mg Fe/g Urea, 0.66 mg Cu/g Urea and 1.06 mg Si/g Urea. Application of such urea @ 200 Kg/ha will supply, 440 g Zn, 220 g Fe, 132 g Cu and 212 g Si along with 87.68 kg N/ha to the crops.

9.10   RISKS INVOLVED WHILE HANDLING NPS

Despite myriad reports available on research strategies and position papers on promises and perils of nanomaterials, basic information regarding the health and environmental risks of NPs is largely lacking. Inhalation of NPs has been associated with oxidative stress, inflammation, fibrosis, multifocal granulomatous inflammation and pneumonia, free radical production, reduced cell viability, induction of apoptosis, and subpleural fibrosis, some of which exhibit dose-dependent behavior. NPs can enter cells by diffusion through cell wall membrane as well as through endocytosis and adhesion. The size distribution of NPs is very important in terms of toxicity in biological system. For example, 20-nm NPs were found to deposit mostly in the alveolar region, whereas, 5 to 10 nm particles deposited in tracheobronchial region, and particles smaller that 10 nm accumulate mostly in the upper respiratory tract. More research efforts are needed to focus on understanding, (1) the form, route, and mass of NPs entering in the human body, (2) the transformation and ultimate fate of NPs inside human body, (3) the transport, distribution and bio-availability, and (4) biochemical response to NPs and their expression.

KEYWORDS

- **Biosafety**
- **Coating**
- **Nano particles**
- **Plant**
- **Stability**
- **Zeta potential**

REFERENCES

Adhikari, Tapan, Kundu, S., Biswas, A. K., Ajay & Subba Rao, A. (2009). Stability of Plant Nutrient Containing Nano-Particles in Aqueous System and their Assimilation by Plants. National Seminar on Developments in Soil Science-2009. Platinum Jubilee Symposium on Soil Science in Meeting the Challenges of Food Security and Environmental Quality. Dec. 22–25, ISSS, New Delhi, 62.

American Society of Testing and Materials. (1985). Zeta Potential of Colloids in Water and Waste Water, ASTM Standard D 4187–82 ASTM, West Conshohocken, PA.

Baalousha, M., Manciulea, A., Cumberland, S., Kendall, K., & Lead, J. R. (2008). Aggregation of Surface Properties of Iron oxide Nano-particles, Influence of pH and Natural Organic Matter, *Environ. Toxicol. Chem.*, 27, 1875–1882.

Banfield, J. F., & Zhang, H. (2001). Nanoparticles in the Environment, In "Nanoparticles and the Environment." (Banfield, J. F., & Navrotsky, A. Eds), 1–58, Mineralogical Society of America, Washington DC Chapter 1.

Banfield, J., & Navrotsky, A. (2001). Nanoparticles and Environment.Reviews in Mineralogy and Geochemistry, 44, Mineral. Soc. of Am. Washington, DC.

Bargar, J. R., Bernier-Latmani, R., Giammar, D. E., & Tebo, B. M. (2008). Biogenic Uraninite Nanoparticles and their importance for Uranium Remediation. Elements (Chantilly. VA, U.S.), 4, 407–412.

Becker, L., Rietmeijer, F. J. M. (ed.,) (2006) Natural Fullerenes and Related Structure of Elemental Carbon, Springer, Dordrecst, Netherland, 6, 95–121.

Blango, M. G., &. Mulvey, M. A. (2009). Bacterial landlines Contact-dependent Signaling in Bacterial Populations, Curr. Opin. Microbiol, 12, 177–181.

Bose, S., Hochella, M. F., Gorby, Y. A., Kennedy, D. W., McCready, D. E., Madden, A. S., & Lower, B. H. (2009). Bioreduction of Hematite Nanoparticles by the Dissimilatory Iron Reducing Bacterium Sbewanella oneidensis MR-1, Geochim Cosmochim Acta, 73, 962–976.

BSI British Standard Institute. (2005). Vocabulary-Nanoparticles, Publicly available Specification (PAS) 71, 25. Department of Trade Industries and British Standard Institution London, UK.

Buffle, J. (2006). The Key Role of Environmental Colloids/Nanoparticles for them Sustainability of life, Environ. Chem., 3, 155e158.

Canas, J. E., Long, M. Q., Nations, S, Vadan, R., Dai, L., Luo, M. X., Ambikapathi, R., Lee, E. H., & Olszyk, D. (2008). Effects of Functionalized and Nonfunctionalized Single-walled Carbon, Nanotubes on Root Elongation of Select Crop Species, Environ. Toxicol. Chem., 27, 1922–1931.

Chadwick, O. A., Derry, L. A., Vitousek, P. M., Huebtert, B. J., &. Hedin, L. O. (1999). Changing Source of Nutrient During Four Million Years of Ecosystem Development, Nature, 397, 491–497.

Darlington, T. K., Neigh, A. M., Spencer, M. T., Nguyen, O. T., & Oldenburg, S. J. (2009). Nanoparticles Characteristics Affecting Environmental Fate and Transport through Soil, Environ. Toxicol. Chem., 28, 1191–1199.

Eichert, T, Kurtz, A., Steiner, U., & Goldbach, H. E. Size Exclusion Limits and Lateral Heterogeneity of the Stomatal Folier Uptake Pathway for Aqueous Solutes and Water Suspended Nano-particles, Physiol. Plant. 134, 151–160.

EPA nanotechnology White paper. U. S. Environmental Protection Agency Report EPA 100/B-07/001, Washington DC 20460, USA, (2007).

Eriksson, E. (1952). Cation-exchange Equilibria on Clay Minerals, Soil Sci., 74, 103–113.

Feitosa-Felizzola, J., Hanna, K., & Chiron, S. (2009). Adsorption and Transformation of Selected Human-used Microlide Antibacterial agents with iron (III) and manganese (IV) oxide, Environ. Pollut 157, 1317–1322.

Fleischer, M. A., Neill, O., & Ehwald, R. (1999). The Pore Size of Non-graminaceous Plant Cell Wall is rapidly decreased by Borate Ester Cross-Linking of the Pectic Polysaccharide Rhamnogalacturon II, Plant physiol., 121, 829–838.

Ghosh, S., Mashayekhi, H., Pan, B., Bhowmik, P., & Xing, B. S. (2008). Colloidal Behavior of Aluminum Oxide Nano-particles as Affected by pH and Natural Organic Matter, Langmuir, 24, 12385–12391.

Gieseking, J. E. (1939). The Mechanism of Cation Exchange in the Montmorillonite-Beidellite-Nontronite Type of Clay Minerals. Soil Sci., 47, 1–13.

Gilbert, B., & Banfield, J. F. (2005). Molecular-scale Processes Involving Nanoparticulate minerals in Biogeochemical Systems. Rev. Mineral Geochem, 59, 109–155.

Gorby, Y. A., Yanina, S., McLean, J. S., Rosso, K. M., Moyles, D., Dohnalkova, A., Beveridge, T. J., Chang, I. S., Kim, B. H., Culley, D. E., Reed, S. B., Romine, M. F., Saffarini, D. A., Hill, E. A., Shi. L., Elias, D. A., Kenned, D. W., Pinchuk, G., watanabe, K., Ishii, S., Logan, B., Nealson, K. H., & Fredrickson, J. K. (2006). Electrically Conductive Bacterial Nanowires Produced by Sbewanella oneidensis Strain MR-1 and other Microorganisms. Proc. Natl. Acad. Sci. USA 103:11358–11363. Gieseking, J. E. (1939) the mechanism of Cation Exchange in the Montmorillonite-beidellite-nontronite type of Clay Minerals. *Soil Sci.* 47, 1–13.

Grieco, W. J., Howard, J. B., Rainey, L. C., &. Vander, J. B. (2000). Sande, Fullerenic Carbon in Combustion Generated Soot *Carbon*, 38, 597–614.

Griffitt, R. J., Luo, J., Gao, J., Bonzongo, J. C., & Barber, D. S. (2008). Effects of Particles Composition and Species on Toxicity of Metallic Nanomaterials in Aquatic Organisms. *Toxicon. Chem*, 27, 1972–1978.

Handy, R. D., Owen, R., & Valsami-Jones, E. (2008). The ecotoxicology of Nano-particles and nano-materials, Current Status, Knowledge Gaps, Challenges and Future Needs, *Ecotoxicology*, 17, 315–325.

Hochella, M. F. (2002). Nanoscience and Technology. The Next Revolution in the Earth Sciences. *jr Earth Planet.Sci.* Lett. 203,593–605.

Hochella, M. F., Jr. (2008). Nanogeoscience from Origins to Cutting Edge Applications, Elements (Chantilly, VA, U.S.), 4, 373–379.

Hochella. M. F. (2008). *jr Element (Chantilly, VA, U.S.)* 4, 373–379.

Illes, E., & Tombacz, E. (2006). The Effects of Humic Acid Adsorption on pH Dependent Surface Charging and Aggregation of Magnetic Nano-Particles, *J. Colloid Interface Sci.*, 295, 115–123.

Jia, G., et al. (2005). Cytotoxicity of Carbon Nano-Materials, Single-wall Nano-tube, Multi-Wall Nano-tube and Fullerene, *Environ. Sci.Technol*, 44, 1036–1042.

Ketzel, & Berkowicz, (2004). *Atmos.Environ.* 38, 2639–2656.

Lee, W. M., An, Y. J., Yoon, H., & Kweon, H. S. (2008). Toxicity and Bioavailability of Copper Nanoparticles to the Terrestrial Plants Mung bean (Phaseolus radistus) and wheat (Triticum aestivum), Plant Agar Test for Waterinsoluble Nanoparticles. Environ. Toxicol. Chem., 27, 1915–1921.

Lee, W. M., An, Y. J., Yoon, H., Kweon, H. S. (2008). Toxicity and Bioavailability of Copper Nanoparticles to the Terrestrial Plants Mung Beans (Phaseolus radiates) and Wheat (Triticum aestivum), Plant Agar Test for Water-insoluble nanoparticles, *Environ Toxicol Chem.,* 27, 1915–1921.

Lin, D. H., & Xing, B. S. (2008). *Environ. Sci. Technol.,* 42, 5580–5585.

Lin, D. H., & Xing, B. S. (2008). Root Uptake and Phytotoxicity of ZnO nanoparticles. *Environ. Sci. Technol.*, 42, 5917–5923.

Lin, D. H., & Xing, B. S. (2008). Root Uptake and Phytotoxicity of ZnO Nanoparticles, Environ. Sci. Technol., 42, 5580–5585.

Lin, S., Reppert, J., Hu, Q., Hudson, J. S., Reid, M. L., Ratnikova, T. A., Rao, H., Luo, A. M., & P. C.

Mahendra, S., Zhu, H., Colvin, V. L., Alvarez, P. J. J. (2008). Quantum Dot Weathering results in Microbial Toxicity, *Environ Sci Technol.*, 42, 9424–9430.

Manceau, A., Nagy, K. L., Marcus, M. A., Lanson, M., Geoffroy, N., Jacquet, T., & Kirpicht-chikova, T. (2008). Formation of Metallic Copper Nanoparticles at the Soil Root Interface, *Environ. Sci. Technol.*, 42, 1766–1772.

Maurice, P. A. & Hochella, M. F. (2008). Nanoscale Particles and processes. A New Dimension in Soil Science Adv. Agron., 100, 123–153.

Moore, M. N. (2006). Do Nano-particles Present Ecotoxicological Risks for the Health of the Aquatic Environment, *Environ. Int.*, 32, 968–976.

Navarro, E., Baun, A., Behra, R., Hartmann, N. B., Filser, J., Miao, A. I., Quigg, A., Santschi, P. H., & Sigg, R. (2008). Environmental Behaviour and Ecotoxicity of Engineered Nanoparticles to Algae, Plants and Fungi, *Ecotoxicology* 17, 372–386.

Necula, B. S., Apachitei, I., Fratila-Apachitei, L. E., Teodosiu, C., & Duszczyk, J. (2007). Stability of Nano/microsized Particles in Deionized Water and Electroless Nickel Solutions, *J. Colloid Interface Sci.*, 314, 514–522.

Seinfeld,J. H., & Pandis, S. N. (2006). *Atmospheric chemistry and physics, From Air Pollution to Climate Change, John Wiley & Sons, Hoboken, N. J.*

Slaveykova, V. I., Startchev, K., & Roberts, J. (2009). Amine and Carboxyl Quantum Dots Affect Membrane Integrity of Bacterium *Cupriavidus metallidurans* CH34, *Environ. Sci. Technol*, 43, 5117–5122.

Theng, B. K. G., & Yuan, G. (2008).Nanoparticles in the Soil Environment.Elements (Chantilly, VA, U.S.), 4, 395–400.

Tilney, L. G., Cooke, T. J., Connelly, P. S., & Tilney, M. S. (1991). *J. Cell Biol.,* 112, 739–747.

Tiwari, A. J., & Marr, L. C. (2010). *J. Environ. Qual.* 39, 1883–1895.

Uzu, G., Sobanska, S., Sarret, G., Munoz, M., & Dumat, C. Foliar lead uptake by lettuce exposed to atmospheric pollution, *Environ. Sci. Technol.* 44:1036–1042

Van Dongen, J. T., Ammerlaan, A. M. H., Wouterlood, M., Van Aelst, A. C. V., & Borstlap, A. C. (2003). Structure of the Developing Pea Seed Coat and the Post-phloem Transport Pathway of Nutrients, *Annals of Botany-London,* 91, 729–737.

Wang, H. H., Wick, R. L., & Xing, B. S. (2009). Toxicity of nanoparticulate and Bulk Zno, Al_2O_5 to the Nematode *Caenorhabditis elegans. Environ. Pollut.*, 157, 1171–1177.

Wierzbicka, M., & Obidzinska, J. (1998). The Effect of Lead on Seed Imbibitions and Germination in Different Plant Species, *Plant Sci.,* 137, 155–171.

Wilson, M. R., Lightbody, J. H., Donaldson, K., Sales J., Stone V. (2002). Interaction between Ultrafine Particles Transition Metals in Vivo and in Vitro, *Toxical Appl Pharm,* 184, 172–179.

Xia, T., Kovochich, M., Liong, M., Madler, L., Gilbert, B., Shi, H. B., Yeh, J. I., Zink, J. I., & Nel, A. E. (2008). Comparison of the Mechanism of Toxicity of Zinc Oxide and Cerium Oxide Nano-particles Based on Dissolution and Oxidative Stress Properties, *ACS Nano* 2, 2121–2134.

Zhang, Y., Chen, Y. S., Westerhoff, P., Hristovski, K., & Crittenden, J. C. (2008). Stability of Commercial Metal Oxide Nano-particles in Water, *Water Res.* 42, 2204–2212.

Zhu, H., Han, J., Xiao, J. Q., & Jin, Y. (2008). Uptake, Translation and Accumulation of Manufactured Iron oxide Nano-particles by Pumpkin Plants, *J Environ Monitor,* 10, 713–717.

CHAPTER 10

PHYTOPATHOSYSTEM MODIFICATION IN RESPONSE TO CLIMATE CHANGE

LAJJA VATI[1] and ABHIJEET GHATAK[2]

[1]Plant Pathology, G.B. Pant University of Agriculture and Technology, Pantnagar, U.S. Nagar, Uttarakhand (263 145) India

[2]Plant Pathology, Bihar Agricultural University, Sabour, Bhagalpur, Bihar (813 210) India; Email: ghatak11@gmail.com

CONTENTS

10.1   INTRODUCTION

Multidimensional effect of climate change pays paramount impact on disease development on agricultural crops. The kind of pathogenic species and their growth trend determine the magnitude of this impact. After a series of 'breath-less efforts' mankind has now recognized that change in climate is drastically affecting plants covering the globe by developing diseases in new unwanted condition developed through air, water and soil pollution. This chapter focuses on various air pollutants influencing quite a few phytopathosystems. The most possible impact of change in atmospheric concentration (including gases, water droplets, etc.) can probably be reflected through Fig. 10.1. Change in air composition is generated by means of changes in different strata of the atmosphere and gas emission (Figs. 10.1 and 10.3). Similarly, introduction of exotic species (e.g., *Phytophthora infestans*, *Puccinia striiformis* f. sp. *graminis*) traveling a long-distance and urbanization helping the foreign member to establish in a new area (Bradley et al., 2012; Gurr et al., 2011; Matyssek et al., 2012; Régnière, 2012). The newly developed composition of air is making suitable those neglected pathogen that are now contributing in development of major diseases (Prospero et al., 2009). Such nonimportant pathogens were out of the top-list category of pathogen due to noneconomical feature under agricultural perspective. McKinnon et al. (2012) made a comprehensive study on impact of climate change on the world economy. The effect of change in climate contributed to death of almost 400 thousand people per annum, which is ultimately of a cost for more than US\$ 1.2 trillion.

The weak pathogens are adopting themselves in such a way that they withstand to this changing condition which can be found with studies on increased aggressiveness (Eastburn et al., 2011; Milus et al., 2009). Overwhelming the changes of concentration of carbon dioxide (CO_2), methane (CH_4), nitrous oxide (N_2O) and ozone (O_3), increasing temperature, uncertain precipitation are managed through the modification in host and their structure, improving the microclimate, nutrition balance in soil and ecological engineering of the planted crop in a manner to achieve noteworthy impact on crop. Although, climate is recognized on accumulated environmental summaries, that is, prevailing in a province or country or continental (a fairly large area) for over 30 years, but change in it is however, the result of a study adopting over the last 100 years entail about the accelerated temperature and gaseous concentration (http://www.esrl.noaa.

gov/gmd/ccgg/trends). Over this period, 0.74°C increased in global mean temperature and an increase in atmospheric CO_2 concentration from 280 ppm in 1750 to 400 ppm in March, 2014 was noticed. The effect of CO_2 concentration on phytopathogens can be seen in Table 1. These changes are contributing in development of a new scenario that is associated to devastating impact on plant diseases (Fig. 10.1).

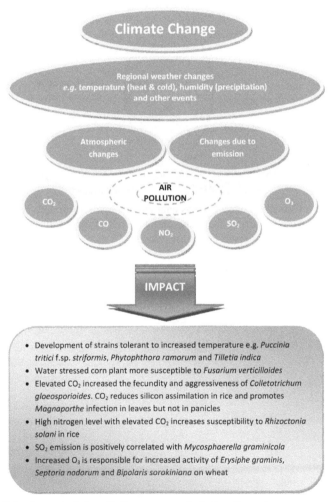

FIGURE 10.1 Impact of air pollution on several phytopathosystems.

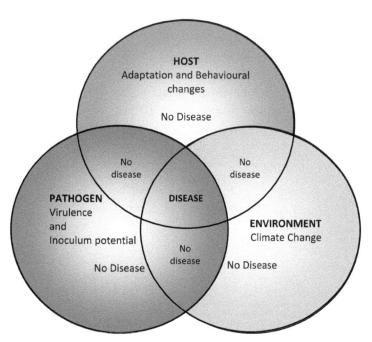

FIGURE 10.2 Climatic features such as temperature, humidity and leaf surface wetness are important drivers of disease, and inappropriate levels of these features for a particular disease may be the limiting factor in disease risk.

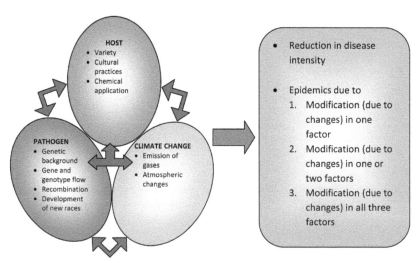

FIGURE 10.3 Result of an impact due to interaction of host, pathogen under scenario of climate change.

Interaction between host and pathogen is significantly affected with change in the gaseous composition of air that ultimately determines the severity of the disease (Eastburn et al., 2011). The gaseous composition may contribute to increase or decrease the level of disease for a said host-parasite interaction. For example, increased concentration of CO_2 lead to reduced epidemics under a wheat-pathosystem (Semenov, 2009) but increased concentration O_3 enlarges the epidemics for a phyto-pathosystem governed by *Botrytis cinerea* (Eastburn et al., 2011) which is contradictory to a report where increased concentration of O_3 was claimed to inhibit sporulation and germination in *B. cinerea* (Krause and Weidensaul, 1978). Thin-walled fungi like *Rhizopus, Penicillium, Aspergillus, Trichoderma*, etc. were susceptible to O_3 (Hibben and Stotzky, 1969), however, spores of *Alternaria* and *Stemphylium* spp. did not affect with increased O_3 concentration having the thick-walled spore (Rich and Tomlinson, 1968). The concentration of SO_2 has increased greatly over centuries, making this gas as one of the most important air pollutant. This gas affects a phytopathosystem in other way: it influences a disease by interfere the metabolic activities of the plant. This gas is played a role as stimulus in production of 'stress ethylene' (Pant, 2000), which finally leads to mango malformation—a mystery. Beside that the increased concentration of this noxious gas has been associated with black-tip in mango (Ranjan and Jha, 1940). These two problems causing reduced production of mango in many mango-growing pouches of Indian subcontinent.

Genes conferring resistance (R) are in general not effective at higher temperature, for example, most R-genes of *P. recondita*-wheat pathosystem (Kolmer, 1996). However, R-gene against wheat stripe rust (Yr36) is effective at 25 to 35°C but ineffective at lower temperatures (15°C) and thus become susceptible to *P. striiformis* f. sp. *tritici* infection (Uauy et al., 2005). Deviation in temperature determines the virulence pattern of the pathogen, has been proved in many phytopathosystems (Milus et al., 2009; Webb et al., 2010). Isolates of *P. striiformis* f. sp. *graminis* collected after year 2000 showed adaptation to a regime of warmer temperature and causes significantly enlarged amount of epidemics in eastern Australia and United States (Milus et al., 2009). In other quantitative analyzes, the post-2000 isolates were found more aggressive for different monocyclic processes. The changing pattern of temperature and future prediction is given in Fig. 10.4.

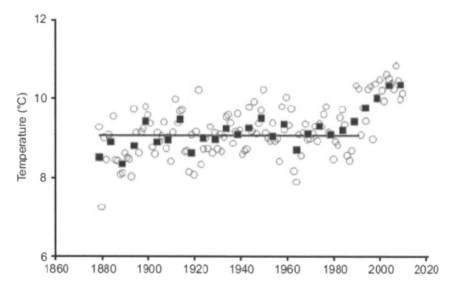

FIGURE 10.4 From 1878 to 2009, change in annual mean (o) and 5-year mean (■) air temperatures at Rothamsted, UK. Horizontal line designates the mean air temperature for the period 1878–1990. Adopted from http://www.rothamsted.ac.uk/

Variation in precipitation affects relative humidity (RH) in an area. The RH determines level of disease intensity for many phytopathosystems. *Magnaporthe oryzae* (Ou, 1985) requires over 89% RH whereas 70% RH is sufficient for *Alternaria solani* (Escuredo et al., 2010) to infect foliage of rice and potato, respectively. However, spore production shows a sigmoid trend in most of the phytopathosystems (Berger et al., 1997; Castano et al., 1989; Rouse et al., 1980; Villareal et al., 1981). The production of spore increases rapidly and then decline gradually.

10.2 IMPACT OF AIR POLLUTION AND STRATEGY AGAINST CLIMATE CHANGE

Due to over deforestation in the humid tropic for intensive agriculture the global carbon percentage has increased considerably (IPCC, 1996). Various spatio-temporal scales, reasonably changes in climate, are based on three parameters. First, the effect on experience basis, second category comprises with extrapolation from expert knowledge/system, and studies on experimental background, and third-one is solely based on computer models. It is equal to the universal truth that the effect of climate change is omnipresent, and therefore, it is applicable to most of the existing issues like plant disease (in particular – air pollution, Fig. 10.1). Hence, a holistic approach is needed to address many issues that are already appearing in agriculture, horticulture, forestry and conservation of the complete ecology (O'Halloran et al., 2012).

10.3   VARIABLES INFLUENCING PHYTOPATHOSYSTEMS

The development of plant disease is occurred by three important factors: (i) susceptible host, (ii) virulent and compatible pathogen (and vector if desired), and (iii) favorable environment (Vanderplank, 1963). It is important to induce a successful infection that all three of these factors must be at right position, at particular degree and at appropriate time (Fig. 10.2). Modification among one of the mentioned variables leads to evolve an altered phytopathosystem. For example, enhanced silicon content in rice plant reduces the chances to *M. oryzae* infection (Nakataa et al., 2008; Seebold et al., 2004). In other view, the lower epidemics may attain in the presence of the local race (of the pathogen) under uncongenial environment. Life strategies are governed by some behavioral adaptations by the plant pathogens. There are two kinds of strategists found in the world of pathogen: (i) *r* strategist-responsible for compound interest disease and (ii) *K* strategist–responsible for simple interest disease (Zadoks and Schein, 1979). The black (stem) rust pathogen *P. graminis* f. sp. *tritici* belongs to the *r* strategist group particularly at uredial stage. The organization is set up in such a way that the whole biomass is divided into 50% mycelium and 50% spore devoted to dispersal. This fungal pathogen is eligible to reproduce 2000 spores per day and 20,000 spores during its lifespan from a pustule developed with a single spore. Contradictorily the *K* strategists are poor producer of functional dispersal unit and without known functional spore, for example, *Sclerotium rolfsii*. For both of the strategist inoculum potential, genetic shift and recombination are the key determinants that decide virulence of the pathogen (Figs. 10.2 and 10.3). It is a matter of chance that the disease may develop at a condition when host, pathogen and environment interact in a compatible manner.

10.4　PHYTOPATHOGENS: INDICATOR OF CLIMATE CHANGE

A living system always responds to a stimulus. Being a living system the phytopathosystems react to climate deviation. Plant pathogens following very unique life history exhibit a wide range of reaction to climate change. For example, latent infection is found in viral pathosystem and expression of symptom depends on thermo-change (DeBokx and Piron, 1977), therefore, detection of such pathogen is varied according to climate. Likewise, fungal pathogens are completely dependent on RH or dew point, temperature associated with other factors for the infection in plant tissue (Huber and Gillespie, 1992). This means shift in disease risk is totally governed by changes in these environmental factors (Vanderplank, 1963).

Similar trend may be seen for pathogen populace with wide range of genetic variability, which often creates complication plant disease management (Strange, 2005). Adaptation toward pesticide and develop resistance is a major problem and a part of evolution of the phytopathogens (Agrios, 2005). Likewise, the pathogen species may rapidly adapt to climate change, however, this depends on the kind of pathogen (McDonald, 2002). Thus, anticipation for best time of management tactics is difficult to apply against potential pathogen population under climate change scenario. Other effect of climate change is activation of weaker pathogen and qualifies them to be economically important, for example, *Rhizoctonia solani* in rice (Fig. 10.1). The long-distance dispersal through wind (e.g., *Ustilago scitaminea*, *P. infestans*) and water (e.g., *X. oryzae*, members of order peronosporales) is a part of such changes that create epidemic in new areas for an exotic spp. (Chakraborty et al., 2008).

10.5　FUNGAL PHYTOPATHOSYSTEMS INFLUENCED BY CLIMATE CHANGE

10.5.1　*PUCCINIA STRIIFORMIS* F. SP. *TRITICI*

The stripe rust pathogen (*P. striiformis* f. sp. *tritici*) has been recognized as a cool climate favoring pathogen. Recently, the isolates collected after 2000 have shown their increased aggressiveness (e.g., shorter latent period) under higher temperature regime (Milus et al., 2009). The new identified race caused severe losses during 2001 to 2003 in the central United States (Chen, 2005).

10.5.2 FUSARIUM VERTICILLOIDES

Intensity of *F. verticilloides* is increased at drought stress condition, which is determined by the frequency and duration of drought (Paica et al., 2013). Mycotoxin production due to *Aspergillus* is also increased under such circumstances.

10.5.3 CRONARTIUM RIBICOLA

The blister rust of pines is favored cooler climate and spread of basidiospores is driven by moisture, air circulation and temperature (Van Arsdel, 1965). Events related to the frequency of 100% RH and changes in temperature influence the disease cycle of *C. ribicola*.

10.5.4 PHYTOPHTHORA RAMORUM

Larger lesions are developed in spring on coast live oaks infected with *P. ramorum*. The synchrony between host cambium activity (host phenology) and pathogen colonization rate (sporulation of the pathogen) are likely to be affected by climate shifts (Donnelly et al., 2011). Likewise, the climate changes adjust the pattern of susceptibility to *P. ramorum* in ecosystems (but not currently affected) at risk (Brasier and Webber, 2010; Tubby and Webber, 2010).

10.5.5 PHYTOPHTHORA CINNAMOMI

Moisture, pH and temperature influence growth and reproduction of *P. cinnamomi* (soilborne nature). Bergot et al. (2004) predict that raised mean temperature will lead to a potential range expansion of *P. cinnamomi*. Similarly, larger mean winter temperature, shifting of seasonal precipitation from summer into winter supports infection by *P. cinnamomi* in Central Europe.

10.6   BACTERIAL PHYTOPATHOSYSTEMS INFLUENCED BY CLIMATE CHANGE

Continuous cultivation of an agricultural plant spp. (often symbolized as variety) introgressed with mono resistance (R) gene may be ineffective at increased temperature. High temperature reduces effectiveness of R-gene and therefore, makes complicate the R-gene mediated defense mechanism against the phytopathogens. However, an R-gene (*Xa7*) for *Xanthomonas oryzae* pv. *oryzae* in rice performs better under increased temperature but other R-genes are not (Webb *et al.*, 2010). Increased multiplication of *X. campestris* pv. *undulosa* in wheat was noticed under high temperature (25°C) but at lower temperature (15°C) even symptom was not produced (Duveiller and Maraite, 1995). Against a few insect vectors temperature has been reported as one of the factors that alter the patterns of gene expression under defense signaling pathways. Elevated CO_2 concentration does not favor the bacterial multiplication in the host tissue. Lower bacterial count of *X. campestris* pv. *pelargonii* was estimated in leaves of geranium under increased level of CO_2 concentration (Jiao et al., 1999).

Apart from pathogenic spp. of bacteria, few beneficial bacteria act according to the temperature exposure, which influences several phytopathosystems. The bioremediation bacterium *Pseudomonas putida* grows optimally with a wide range of temperature (4 to 30°C) but the lower temperature (below 10°C) helps to express of at least 266 genes involving energy metabolism and assist other cellular functions (Fonseca, 2011). However, a strain of the same bacterial sp. tolerates high temperature (40°C) and performs well under drought areas (Srivastava et al., 2008).

10.7   VIRAL PHYTOPATHOSYSTEMS INFLUENCED BY CLIMATE CHANGE

Viruses are severe in subtropical and tropical region (Thresh, 2006) and in organic cultivation or protected systems (Pappu et al., 2009; Jones, 2004). Unlike other pathogens viruses are reported to infect wild relatives of main host and create most alarming situation to the biodiversity (Jones, 2009). This affects the environment for both micro and macrolevel. Viruses express their symptom under elevated temperature. Higher temperature increases susceptibility to virus infection under artificial inoculation

(Kasannis, 1952; Matthews, 1970). Similarly, elevated mean temperature in a region increases incidence of virus infected plants developed through contact, for example, white clover mosaic virus (Coutts and Jones, 2002), however, the impact of increased temperature may reduce infection probability for other virus, for example, tobacco rattle virus multiplied (Matthews, 1970). The favorable temperature for multiplication of tobacco rattle virus in tobacco is 18 to 22°C but a sharp decline may be noticed at 26°C. Further, in a mixed infection site, the behavior of system may be influenced by increased temperature: systemic infection has found at 31°C in a case when a mixture of brome mosaic virus and tobacco mosaic virus held at 20°C, 25°C and 31°C (Hamilton and Nichols, 1977). Ford and Ross (1962) observed that higher accumulation of potato virus (PV-X) was greatest at 30°C under a mixed infection of tobacco plants with PV-X and PV-Y compared to other lower temperature regime.

10.8 MANAGEMENT OF THE PHYTOPATHOSYSTEM UNDER CHANGING CLIMATE

Augmentation of global food production by 50% up to 2050 is necessarily required to save the multiplying human population (Chakraborty and Newton, 2011). There may be more complex situation to retain the potential yield under ever-enlarging warmer regions (heat stress) provided by drought, evolution in microbial world to damage crops (involving phytopathosystems), and soil erosion due to intense precipitation events. Beside these obstructers the task is hard to achieve due to the melting Himalayan glaciers, reasonably increased temperature, which may affect 25% cereal production of the globe. Studies under realistic field locations integrate with simulators for confirmation of multicomponent program will be useful to understand the features of this topic. The possible effects of new composition of the atmosphere in a shifting scenario correlating the biological processes of phytopathogens are then could be realized.

Adjustments are very much needful for plant disease management strategies under climate change (Garrett et al., 2006). Impact of climate change on host-plant resistance is one of the areas that has been most investigated (Boland et al., 2004; Legreve and Duveiller, 2010). Host-plant resistance has been made major contribution to ensure global food security under both climate change and sustainable management systems (Berry et al.,

2008; Legreve and Duveiller, 2010). Partially resistant varieties have often been proved to be a reliable option. This approach reduces the chances of complete elimination of respective pathogen which may in otherwise lead to – "boom and burst cycles". However, such approaches principally rely on the resistance genes. There is always threatening for domination of newly emerged race Ug99 (*P. graminis* f. sp. *tritici*) even the stem rust resistance gene *Sr31* has been effective for over 30 years. The breakdown of the gene *Sr31* will definitely convey serious consequences where other resistant cultivars are not available (Flood, 2010).

The number and virulence status of inoculum are influenced by the climate change or the survival of the pathogen is primarily taken care by the surrounding environment. Therefore, severity of epidemics is regulated by the primary inoculum (infection unit) under changing climate (Chaube and Pundhir, 2005). Management of infection unit is the important key to manage a phytopathosystem (Zadoks and Schein, 1979). Under a monocyclic process the management of a single infection unit and population of infection units are required. However, the polycyclic process could be managed with management of basic infection rate developed from the infection units repeatedly infect the host.

The chemicals being used in plant disease management are most effective under prescribed climatic conditions (Fig. 10.3). Change in the climate may lead to malformed effect of the applied chemical. Schepers (1996) advocates for precipitation in order to improve distribution of fungicide, however, heavy rainfall deplete residue of fungicide on the crop canopy (Neuhaus et al., 1974). Atmospheric changes manipulate the microflora community structure of rhizosphere and phyllosphere. Such manipulation is driven by the physical climate over biocontrol agents (Ayres et al., 1996; Manning and Tiedemann, 1995). Elevated CO_2 interacts with the nitrogen and other essential availabilities in soil to promote population and kind of mycorrhizal fungi that improves root and plant health (Staddon and Fitter, 1998).

The biotic environment is pathogen limiting (Zadoks and Schein, 1979): *Mycosphaerella musicola* (causes sigatoka disease) can infect banana earlier than a phyllosphere microflora build up, that is, within one or two days after the leaves become unfolded. The above discussed management strategies (resistance gene(s), pathogen, chemicals and the environment) can improve the phytopathosystems addressing plant health even under climate change. Accordingly, this issue will assist us to choose best and optimized way in plant disease management.

10.9 CONCLUSION

Limited efforts have been attempted on role of climate change on phytopathosystems. Therefore, efficacy of physical, chemical and biological control means must be evaluated. This may include disease-resistant varieties and aspects on changing climate in research priorities. Considering the phytopathosystems climate change is not an isolated issue but such problem is always there with every system (e.g., agriculture) over period. The changing climate is associated with changes in phytopathosystems through responses of the host, the overall pathosystem, and the specific environmental condition. Under this changing scenario the elevated level of CO_2 has been contributed to reduced aggressiveness of a few studied pathogens. Therefore, a significant reduction in disease severity may be observed for such phytopathosystems under rising CO_2 concentration. This is important information that may be plant responses that could be used for breeding programs. However, fecundity of the pathogen may increase (Table 10.1); thus, change in aggressiveness may certainly applicable in order of pathogen evolution under accelerated CO_2 condition. In this regard development of the alternatives of management strategies are imposed.

TABLE 10.1   Effect of CO_2 Concentration or Temperature Variation on Pathogen Biology

Pathogen	Observation	Worker
Colletotrichum gloeosporioides	Increased level of CO_2 delayed germtube growth and increased fecundity	Chakraborty et al. (2000)
Phytophthora parasitica	Increased CO_2 amplified the pathogen biomass	Jwa and Walling (2001)
Puccinia recondita	No effect of CO_2 in leaf area development covered by pustules	Tiedemann and Firsching (2000)
Puccinia striiformis	Fecundity was unchanged under high level of CO_2	Chakraborty et al. (2010)
Puccinia striiformis	Latent period was positively correlated with increased temperature	Milus et al. (2009)

Shift in distribution of host and pathogen is reported as a resultant of climate change (Mina and Sinha, 2008). Without seed treatment of wheat diseases like common bunt (*Tilletia caries*) and Karnal bunt (*T. indica*) become important under changing climatic conditions (Oerke, 2006). Change in disease scenario of chickpea and pigeon pea during last decade has been documented (Pande and Shanna, 2010). Diseases of minor importance like dry root rot of chickpea (*Macrophomina phaseolina*) and Phytophthora blight of pigeon pea (*P. drechsleri* f. sp. *cajani*) have drew the attention of plant pathologist as a potential threat. Although the raised temperature is an indicator of increase in disease severity or it paid importance to the pathogens of minor importance, but quantity of precipitation could act as deciding factor for disease severity and its spread (Woods et al., 2005). Bacterial pathogens like *Ralstonia solanacearum, Burkholderia glumea,* etc. are of serious impact in those areas where temperature-dependent strategies do not work (Kudela, 2009). Complex effect of the changing climate again draws attention to solve or to establish 'new' management strategies. Viruses are more prone to temperature as well as drought stress condition. The severity and incidence of maize dwarf mosaic virus and beet yellows virus were increased observed under drought stress (Clover et al., 1999; Olsen et al., 1990). Similarly, elevated temperature along with increased CO_2 concentration in the atmosphere creates hazardous situation for rice blast (*M. oryzae*), rice sheath blight (*Rhizoctonia solani*) and late blight of potato (*P. infestans*).

Remarkably climate change may alter genetic shift in the populace of the pathogen, rate of their development, improve the resistance and physiology in the host to reply severity of disease. Overall, climate change has the positive, negative, or even neutral impact on the phytopathosystem. It is due to particular nature of the interaction.

KEYWORDS

- **Air Pollution**
- **Climate Change**
- **Disease Management**
- **Disease Triangle**
- **Life Strategy**
- **Phytopathosystem**

REFERENCES

Agrios, G. N. (2005). Plant Pathology, Academic Press, San Diego.

Ayres, P. G., Gunasekera, T. S., Rasanayagam, M. S., & Paul, N. D. (1996). Effects of UV-B Radiation (280–320 nm) on Foliar Saprophytes and Pathogens. In: Frankland, J. C., Magan, N., Gadd, G. M. (Eds.), Fungi and Environmental Change. Cambridge University Press, Cambridge, 32–50.

Berger, R. D., Bergamin Filho, A., & Amorim, L. (1997). Lesion Expansion as an Epidemic Component. Phytopathology, 87, 1005–1013.

Bergot, M., Cloppet, E., Pe´rarnaud, V., De´que´, M., Marcais, B., & Desprez-Loustau, M. L. (2004). Simulation of Potential Range Expansion of Bos L. crop losses caused by Viruses. Crop Protection 1, 263–82.

Berry, P. M., Kindred, D. R., & Paveley, N. D. (2008). Quantifying the Effects of Fungicides and Disease Resistance on Greenhouse Gas Emissions Associated with Wheat Production. Plant Pathology, 57, 1000–1008.

Boland, G. J., Melzer, M. S., Hopkin, A., Higgins, V., & Nassuth, A. (2004). Climate Change and Plant Diseases in Ontario, Canadian Journal of Plant Pathology, 26, 335–50.

Bradley, B. A., Blumenthal, D. M., Early, R., Grosholz, E. D., Lawler, J. J., Miller, L. P., et al. (2012). Global Change, Global Trade, and the Next Wave of Plant Invasions. Frontiers in Ecology and the Environment. doi: 10.1890/110145.

Brasier, C., & Webber, J. (2010). Sudden Larch Death. Nature 466, 824–825.

Castano, J., MacKenzie, D. R., & Nelson, R. R. (1989). Components Analysis of Rate Non-specific Resistance to Blast Disease of Rice Caused by Pyiricularia oryzae. J. Phytopathol, 127, 89–99.

Chakraborty, S., & Newton, A. C. (2011). Climate Change, Plant Diseases and food security, An Overview, Plant Pathology, 60, 2–14.

Chakraborty, S., Luck, J., Hollaway, G., Fitzgerald, G., & White, N. (2010). Rust-proofing Wheat for a Changing Climate. BGRI. (2010) Technical Workshop, (30–31 May 2010). St Petersburg, Russia. (http://www.globalrust.org/db/attachments/bgriiwc/21/2/10-Chakraborty-A4-ca-embargo.pdf).

Chakraborty, S., Luck, J., Hollaway, G., Freeman, A., Norton, R., Garrett, K.A., et al. (2008). Impacts of Global Change on Diseases of Agricultural Crops and Forest Trees CAB Reviews: Perspectives in Agriculture, Veterinary Science, Nutrition and Natural Resources 3, 054.

Chakraborty, S., Pangga, I. B., Lupton, J., Hart, L., Room, P. M., & Yates, D. (2000). Production and Dispersal of Colletotrichum, gloeosporioides spores on Stylosanthes scabra under elevated CO_2. Environmental Pollution, 108, 381–387.

Chaube, H. S., & Pundhir, V. S. (2005). Crop Diseases and their Management. Prentice-Hall of India, New Delhi.

Chen, X. M. (2005). Epidemiology and Control of Stripe Rust, [Puccinia striiformis f. sp. tritici] on wheat. Can. J. Plant Pathol. 27, 314–337.

Clover, G. R. G., Smith, H. G., Azam-Ali, S. N., & Jaggard, K. W. (1999). The Effects of Drought on Sugar Beet Growth in Isolation and in Combination with Beet Yellows Virus Infection, J. Agric. Sci. 133, 251–261.

DeBokx, J. A., & Piron, P. G. M. (1977). Effect of Temperature on Symptom Expression and Relative Virus Concentration in Potato Plants Infected with Potato Virus Y n and Potato Virus, Y O. Potato Research, 20(3), 207–214.

Donnelly, A., Caffarra, A., & O'Neill, B. F. (2011). A Review of Climate-Driven Mismatches Between Interdependent Phenophases in Terrestrial and Aquatic Ecosystems. International Journal of Biometeorology, 55, 805–817.

Duveiller, E., & Maraite, H. (1995). Effect of temperature and Air Humidity on Multiplication of *Xanthomonas campestris* pv. *undulosa* and Symptom Expression in Susceptible and Field-Tolerant Wheat Genotypes, Journal of Phytopathology, 143, 227–232.

Eastburn, D. M., McElrone, A. J., & Bilgin, D. D. (2011). Influence of Atmospheric and Climatic Change on Plant-Pathogen Interactions. Plant Pathology, 60, 54–69.

Escuredo, O., Seijo, M. C., Fernández-González, M., & Iglesias, I. (2010). Effects of Meteorological Factors n The Levels of Alternaria Spores on a Potato Crop, Int J Biometeorol. DOI: 10.1007/s00484-010-0330-4.

Flood, J. (2010). The Importance of Plant Health to Food Security, Food Security, 2, 215–31.

Fonseca, P., Moreno, R., & Rojo, F. (2011). Growth of *Pseudomonas putida* at Low Temperature, Global Transcriptomic and Proteomic Analyzes. Environmental Microbiology Reports. doi:10.1111/j.1758–2229.2010.00229x.

Ford, R. E., & Ross, A. F. (1962). Effect of Temperature on the Interaction of Potato Viruses X and Y in Inoculated Tobacco Leaves, Phytopathology, 52, 71–77.

Garrett, K. A., Dendy, S. P., Frank, E. E., Rouse, M. N., & Travers, S. E. (2006). Climate Change Effects on Plant Disease, Genomes to Ecosystems, Annual Review of Phytopathology, 44, 489–509.

Gurr, S., Samalova, M., & Fisher, M. (2011). The Rise and Rise of Emerging Infectious Fungi Challenges Food Security and Ecosystem Health, Fungal Biology Reviews, 25, 181–188.

Hamilton, R. I., & Nichols, C. (1977). The Influence of Bromegrass Mosaic Virus on the Replication of Tobacco Mosaic Virus in *Hordeum Vulgare*. Phytopathology, 67, 484–489.

Huber, L., & Gillespie, T. J. (1992). Modeling Leaf Wetness in Relation to Plant Disease Epidemiology. Ann. Rev. Phytopathol, 30, 553–577.

IPCC (1996). Climate Change (1995), The Science of Climate Change. Houghton, J. T., Meira Filho, L. G., Callander, B. A., Harris, N., Kattenberg A., Maskell, K. (Eds.), Contribution of Working Group I to the Second Assessment Report of the Intergovernmental Panel on Climate Change. Cambridge University Press, Cambridge.

Jiao, J., Goodwin, P., & Grodzinski, B. (1999). Inhibition of Photosynthesis and Export in Geranium Grown at two CO_2 Levels and Infected with *Xanthomonas campestris* pv. *pelargonii*. Plant, Cell and Environment, 22, 15–25.

Jones, R. A. C. (2004). Using Epidemiological Information to Develop Effective Integrated Virus Disease Management Strategies, Virus Research, 100, 5–30.

Jones, R. A. C. (2009). Plant Virus Emergence and Evolution, Origins, New Encounter Scenarios, Factors Driving Emergence, Effects of Changing World Conditions, and Prospects for Control, Virus Research, 141, 113–130.

Jwa, N. S., & Walling, L. L. (2001). Influence of Elevated CO_2 Concentration on Disease Development in Tomato. New Phytologist, 149, 509–518.

Kolmer, J. A. (1996). Genetics of Resistance to Wheat Leaf Rust, Annual Review of Phytopathology. 34, 435–455.

Krause, C. R., & Weidensaul, T. C. (1978). Effects of Ozone on the Sporulation, Germination, and Pathogenicity of *Botrytis cinerea*, Phytopathology, 68, 195–198.

Kudela, V. (2009). Potential Impact of Climate Change on Geographic Distribution of Plant Pathogenic Bacteria in Central Europe. Plant Prot. Sci., 45, S27–S32.

Legreve, A., & Duveiller, E. (2010). Preventing Potential Diseases and Pest Epidemics Under a Changing climate, In: Reynolds, M. P., ed. Climate Change and Crop Production, Wallingford, UK CABI Publishing, 50–70.

Manning, W. J., & Tiedemann, A. V. (1995). Climate change: potential effects of increased atmospheric carbon dioxide (CO_2), ozone (O_3), and ultraviolet-B (UVB) radiation on plant diseases. Environmental Pollution 88, 219–245.

Matthews, R. E. F. (1970). Plant Virology, Academic Press, New York.

Matyssek, R., Wieser, G., Calfapietra, C., deVries, W., Dizengremel, P., Ernst, D., et al. (2012). Forests under Climate Change and Air Pollution, Gaps in Understanding and Future Directions, for Research. Environmental Pollution, 160, 57–65.

McDonald, B. A., & Linde, C. (2002). Pathogen Population Genetics, Evolutionary Potential, and Durable Resistance Ann. Rev. Phytopathol, 40, 349–379.

Milus, E. A., Kristensen, K., & Hovmøller, M. S. (2009). Evidence for Increased Aggressiveness in a Recent Widespread Strain of Puccinia Striiformis f. sp. tritici Causing Stripe Rust of Wheat, Phytopathology, 99, 89–94.

Mina, U., & Sinha, P. (2008). Effects of Climate Change on Plant Pathogens. Environ. News 14(4), 6–10.

Nakataa, Y., Uenoa, M., Kiharaa, J., Ichiib, M., Taketac, S., & Arasea, S. (2008). Rice Blast Disease and Susceptibility to Pests in a Silicon Uptake-Deficient Mutant lsi1 of rice. Crop Protection, 27, 865–868.

Neuhaus, W., Stachewicz, H. & Dunsing, M. (1974). (English summary) UÈ ber den Einfluss von Niederschlaegen auf die Biologische Wirkung von Fungiziden zur Phytophthora-Bekampfung. Nachrichtenbl. Pflanzensch. DDR. 28, 149–153.

O'Halloran, T. L., Law, B. E., Goulden, M. L., Wang, Z., Barr, J. G., Schaaf, C., et al. (2012). Radiative Forcing of Natural Forest Disturbances, Global Change Biology, doi: 10.1111/j.1365–2486.2011.02577x.

Oerke, E. C. (2006). Crop Losses to Pests. J. Agric. Sci. 144, 31–43.

Olsen, A. J., Pataky, J. K., D'arcy, C. J., & Ford, R. E. (1990). Effects of Drought Stress and Infection by Maize Dwarf Mosaic Virus in Sweet Corn. Plant Dis. 74, 147–151.

Ou, S. H. (1985). Rice Diseases, 2nd ed. Commonwealth Mycological Institute, Kew, U.K.

Paica, A., Enc, C. I., & Tefan, L. A. (2013). Fungal Biodiversity and Climate Change on Corn, a Key Tool in Building an Innovative and Sustainable Agriculture on Dobrogea Area. Scientific Papers, Series A. Agronomy LVI, 406–411.

Pande, S., & Shanna, M. (2010). Climate Change, Potential Impact on Chickpea and Pigeon Pea Diseases in the Rainfed Semi-Arid Tropics (SAT). In Proceedings of the 5th International Food Legumes Research Conference (IFLRC V) and 7th European Conference on Grain Legumes (AEP VII), Antalya, Turkey.

Pant, R. C. (2000). Is 'Stress ethylene' the cause of mango (*Mangifera indica* L.) malformation? Physiol. Mol. Biol. Plant. 6, 8–14.

Pappu, H. R., Jones, R. A. C., & Jain, R. K. (2009). Global Status of Tospovirus Epidemics in Diverse Cropping Systems, Successes Achieved and Challenges ahead. Virus Research, 141, 219–36.

Prospero, S., Grunwald, N. J., Winton, L. M., & Hansen, E. D. M. (2009). Migration Patterns of the Emerging Plant Pathogen *Phytophthora ramorum* on the West Coast of the United States of America. *Phytopathol*, 99, 739–749.

Ranjan, S., & Jha, V. R. (1940). The Effect of Ethylene and Sulphur Dioxide on the Fruits of *Mangifera indica*. Proceedings of the Indian Academy of Sciences, Section B, 11(6), 267–288.

Régnière, J. (2012). Invasive Species, Climate Change and Forest Health, In Schlichter, T., & Montes, L. (Eds.), Forests in development, a Vital Balance (27–37), Berlin, Springer, doi: 10.1007/978–94–007–2576–8_3.

Rouse, D. I., Nelson, R. R., MacKenzie, D. R., & Armitage, C. R. (1980). Components of Rate-Reducing Resistance in Seedlings of Four Wheat Cultivars and Parasitic Fitness in Six Isolates of *Erysiphe graminis* f. sp. *tritici*. Phytopathology, 70, 1097–1100.

Schepers, H. T. A. M. (1996). Effect of Rain on Efficiency of Fungicides on Potato against *Phytophthora infestans*. Potato Research, 39, 541–550.

Seebold, K. W. Jr., Datnoff, L. E., Correa-Victoria, F. J., Kucharek, T. A., & Snyder, G. H. (2004). Effects of Silicon and Fungicides on the Control of Leaf and Neck Blast in Upland Rice, Plant Dis. 88, 253–258.

Semenov, M. A. (2009). Impacts of Climate Change on Wheat in England and Wales, Journal of the Royal Society Interface, 6, 343–50.

Srivastava, S., Yadav, A., Seem, K., Mishra, S., Chaudhary, V., & Nautiyal, C. S. (2008). Effect of High Temperature on *Pseudomonas putida* NBRI0987 Biofilm Formation and Expression of Stress Sigma Factor Rpos, Current Microbiology, 56(5), 453–457.

Staddon, P. L., & Fitter, A. H. (1998). Does Elevated Carbon dioxide Affect Arbuscular Mycorrhizas? Trends in Ecology and Evolution, 13, 455–458.

Thresh, J. M. (2006). Plant Virus Epidemiology, The Concept of Host Genetic Vulnerability, Advances in Virus Research, 67, 89–125.

Tiedemann, A. V., & Firsching, K. H. (2000). Interactive Effects of Elevated Ozone and Carbon Dioxide on Growth and Yield of Leaf Rustinfected Versus Non-Infected Wheat, Environmental Pollution, 108, 357–363.

Tubby, K. V., & Webber, J. F. (2010). Pests and Diseases Threatening Urban Trees under a Changing Climate, Forestry, 83, 451–459.

Uauy, C., Brevis, J. C., Chen, X., Khan, I., Jackson, L., Chicaiza, O., Distelfeld, A., Fahima, T., & Dubcovsky, J. (2005). High-temperature Adult-Plant (HTAP) Stripe Rust Resistance Gene Yr36 from *Triticum turgidum* ssp. *dicoccoides* is Closely Linked to the Grain Protein Content Locus Gpc-B1. Theoretical and Applied Genetics, 112, 97–105.

Van Arsdel, E. P. (1965). Micrometeorology and Plant Disease Epidemiology, Phytopathology, 55, 945–950.

Vanderplank, J. E. (1963). Plant Diseases, Epidemics and Control. Academic Press, New York.

Villareal, R. L., Nelson, R. R., MacKenzie, D. R., & Coffman, W. R. (1981). Some Components of Slow-Blasting Resistance, in Rice. Phytopathology, 71, 608–611.

Webb, K. M., Ona, I., Bai, J., Garrett, K. A., Mew, T., Vera Cruz, C. M., & Leach, J. E. (2010). A benefit of High Temperature, Increased Effectiveness of a Rice Bacterial Blight Disease Resistance Gene, New Phytologist, 185, 568–576.

Woods, A., Coates, K. D., & Hamann, A. (2005). Is an Unprecedented Dothistroma Needle Blight Epidemic Related to Climate Change? BioScience, 55, 761–769.

Zadoks, J. C., & Schein, R. D. (1979). Epidemiology and Plant Disease Management, Oxford University Press, New York.

CHAPTER 11

SOIL FERTILITY DYNAMICS VIS-À-VIS CLIMATE CHANGE IN CITRUS

A. K. SRIVASTAVA

Principal Scientist (Soil Science) National Research Centre for Citrus, Nagpur 440 010, Maharashtra, India; E-mail: aksrivas2007@gmail.com

CONTENTS

ABSTRACT

Climate change is now an established phenomenon, irrespective of weather or not, it is anthropogenic or nonanthropogenic in nature. The story of climate change is an incomplete exercise unless the dynamics of soil fertility is addressed. Soil fertility is the basis of all life, its origin and the place of its continuous renewal compels us to see dynamic changes taking place in soil fertility via climate change. In addition to food security, nutritional security through fruit crops has become a core agenda of extreme scientific debate. Of them, citrus holds a place of prominence to fulfill these objectives on one hand, and on the other hand, makes it highly imperative to undertake an incisive analysis as how climate change is dictating the performance of citrus through soil fertility changes. Elevated atmospheric CO_2 concentration and consequent rise in temperature will trigger the simultaneous increase in soil temperature, with the result, dynamics of microbial community structure and diversity both will be influenced, and hence, available pool of nutrients in soil to varying magnitude as per available limited studies. These changes need to be mitigated through some soil carbon sequestering techniques like long-term organic manuring, use of microbial consortium, integrated nutrient management instead of exclusive chemical fertilizers, etc. With more databases accruing through researches on these issues, complexities involved in unraveling the soil fertility-climate change nexus will be eased out in the larger benefit of citriculture in the years to come. Since citrus is globally grown across 153 countries, in the context of uncertainties associated with global climate change (direction, rate, seasonal and geographical distribution), such exercise, however, needs appropriate accelerated efforts.

11.1 INTRODUCTION

An estimate states over 900 million people in the world are undernourished. It has been discussed that malnutrition alone is accountable for 3.5 million deaths per annum (Srivastava, 2012). Plant nutrition in response to soil fertility is a intricate process that has developed over the course of plant evolution with the discovery of fundamental importance of plant nutrition; only second to the invention of photosynthesis as an efficient medium to strengthen plant defense system. Horticulture is pondered as a rising field in Indian economy where fruit crops the highest promise. Horticulture contribution to the GDP is 31% and 54% to the export in agriculture contributed merely through 9% of area. Citrus is one of the most important fruit crops of India grown in 9.27 lakh ha with a production of 86.95 lakh tons and average productivity of 9.23 tons ha^{-1} (Indian Horticulture Database, 2012).

There are two major agencies, which are highly devoted to study the soil fertility and climate change. The first one is the Tropical Soil Biology and Fertility (TSBF) initiated in 1984 by International Union of Biological Sciences and NESCO to research on the role of biological processes in maintenance of soil fertility. Whereas the second one is, the International Geosphere-Biosphere Program (IGBP) aims to describe and understand the interactive physical, chemical, and biological processes that regulate total earth system.

Recently, Inter-Governmental Panel of Climate Change (IPCC) has reconfirmed that the global atmospheric concentrations of carbon dioxide, methane and nitrous oxide, greenhouse gases (GHGs), have increased markedly as a result of human activities since 17th century and presently huge preindustrial values determined from ice cores across many thousands of years (Drake et al., 1997). The increment in GHGs raised the temperature increase in the climate by 0.74°C between 1906 and 2005. An estimation of 12 years (1995–2006), considering the temperature of the environment, revealed that 11 years ranked among the 12 hottest years in the instrumental record of global surface temperature. IPCC has opted that the increase in temperature by the end of this century may be ranged between 2 and 4.5°C.

It is proposed that rainfall over India will rise by 15–40%, whereas the mean annual temperature will increase by 3–6°C by the end of 21st century. Therefore, such changes in terms of climate change could affect citrus crops through its direct and indirect effects.

11.2   CLIMATE ANALYSIS FOR CITRUS

Climate is the most important component of commercial citriculture, which determines the difference in growth, yield and quality due to differential behavior of citrus in relation to climate-soil nexus (Kurihara, 1969; Reuther, 1973). However, citrus is well adapted to a wide range of climate from humid tropical to desert (Levy and Syvertsen, 1981). Few attempts in the past, have been made to characterize climate under which citrus is grown world over. Based on mean annual temperature and consequently the diurnal changes, Reuther (1973) divided climate of world citrus into three distinct thermal environments, namely, subtropical zone with low diurnal variation varying between 4.7°C and 10.8°C in cooler and hotter months, respectively covering Ocho island in Hiroshima representing humid climate, with well distributed rainfall; tropical zone characterized by absence of seasonal change in temperature represented by Colombia with mean maximum temperature of 32.8□C in the coolest month and 34.6°C in the hottest month; and subtropical zone with moderate diurnal variation represented by San Joaquin valley of California having semiarid climate with diurnal variation up to 12°C and 22°C in coolest and hottest month, respectively.

While, according to Yelenosky (1977), climatically five citrus growing regions of the world are delineated as: moist marine areas, such as southern Japan, the black sea coast, Adriatic sea coast of Yugoslavia, parts of Turkey, and New Zealand; subtropical, Mediterranean climate is typified Spain, Israel, Italy, Turkey, parts of Australia, Lebanon, Greece, and the coastal areas of California; subtropical, arid areas such as the U.S. desert regions of southern California, Arizona and possibly the Rio Grande valley of Texas; subtropical, moist areas include Florida (USA) parts of China, India, South Africa, Brazil, and Argentina; and tropical, moist areas such as Hawaii, parts of Mexico, Central America, and others include those where temperature lower than 18.3°C rarely occur, and rainfall is usually heavy.

Of late, analysis on climate requirement undertaken by Jackson and Looney (1999) revealed that ideal growth of citrus cultivars such as sweet orange takes place between 13 and 40°C. While, lemons are more tolerant to cool summer temperatures than sweet orange, and do not grow well under tropical climate. Grapefruits prefer tropical and warm tropical climate. However, not many cultivar specific climatic norms are available. Earlier Pehrson (1976) presented than annual maturity ratios (TSS:acid) of navel oranges for 22 years and the mean monthly temperatures for years in which maturity advanced or delayed, each average over five years. Advanced maturity is associated during the year of high spring temperature (number of degree, days over 23°C).

Jones and Cree (1965) reported that high temperature during the June drop period (May to June) can be very detrimental to the crop size of navels. Later, Jackson and Hamer (1980) showed that three climatic factors accounted for 64% variation in yield of Cox orange between seasons. Moss and Muirhead (1971) indicated a negative effect of high September temperature on yield of Navel Orange in Australia, but had a positive effect of high temperature during November. Models have been developed through long-term field experiments in Western Georgia to separate effect of fertilizer from climate and other incidental factors (Tsanava et al., 1989). At low altitude in subtropical California, Valencia matures in about 9.5 months of anthesis in the hottest and about 14 months in the coolest climate zone (Monselise, 1981). Quantification analysis regarding the contribution of environments, soil and management factors on yield of Satsuma mandarin revealed that yield is more affected by physiographic environment than either climate or even fertilizer application (Egashira et al., 1990).

Tubelis and Salibe (1988) observed that production is correlated with the orchard age and total rainfall received in 16 months, before picking season the relationship (HA/CA $= -18.91 + 20.35I + 0.8190P8 - 1.3788P9 + 0.5805P10 - 0.149615$, where HA/CA stand for production express in kg plant^{-1}; I, orchard's age expressed in years; P8, P9, P10, and P15, total rainfall received, respectively during the month of August, September, October, and March of the year before picking) showed the best determination coefficient up to 98% up to year to year variation in yield, having a good yield forecast ability, with substantial advance at blooming. Haggag and Maksoud (1996) later developed a mathematical model for yield forecasting in navel orange in Egypt. Similar to above models, various other mathematical models were developed for predicting temperature requirement for Texas citrus (Chance and Rathewell, 1979) and California Citrus (Pehrson, 1966).

11.3   CLIMATE CHANGE AND CITRUS

Recent concerns about the "greenhouse effect" and damage to the ozone layer have resulted in more concerted studies on the quantities, kinds, distributions, and behavior of climate in the different systems (Johnson and Kerns, 1991). India is one of the 27 countries which are more likely to be affected by the impact of climate change with citriculture no exception. Rising carbon dioxide levels due to anthropogenic greenhouse gases may have both, detrimental and beneficial effect on citrus productivity through dynamic changes in soil fertility. The overall effects of climate change on citrus will depend on balance of these effects. Climate change is not local but global, however, the variability in climate change in different citrus growing belts is different. Hence, the issues of climate change on citriculture need to be handled at local level in order to harness an ensured sustainability in quality production. However, it is not easy for a citrus orchardist to respond promptly in order to change crop species or variety under climate change situation. This needs a thorough understanding of the near future patterns of climate change at a regional level and to act accordingly. While the future climate change will dictate the choice of different combat technologies, the present day situation is to be handled carefully for meeting the current demand of citrus fruits. This not only demands for deeper study on climate change scenario but also on its impact on the factor productivity. Therefore, the challenges ahead are to have sustainability and competitiveness in addition to achieve target production in the environment of declining land and water resources coupled with threat of climate change.

With climate change, soil degradation is more likely to occur, and soil fertility would probably be affected. However, because the ratio of carbon-to-nitrogen is a constant, a doubling of soil carbon is likely to imply a higher sink capacity of nitrogen in soil, thus providing better opportunities for soil fertility improvements. On the other hand, high precipitation will lead of higher magnitude of hydration of soil, thereby releasing some of the nutrients, especially the swell-shrink soils (smectite rich soils), for example, K extracted through 1 N neutral NH_4OAc which was highest order during July August (311.6 mg kg^{-1}) compared to April May (148.2 mg kg^{-1}) according to our earlier studies (Srivastava, 2011), and adopting site specific nutrient management could restore a greater amount of nonexchangeable-K compared to conventional fertilizer practices.

11.3.1 FRUIT YIELD AND QUALITY

Higher temperatures can increase the capacity of air to absorb water vapor and consequently, generate a higher demand for water. Higher evapotranspiration indices could lower or deplete the water reservoir in soils, creating water stress in plants during dry seasons. For example, water stress is of great concern in fruit production, because trees are not irrigated in many production areas around the world. It is well documented that water stress not only reduces crop productivity but also tends to accelerate fruit ripening (Henson, 2008). Other investigators forecast for the near future that rising air temperature could induce more frequent occurrence of extreme drought, flooding or heat waves than in the past (Assad et al., 2004).

Carbon dioxide (CO_2), also known as the most important greenhouse gas, and ozone (O_3) concentrations in the atmosphere are changing during the last decade and are affecting many aspects of fruit and vegetable crops production around the globe (Felzer et al., 2007; Lloyd and Farquhar, 2008). Carbon dioxide concentrations are increasing in the atmosphere during the last decades (Mearns, 2000). The current atmospheric CO_2 concentration is higher than at any time in the past 420,000 years (Petit et al., 1999). Exposure to elevated temperatures can cause morphological, anatomical, physiological, and, ultimately, biochemical changes in plant tissues and, as a consequence, can affect growth and development of different plant organs. These events can cause drastic reductions in commercial yield. However, by understanding plant tissues physiological response to high temperatures, mechanisms of heat tolerances and possible strategies to improve yield, it is possible to predict reactions that will take place in the different steps of fruit and vegetable crops production, harvest and postharvest (Kays, 1997).

In citrus, climate risk is mainly influenced by temperature in addition to elevated CO_2. Woolf and Ferguson (2000) reported that the high temperatures, both in terms of diurnal fluctuations and long-term exposure, can result in differences in internal quality properties such as sugar contents, tissue firmness, and oil levels. Singh et al. (1998) found that the Kinnow fruit showed sigmoid curve with three distinct phases from 15th May to 15th July; 15th August to 15th November and 29th November to 8th February. The TSS content as sugars increased while juice acid decreased during the entire period of fruit growth, whereas ascorbic acid increased initially and decreased with the advancement of the fruits development distinctly in subtropical climatic condition. Tao et al. (2003) elaborated the effect of light as environmental signal on stimulating carotenoid synthesis, especially the accumulation of β-cryptoxanthin in citrus fruit peel. Other

studies (Wen et al., 2001; Idso et al., 2002) showed that the activity vari-
ance of enzymes is related to organic acid metabolism during development
of Robertson Navel orange fruit.

Rosenzweig et al. (1995) investigated **the** potential impacts of global
climate change on fruit yield in the US were through simulations of citrus. Simulated treatments included combinations of three increased temperature regimes (+1.5, +2.5 and +5.0°C), and three levels of atmospheric
carbon dioxide (440, 530, and 600 ppm) in addition to control runs representing current climatic conditions. **Downton et al.** (1987) recorded the
response to elevated CO_2 of 3-year-old fruiting Valencia orange scions
(*Citrus sinensis* Osbeck) on citrange rootstock (*C. sinensis   Poncirus tri-
foliata* (L.) Raf). Fruit yield from the CO_2 enriched trees did not differ
from the controls besides soluble solids content, dry weight, seed number or rind thickness. The progression of fruit coloration was more rapid
for the CO_2 enriched trees. These results indicated that fruit yield will
increased as global levels of CO_2 continued to rise, at least in those species that experience source limitation during fruit development (Moretti
et al., 2010).

11.4   SOIL FERTILITY DYNAMICS

Dependence on curiosity about soil, exploring the diversity and dynamics of this resource continues to yield fresh discoveries and insights. New
avenues of soil research are coupled by a need to understand soil in context of climate change, greenhouse gases and carbon sequestration. The
major challenge is to reverse the side of nutrient loss and increase the
soil stocks through recapitalization initiatives. In recent years, there has
been an explosion of insight into genetic ecological and biogeochemical
dynamics of C-cycling processes in soils (Varallyay, 2010). These include:
plant and rhizosphere effects on soil respiration, the effects of microbial
communities on productivity and soil-C turnover, and interactions among
element cycles (role of organic nitrogen as a C and N source to microbes).
Fertilizers are the biggest source of greenhouse gases with reference to
horticulture. Large amounts of CO_2 and N_2O are released into atmosphere
during the production of fertilizers; offsetting the amount of carbon sequestered by trees. Nitrous oxide has more adverse effect than CO_2 over
climate change since each ton of N_2O is equivalent to 300 tons of CO_2. The

transport of fertilizer adds to the footprint as a ton of fertilizer transported over 1 km releases about 1 kg of CO_2.

Soil fertility is most essential for sustaining productivity and integrally depends on a complex network of soil structure and balanced water, oxygen and nutrients availability. Soil fertility is influenced by crop cultivars, growth promoting microbes and inorganic, organic absorber stabilized slow release fertilizers. In developing soil fertilities and a sustaining human subsistence an optimal crop selection and growth management, a skilled handling of growth promoting microbes, and a flawless application of fertilizers are potential strategies.

The sensitivity of soil carbon to warming is a major uncertainty to projections of carbon dioxide and climate. According to Hari Eswaran et al. (1993), the C stored in soils is nearly three times that in the above ground biomass and approximately double that in the atmosphere. Globally, 1576 Pg of C is stored in soils, with ~506 Pg (32%) of this in soils of the tropics. It is also estimated that ~40% of the C in soils of the tropics is in forest soils. Experimental studies indicate increased soil organic carbon (SOC) decomposition at higher temperatures, resulting in increased CO_2 emissions from soils. But recent evidences favor against this theory. The initially increased CO_2 efflux returns to the prewarming rates within 1–3 years, and apparent carbon pool turnover times are insensitive to temperature with non-labile SOC showing more sensitivity to temperature than labile SOC (Knorr et al., 2005). Significantly more carbon is stored in the world's soils than in present in atmosphere. Disagreements exist, regarding the effects of climate change on global soil carbon stocks. Despite much research, a consensus has not yet emerged on the temperature sensitivity of soil carbon decomposition due to differential kinetic properties of diverse soil organic compounds (Davidson and Janssen, 2006)

Soil fertility depletion is the fundamental biophysical cause declining productivity (Pawlson, 2005). Citrus-based farming system has high potential for sequestering carbon for mitigation of climate change. The perennial nature of citrus trees act as carbon sinks by sequestering the atmospheric carbon. Modifying fertilizer application to enhance nutrient availability, use of soil amendments to improve soil fertility providing irrigation at critical growth stages and conservation of soil moisture are important interventions. Considerable studies have into questions of just how climate change will affect crop and soil productivity. However, the problem of predicting the future course of citrus cultivation in a changing world is confounded by the fundamental complexity arising on account of interplay of set of dynamic factors specific to each citrus belt. Experimental studies of the long-term effects of CO_2 in more realistic field settings have not yet been done on a comprehensive scale (Rosenzweig and Hillel, 1993).

In high-N ecosystem, the maximum amount of C that can be decomposed may be less than in low-N ecosystem, resulting in greater C storage in high N ecosystem (De Deyn et al., 2008), Similarly, atmospheric deposition of N can retard rates of soil organic matter (SOM) decomposition by reducing the production of lignolytic and cellulolytic soil enzymes; however, the effect of enhanced soil fertility on decomposition processes may be ecosystem specific (Waldrop et al., 2004). Differences in C sequestration responses to N addition may depend on whether the largest effect of N in soils is to increase nonbiological formation of recalcitrant soil organic matter (reducing decomposability) or to increase the growth and metabolism of soil decomposers (N stimulation of decomposition).

It is contemplated that global warming may significantly alter soil composition at the molecular level, and that such changes could have a major impact on atmospheric levels of CO_2. Global warming may change present day decomposition patterns by altering the soil microbial communities and activities, thus changing the overall flow of carbon into and out of the soil, and affecting the soil fertility as well. The implication of the increased degradation of lignins is that less carbon remains in soil solid phase, and more CO_2 is released from soil into the atmosphere (Lance Frazer, 2009). To understand the soil-climate interactions better, we need more soil research to focus the molecular level within an eye toward predicting both short- and long-range changes in system. In this context, the studies pertaining to changes in nutrient pool of citrus rhizosphere, microbial communities and soil carbon partitioning are highly imperative in response to elevated CO_2 and temperature.

11.5   CLIMATE VERSUS SOIL MICROBIAL DYNAMICS

Soils are of particular importance in atmospheric CO_2 budget for number of reasons. At the global scale, rates of soil CO_2 efflux correlated significantly with temperature and precipitation, however, did not correlate well with soil carbon pools, soil nitrogen pools or soil C:N ratio (Raich and Potter, 1995).

11.5.1 SOIL MICROBIAL POPULATION

Variation in rhizosphere soil temperature under two different systems of nutrient management showed that soil temperature irrespective of timing, was invariably lower under INM-treated compared to IF-treated trees (Table 11.1). During 4-month study, soil temperature within rhizosphere of INM-treated trees was lower by 1–3°C over conventionally IF-treated trees rhizosphere. Variation in bacterial count (BC) and fungal count (FC) both, displayed large variation between two timings of observation and between two nutrient management systems. From January to April, with the rise in both morning and afternoon temperature, from 18.2 to 24.4°C and from 22.4 to 37.6°C under IF, the BC reduced from 108×10^4 to 54×10^4 cfu g^{-1}, and from 81×10^4 to 31×10^4 cfu g^{-1}, respectively. While, FC reduced from 69×10^3 to 18×10^3 cfu g^{-1} and from 52×10^3 to as low as 10×10^3 under IF with morning and afternoon temperature variation of 18–24.4°C and 22.4–37.6°C, respectively. On the other hand, INM-treated plants rhizosphere maintained a much high population of bacteria in morning (168×10^4-111 $*10^4$ cfu g^{-1}) and afternoon ($142 \times 10^4 - 76 \times 10^4$ cfu g^{-1}), although, the count reduced during afternoon but distinctly higher than IF-treated plants rhizosphere. The FC recorded likewise on the higher side, which reduced from $81 \times 10^3 \times 48 \times 10^3$ with rise in morning temperature from 18.2 to 24.4°C and from 72×10^3 to 28×10^3 cfu g^{-1} with rise in afternoon temperature from 21.4 to 34.3°C (Table 11.1).

11.5.2 RHIZOSPHERE MICRONUTRIENTS AVAILABILITY

Availability of micronutrients within the rhizosphere was observed to be influenced by variation in soil temperature under both IF-and INM-treated plants rhizosphere (Table 11.1). Interestingly, DTPA-Cu and DTPA-Zn were not so much influenced by variation in soil temperature compared to availability of DTPA-Fe and DTPA-Mn. Under IF-treated plants rhizosphere, rise in morning temperature from 18.2 to 24.4°C was associated in reduction in DTPA-Fe from 10.1 to 9.0 mg kg^{-1} and DTPA-Mm from 8.2 to 7.0 mg kg^{-1}, without any distinct change in DTPA-Cu and DTPA-Zn.

TABLE 11.1   Change in Soil Microbial Biomass and Available Supply of Micronutrients in Response to Changing Soil Temperature

Time of Sampling	Soil temp. (°C) (0–15 cm depth)		SMB (cfu g⁻¹ soil)				Micronutrients (mg kg⁻¹)							
			IF		INM		DTPA-Fe		DTPA-Mn		DTPA-Cu		DTPA-Zn	
	IF	INM	BC (×10⁴)	FC (×10³)	BC (×10⁴)	FC (×10³)	IF	INM	IF	INM	IF	INM	IF	INM
Jan. 2012														
Morning	18.2	17.2	108	69	168	81	10.1	14.2	8.2	11.4	1.20	1.80	0.82	1.04
Afternoon	22.4	21.4	81	52	142	72	8.1	13.2	7.2	10.9	1.10	1.76	0.81	1.06
Feb., 2012														
Morning	18.9	17.8	91	61	148	72	9.6	13.8	8.0	11.2	1.18	1.76	0.79	1.02
Afternoon	24.2	23.1	72	48	131	61	8.0	13.2	6.9	10.8	1.10	1.70	0.80	1.02
Mar. 2012														
Morning	20.2	19.3	78	51	129	52	9.2	13.3	7.6	11.0	1.20	1.82	0.80	1.04
Afternoon	31.2	29.4	60	39	116	42	7.6	12.8	6.2	10.2	1.18	1.76	0.79	1.06
Apr. 2012														
Morning	24.4	22.3	54	18	111	48	9.0	13.0	7.0	10.8	1.12	1.80	0.82	1.00
Afternoon	37.6	34.3	31	10	76	28	6.9	12.5	5.8	9.6	1.18	1.79	0.80	1.04

SMB, IF, INM, BC and FC stand for soil microbial biomass, inorganic fertilizers, integrated nutrient management, bacterial count and fungal count, respectively.

Source Srivastava (2013, unpublished data).

While afternoon temperature rise from 22.4 to 37.6°C induced reduction in DTPA-Fe from 8.1 to 6.9 mg kg^{-1}, DTPA-Mn from 7.2 to 5.8 mg kg^{-1} with marginal rise in DTPA-Cu from 1.10 to 1.18 mg kg^{-1}, without much change in DTPA-Zn (0.81 to 0.80 mg kg^{-1}). Under INM-treated plants rhizosphere, these changes were of comparatively lower order. There was a reduction in DTPA-Fe form 14.2 to 13.0 mg kg^{-1}, DTPA-Mn from 11.4 to 10.8 mg kg^{-1} without much change in either DTPA-Cu or DTPA-Zn with rise in morning temperature from 17.2 to 22.3°C. While under same INM-treated trees rhizosphere, afternoon rise in soil temperature from 21.4 to 34.3°C, brought reduction in DTPA-Fe from 13.2 to 12.5 mg kg^{-1}, DTPA-Mn from 10.9 to 9.6 mg kg^{-1} with just marginal increase in DTPA-Cu (1.76–1.79 mg kg^{-1}) and just marginal reduction in DTPA-Zn (1.06 to 1.04 mg kg^{-1}). These preliminary observations provide some database support to the fact that differential index of microbial and nutrient availability is maintained within the plants rhizosphere.

Soil bacteria count and fungal count exhibited direct relation with air temperature (maximum temperature and minimum temperature (Table 11.2). The highest bacterial count and fungal count (46–60   10^4 cfu g^{-1} soil and 18–24   10^4) was observed in the air temperature (maximum) range of 32.3–38.4°C and soil temperature range of (29.2–29.5°C) during September–October which bacterial and fungal count further reduced drastically (26–27   10^4 cfu g^{-1} bacterial count and 13–15   10^4 cfu g^{-1} soil fungal count) with rise in soil temperature up to 34.5°C in March at Site 1 (Pipla Kinkhede, Nagpur). While at Site 2 (Jarud, Warud, Amravati), almost similar pattern was observed, although maximum air temperature was on the comparatively lower side (26.8–28.2°C) but relative humidity was almost the same during September-October. The highest bacterial count and fungal count (28–47 × 10^4 cfu g^{-1} soil and 16–23 × 10^4 cfu g^{-1} soil) was observed during September–October. Again, during February–March, their no change in either maximum air temperature (27.8–33.8°C) or soil temperature (30.1–30.7°C), the bacterial count and fungal count (27–30 × 10^4 cfu g^{-1} soil and 13–14 × 10^4 cfu g^{-1} soil) showed no further substantial decline during February to March.

Efforts were further made to study the dynamics of six major microbial species (*Bacillus mycoides, Pseudomonas fluorescens, Bacillus polymyxa, Azotobacter chroococcum, Trichoderma harzianum* and *Bacillus licheniformis)* in relation to different soil and climatic factors with a purpose to identify optimal conditions for different species and dominant species

TABLE 11.2 Soil Microbial Biomass and Soil Microbial Dynamics (Soil Microbial Community and Structure) In Relation to Different Soil Climatic Parameters

Sr. no.	Month	Soil microbial biomass(cfu g⁻¹ soil) (×10⁴)		Soil microbial dynamics (×10³)					
		BC	FC	Bm	Pf	Bp	Az	Th	Bl
Location : Pipla, Kinkhede (21°19×19.8"N longitude; 78°52×6.8" E latitude), Kalmeshwar, Nagpur									
1.	Sept. (2012)	46	18	09	08	10	15	08	08
2.	Oct. (2012)	60	24	08	06	11	13	10	10
3.	Nov. (2012)	24	18	07	05	08	14	7	08
4.	Dec. (2012)	40	20	06	09	10	24	22	12
5.	Jan. (2013)	29	17	08	10	04	16	15	09
6.	Feb. (2013)	26	15	02	08	11	12	11	06
7.	Mar. (2013)	27	13	02	05	10	13	8	05
Location : Jarud, (21°27× 32.2" N longitude; 78°12× 53.8" E latitude), Warud, Amravati									
1.	Sept. (2012)	28	16	08	09	07	13	08	10
2.	Oct. (2012)	47	23	06	11	12	16	11	11
3.	Nov. (2012)	22	26	04	08	05	11	04	10
4.	Dec. (2012)	26	15	05	12	25	18	18	12
5.	Jan. (2013)	33	22	05	10	03	12	03	08
6.	Feb. (2013)	27	14	01	17	03	15	09	05

BC and FC stand for bacterial count and fungal count, respectively.
Bm, for *Bacillus mycoides* on nutrient agar; Pf, *Pseudomonas fluorescens* on *Pseudomonas* isolation agar; Bp, *Bacillus polymyxa* on starch agar; Az, *Azotobacter chroococcum* on *Azotobacter* agar; Th, *Trichoderma harzianum* on Potato dextrose agar; and Bl, *Bacillus licheniformis* on nutrient agar.

Source: Srivastava (2013, unpublished data).

governing the soil fertility dynamics (Table 11.2). The maximum number of colonies of *Bacillus mycoides* ($8–9 \times 10^3$ cfu g^{-1} soil) was observed during September month in a air temperature range of 21.7–32.3°C and soil temperature of 27.6–29.5°C with a relative humidity of 82.7–88.7%, irrespective of orchard site, though both the orchard sites different considerably in the agropedolgical setup. While maximum colonies of *Pseudomonas fluorescens* (10×10^3 cfu g^{-1} soil) was observed during January month with air temperature range of 11.2–27.3°C and soil temperature of 23.0–23.5°C. The maxima for colonies of *Bacillus polymyxa* ($10–25 \times 10^3$ cfu g^{-1}) and *Bacillus licheniformis*, (12×10^3 cfu g^{-1}) was observed during December month having air temperature range of 11.7–35.0°C and soil temperature range of 19.2–29.6°C with relative humidity of 53.6–78.8%. Similarly, during the same December month, the maximum colony of *Azotobacter chroococcum* ($18–24 \times 10^3$ cfu g^{-1} soil) and *Trichoderma harzianum* ($18–22 \times 10^3$ cfu g^{-1} soil) was observed. But, with regard to *Trichoderma harzianum*, at site 1 December witnessed the maximum colony (22×10^3 cfu g^{-1} soil). Interestingly, the colonies of most of the microbial species declined sharply towards March witnessing higher air as well a soil temperature with substantially much reduced relative humidity.

11.6   CLIMATE VERSUS NUTRIENT DYNAMICS

Climate induced variation in soil temperature has shown strong influence on nutrient dynamics. The soil properties such as soil pH and EC, remained unaffected by any changes in climatic features. However, dynamics of nutrient availability, both macro as well as micronutrients were considerably unaffected by maximum air temperature, relative humidity and soil temperature (Table 11.3). KMnO$_4$-N in soil at both the sites up to November month showed no changes, but thereafter, it increased to 128.8 mg kg^{-1} at Pipla and 126.3 mg kg^{-1} at Jarud with the rise in soil temperature from 19.2 to 34.5°C and from 29.6 to 30.7°C, respectively, corresponding to reduction in NH$_4$OAc-K from 202.3–191.4 mg kg^{-1} and from 198.7 to 176.1 mg kg^{-1} although with a any distinct relation between these climate-soil parameters versus Olsen-P. On the other hand, with regard to DTPA-extractable micronutrients, except DTPA-Fe, none of the micronutrients like Mn, Cu and Zn displayed any visible relation with any of the soil-climatic parameters.

TABLE 11.3 Nutrient Availability in Relation to Different Soil Climatic Parameters at two Different Contrasting Nagpur Mandarin Orchards Established on Haplustert at Pipla (Kinkhede), Nagpur and Jarud, Amravati, Maharashtra

Sr. no	Month	Soil pH	Soil EC (dSm⁻¹)	Macronutrients (mg kg⁻¹)			Micronutrients			
				N	P	K	Fe	Mn	Cu	Zn
Location : Pipla, Kinkhede (21°19×19.8"N longitude; 78°52×6.8" E latitude), Kalmeshwar, Nagpur										
1.	Sept. (2012)	7.21	0.19	109.4	10.7	182.4	14.2	10.8	2.2	1.05
2.	Oct. (2012)	7.30	0.14	110.8	12.0	198.2	14.4	12.4	2.2	0.98
3.	Nov. (2012)	7.48	0.22	109.4	10.1	208.8	11.5	6.5	2.1	1.03
4.	Dec. (2012)	7.50	0.20	120.4	12.5	202.3	11.5	10.3	2.2	1.00
5.	Jan. (2013)	7.43	0.17	123.7	9.2	208.7	14.3	10.3	2.1	1.05
6.	Feb. (2013)	7.34	0.18	126.0	11.9	194.8	14.6	10.5	2.2	1.05
7.	Mar. (2013)	7.42	0.15	128.8	9.9	191.4	13.6	10.4	2.2	1.06
Location : Jarud, (21° 27× 32.2" N longitude; 78°12× 53.8" E latitude), Warud, Amravati										
1.	Sept. (2012)	7.61	0.19	114.2	12.2	190.9	14.5	10.8	2.1	1.08
2.	Oct. (2012)	7.29	0.14	117.3	10.6	194.6	14.5	10.7	2.2	1.06
3.	Nov. (2012)	7.86	0.10	112.3	10.7	198.7	6.4	10.4	2.3	1.76
4.	Dec. (2012)	7.74	0.11	117.0	10.9	196.8	8.2	11.1	2.8	1.34
5.	Jan. (2013)	7.48	0.11	118.1	10.9	204.1	13.1	9.9	2.2	1.08
6.	Feb. (2013)	7.50	0.11	119.0	11.9	184.8	14.6	10.5	2.1	1.05
7.	Mar. (2013)	7.10	0.11	126.3	12.4	176.1	14.7	10.9	2.1	1.02

Source: Srivastava (2013, unpublished data).

11.7 COMBATING CLIMATE CHANGE INDUCED SOIL FERTILITY CHANGES

Areas covered with citrus as perennial crop are not easy to be replaced with some other crops under climate change situation. A thorough understanding on climate change is needed to develop some mitigation technologies in the wake of crop responses via C_3 (Plants that fix carbon dioxide through C_3 enzyme route as ribulose 1,5-diphosphate carboxylase have greater potential for response to increased atmospheric levels of carbon dioxide) or C_4 (plants that fix carbon dioxide through C_4 enzyme route as phosphoenolpyruvate carboxylase), thereby, earning some carbon credits and enhancing the income through trading of carbon (Farquhar et al., 1980; Furbank and Hatch, 1987). In this regard, some production practices are needed to undergo some adjustments to compensate climate change based on practices facilitating improvised carbon sequestration as an effective mitigation option since citrus trees act as strong carbon sinks (Paustian et al., 1998).

Carbon sequestration (recognized by the Intergovernmental Panel on Climate Changes and the European Commission as one of the possible measures through which greenhouse gas emissions can be mitigated) is the process through which agricultural practices remove carbon from the atmosphere. Sequestration slows the rate of climate change by enhancing carbon storage in trees architectural framework and soils (Johnson and Kerns, 1991; Paustian et al., 1998).

11.7.1 ADJUSTING FERTILIZER SCHEDULING

It is our experience that the fertilizer schedule, which we developed in 1990s, is no longer so responsive, and when the same fertilizer schedule was evaluated against other options during 2006–2013, some interesting results were obtained without altering the total amount of fertilizer (Srivastava, 2013). The fruit yield significantly influenced by different treatments having varying schedules of fertilization (Table 11.4). The highest fruit yield of 61.10 kg tree^{-1}(16.92 tons ha^{-1}) obtained with T_2 was significantly higher to treatment T_1(48.12 kg tree^{-1} or 13.32 tons ha^{-1}), T_3(52.32 kg tree^{-1} or 14.49 tons ha^{-1}) and T_4(58.30 kg tree^{-1} or 16.14 tons ha^{-1}). Thus, without changing the dose of fertilizer, just scheduling across

TABLE 11.4 Response of Different Fertilizer Scheduling N Fruit Yield, Quality and Leaf Nutrient Composition (Pooled Data 2006–2013)

Treatments	Fruit yield (kg tree^{-1})	Fruit quality (%)			Leaf nutrient composition							
					Macronutrients (%)			Micronutrients (ppm)			
		Juice	TSS	Acidity	N	P	K	Fe	Mn	Cu	Zn
T_1	48.12 (13.32)	45.10	8.52	0.87	2.18	0.09	1.38	44.2	32.4	6.4	20.3
T_2	61.10 (16.92)	47.78	9.92	0.78	2.38	0.13	1.56	59.6	39.0	5.8	24.8
T_3	52.32 (14.49)	46.94	9.24	0.82	2.22	0.10	1.32	54.2	36.2	6.9	23.1
T_4	58.30(16.14)	46.50	9.10	0.94	2.16	0.12	1.20	47.9	33.3	6.2	21.4
LSD($P=0.05$)	2.60 (–)	0.44	0.08	0.04	0.08	0.008	0.11	1.8	1.1	NS	0.96

T_1 : 100% N and 50% K up to stage IV; 100% P up to stage III; 50% K divided at V and VI.
T_2 : 80% N, 100% P and 50% K up to stage IV; 20% N and 50% K divided at V and VI.
T_3 : 100% N, 100% P and 30% K up to IV; 70% K divided at V and VI.
T_4 : 67% N and 100% P up to IV; 33% N and 100% K at V.
Figures in parenthesis indicate the fruit yield in tons/ha.
Source: Srivastava (2013).

growth stages brought so much variation in ultimate productivity. This is where reorientation of fertilizer application synchronizing with nutrient demand holds key to fertilizer use efficiency.

All the three fruit quality parameters viz., juice content, total soluble solids (TSS) and acidity (Table 11.4) were significantly influenced with treatments involving different fertilizer scheduling. The treatment T_2 registered highest juice content (47.78%) and TSS (9.92%) and lowest acidity (0.78%). While rest of the treatments produced a significantly inferior responses viz., T_1(45.10% juice content, 8.52% TSS and 0.87% juice acidity), T_3(46.94% juice content, 9.24% TSS and 0.82% juice acidity) and T_4(46.50% juice content, 9.10% TSS and 0.94% juice acidity). These observations strongly supported the notion that correct fertilizer scheduling could dictate the fruit quality development to varying proportions, out of which the proportionate application of K to the ratio of N and P has most profound effect. Such an empirical relationship between different nutrients needs to be befittingly used in order to harness elevated orchard productivity.

The treatment T_2(2.38% N, 0.13% P, 1.56% K, 59.6 ppm Fe, 39.0 ppm Mn and 24.8 ppm Zn) maintained the highest concentration of both macronutrients as well as micronutrients compared to rest of other treatments like T_1(2.18% N, 0.09% P, 1.38% K, 44.2 ppm Fe, 32.4 ppm Mn and 20.3 ppm Zn), T_3(2.20% N, 0.10% P, 1.32% K, 54.2 ppm Fe, 36.2 ppm Mn and 23.1 ppm Zn) and T_4(2.16% N, 0.12% P, 1.20% K, 47.9 ppm Fe, 33.3 ppm Mn and 21.4 ppm Zn). Such magnitude of changes in leaf nutrient composition suggested that in order to maintain the maximum fertilizer use efficiency, scheduling of fertilization holds a key issue, without altering the quantum of fertilizers to be synchronized across crop growth stages (Table 11.4).

11.7.2 ORGANIC MANURING

Compositing can contribute in a positive way to the twin objectives of restoring soil quality and sequestering carbon in soils. Application of organic manures can lead either to a build-up of soil organic carbon over time or a reduction in the rate at which organic matter is depleted from soils. In either case the overall quality or organic matter in soils will be higher using no organic manure (Srivastava et al., 2002).

Applying organic fertilizers, such as those resulting from composting, could increase the amount of carbon stored in these soils (stable organic fractions) and contribute significantly to the reduction of greenhouse gas emissions as climate change mitigation measure. The Intergovernmental Panel has recognized carbon sequestration in soil on Climate Change and the European Commission as one of the possible measures through which greenhouse gas emissions can be mitigated. One estimate of the potential value of this approach was assumed that 20% of the surface of agricultural land in the EU could be used as a sink for carbon – suggested it could constitute about 8.6% of the total EU emission-reduction objective (Favoino and Hogg, 2008). An increase of just 0.15% in organic carbon in arable soils in a country like Italy would effectively imply the sequestration of the same amount of carbon within soil that is currently released into the atmosphere in a period of one year with fossil fuels. Furthermore, increasing organic matter in soils may cause other greenhouse gas-saving effects, such as improved workably of soils, better water retention, less production and use of mineral fertilizers and reduced release of nitrous oxide.

According to (Favoino and Hogg, 2008), this loss of carbon sink capacity is not permanent. Composting can contribute in a positive way to the twin objectives of restoring soil quality and sequestering carbon in soils. Applications of organic matter (in the form of organic fertilizers) can lead either to a build-up of soil organic carbon over time, or a reduction in the rate at which organic matter in soils will be higher than using no organic fertilizer. The results suggested that soils where manure was added have soil organic carbon levels 1.34% higher than un-amended soils, and 1.13% higher than soils amended with chemical fertilizers, over a 50-year period. This is clearly significant given the evaluations reported above regarding carbon being lost from soils, and the increasing amount of carbon dioxide in the atmosphere. Based on these studies, it is possibility to develop models for studying dynamics of carbon storage compost application.

Response of different organic manuring on soil organic carbon changes was studied (Srivastava, 2012) during 2003–2011. It was observed (Fig. 11.1) that with all the treatments, the organic carbon content continued to increase e.g., from an initial level of 0.50% to 0.62% with FYM (T_1) treatment, 0.51% to 0.79% with vermicompost treatment (T_2), 0.54% to 0.64% with poultry manure (T_3), no change from initial value of 0.52% with green manuring of fodder cowpea (T_4), 0.58% to 0.66% with sunhemp green manuring (T_5) and from 0.54% to 0.56% with inorganic fertilizers (T_6). The maximum carbon loading of 0.28% was observed with vermicompost treatment (T_2) followed by 0.12% with FYM (T_1), 0.10% with poultry manure (T_3), 0.08% with green manuring of sunhemp and only 0.02% with inorganic fertilizers (T_6). These observations warranted that increase in organic carbon content in soil in other words, carbon sequestration operates in cyclic manner. Such cycles differ from one production system to another production system and from one soil type to another soil type.

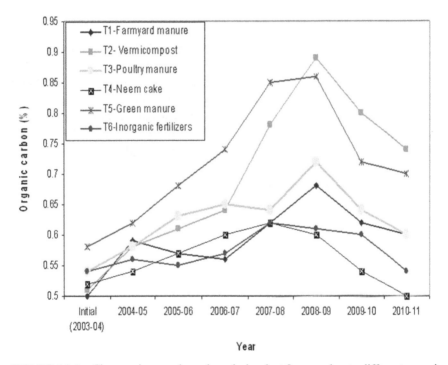

FIGURE 11.1 Changes in organic carbon during last 8 years due to different organic manuring versus inorganic fertilizer treatments.

11.7.3 MULTIPLE INOCULATION THROUGH MICROBIAL CONSORTIUM

Formation of associations with other organisms to promote protection from potentially inhibitory environmental factors where such associations reflect synergistic lifestyles facilitating more effective and efficient growth and biogeochemical cycles than individual populations as a community. Such associations are often called microbial consortium in which members of the consortium maintain metabolic and ecological compatibility for individual niches to exist in the close proximity in soil (Radianingtyas et al., 2003, Lazdunski et al., 2004). Such microbial consortium is more resistant to environmental changes, and can compete much better than single microorganism. As a result different species of microbial consortium inside an ecosystem propagate with different dynamics depending upon their genetic potentiality as well as capacity of adjustment to the microenvironmental conditions giving better yield and quality (Bashan, 1998). If the microorganisms interactions are evaluated and included in such selection process, a microbial consortium may out-perform the results achieved by pure cultures. When different microbial strains are made into an inoculum consortium, each of the constituent stains of the consortium not only out-compete with others for rhizosphere establishments, but complement functionally for plant growth promotion (Wu and Srivastava, 2012). Different microbial components in a microbial consortium should possess:

1. high rhizosphere competence,
2. high competitive saprophytic ability,
3. ease for mass multiplication,
4. safe to environment,
5. broad spectrum of action,
6. excellent and reliable efficacy,
7. compatible with other rhizosphere microbes, and in able to tolerate other abiotic stresses (Date, 2001; Rainey, 1999).

The developed microbial consortium (containing *Bacillus polymyxa, Bacillus mycoides, Pseudomonas fluorescens, Azotobacter chroococcum* and *Trichoderma harzanium*) was evaluated in nursery plants, both on seedlings for 45 days and in buddlings for 124 days, using a total of 354 plants. Out of these 354 plants, 172 plants in 13 replications (each with 4 units) were treated and other 172 plants in 13 replications were kept as untreated control (Table 11.5). The response of microbial consortium on rough lemon seedlings showed a significant increase in various growth parameters (9.59 g root weight, 24.86 g shoot weight, and 11.9 mm stem diameter on per plant basis) over control (2.99 g root weight, 9.08 g shoot weight, and 8.6 mm stem diameter on per plant basis). Similar observations were made on buddlings also. There was a significantly higher growth with microbial consortium treated buddlings (11.76 g root weight, 26.41 g shoot weight, and 28.51 mm stem diameter on per plant basis) compared to untreated control (4.10 g root weight, 10.72 g shoot weight, and 20.20 mm stem diameter). These observations confirmed the effectiveness of developed microbial consortium in growth promoting abilities.

TABLE 11.5   Evaluation of Microbial Consortium in Nursery Plants (Pooled Data of Two Seasons)

Treatments	Root weight (g)	Shoot weight (g)	Stem diameter (mm)
Seedlings (Period : 45 days)			
Control	2.99	9.08	8.61
Treated	9.59	24.86	11.9
$t_p = 0.05$	3.65	5.63	1.43
Buddlings (Period : 124 days)			
Control	4.10	10.72	20.20
Treated	11.76	26.41	28.51
$t_p = 0.05$	2.03	5.635	2.02

Source: Wu and Srivastava (2012).

The inoculation with microbial consortium brought a significant change in available supply of different nutrients in soil and microbial biomass nutrients (Table 11.6). A significantly higher soil fertility status with microbial consortium treated plants (123.4 N – 16.2 P – 13.7 Fe – 10.2 Mn – 0.88 Zn mg kg^{-1}) was observed compared to untreated control (116.2 N – 13.2 P – 8.8 Fe – 6.7 Mn – 0.62 Zn mg kg^{-1}). Similarly, microbial biomass nutrients were higher in the rhizosphere treated with microbial consortium (147.7 mg kg^{-1} C_{mic}, 34.1 mg kg^{-1} N_{mic} and 17.8 mg kg^{-1} P_{mic}) than untreated control (119.8 mg kg^{-1}, C_{mic}, 21.8 mg kg^{-1} N_{mic} and 13.5 mg kg^{-1} P_{mic}). The above observations strongly supported the effectiveness of microbial consortium in improving chemical and biological indices of citrus rhizosphere.

TABLE 11.6   Changes in Soil Fertility indices in Response to Inoculation with Microbial Consortium (Pooled Data of 2 Seasons)

	Soil fertility (mg kg⁻¹)						
	N	P	K	Fe	Mn	Cu	Zn
Control	116.2	13.2	166.7	8.8	6.7	1.12	0.62
Treated	123.4	16.2	169.7	13.7	10.2	1.16	0.88
$t_{P = 0.05}$	3.95	2.0	NS	1.75	1.35	NS	0.12

	Microbial biomass nutrients (mg kg⁻¹)		
	C_{mic}	N_{mic}	P_{mic}
Control	119.8	21.8	13.5
Treated	147.7	34.1	17.8
$t_{P = 0.05}$	9.85	2.5	1.25

Computed on the basis of analysis after 162 days of inoculation
C_{mic}, N_{mic}, and P_{mic} stand for microbial biomass-C, microbial biomass-N and microbial biomass-P, respectively.
Source: Wu and Srivastava (2012).

The treatment combination of ¾P + AM + N was observed the best treatment with reference to better growth and yield of high quality fruits of 'Mosambi' sweet orange suggesting the compatibility of biofertilizers (AZO) and AM inoculation in combination with chemical fertilizers for better growth, yield and fruit quality. Such observations in the long-term are expected to cut down the cost of chemical fertilizers, particularly N and P and building up fertility by maintaining better soil physical conditions (Chonkhe et al., 2000). High efficiency of *Azospirillum* for fixing nitrogen and better mobilization of fixed phosphorus by AM even at high temperatures can make these highly suited for Mosambi sweet orange (Manjunath et al., 1983).

11.7.4   INM AND SOIL CARBON SEQUESTRATION

Citrus by the virtue of its extensive cultivation as a key perennial crop in the world trade, has attracted worldwide investigation from various angles.

The present citrus production trends are characterized by either frequent crop failure or recurrence of alternate on-and-off years, setting unaccountable monetary loss to the industry (Rojas, 1998). In recent years, nutrient additions have been exclusively in favor of mineral fertilizers due to demographic pressure, demands related to life styles and trade involvement. While the quick and substantial response to fruit yield due to mineral fertilizers eclipsed the use of organic manures, the inadequate supply of the latter sources exacerbated this change (Srivastava and Ngullie, 2009; Srivastava, 2009). Differential efficacy of two conventional methods of fertilization (soil versus foliar application) has, although helped in improving the quality citrus, but of late, continuous fertilization has failed to sustain the same yield expectancy on a long-term basis due to depletion of soil carbon stock and consequently, emerged multiple nutrient deficiencies, irrespective of soil type. The menace of multiple nutrient deficiencies is further triggered through increase in air temperature via changes in microbial communities and activities within the rhizosphere in the light of climate change. Such changes will dictate adversely on the orchard's productive life in long run. Gradual shift from purely inorganic to organic fertilizers started gaining wide scale use for enhanced nutrient cycling (Srivastava et al., 2002).

Integrated Nutrient Management (INM) as a dynamic concept of nutrient management considers the economic yield in terms of fruit yield coupled with quality on one hand, and soil physicochemical and microbial prospects on other hand as a marker of resistance against the nutrient mining (arises because of failure to strike a balance between annual nutrient demand versus the quantity of nutrients applied). Soils under citrus differ from other cultivated soils, which remain fallow for 3–6 months every year forcing depletion of SOM (Bhargava, 2002). In contrast, biological oxidation of existing C continues in soil covered under citrus (Srivastava et al., 2002). Multiple nutrient deficiencies are considered to have a triggering effect on potential source of atmospheric CO_2. Soil carbon stock is considered as an important criterion to determine the impact of INM in the longer version of impact assessment (He et al., 1997). The amount of accumulated C within the rhizosphere soil does not continue to increase with time with increasing C outputs. An upper limit of C saturation level occurs, which governs the ultimate limit of soil C sink and rate of C sequestration in mineral soils, independent of C input rate. An understanding of the mechanism involved in C stabilization in soils is needed for controlling and enhancing soil C sequestration (Goh, 2004) under varying modes of nutrient management.

11.7.5 SOIL MICROBIAL POPULATION

An increase in the microbial biomass often goes along with increased nutrient immobilization. Over the years, the concepts of INM and integrated soil management (ISM) have been gaining acceptance, moving away from a more sectoral and inputs driven approach. INM advocates the careful management of nutrient stocks and flows in a way that leads to profitable and sustained production. ISM emphasizes the management of nutrient flows, but also highlights other important aspects of soil complex such as maintaining organic matter content, soil structure, moisture, and microbial biodiversity. Still more attention is needed towards integrated soil biological management as a crucial aspect of soil fertility management since providing protection to citrus rhizosphere against the nutrient depletion is of utmost importance for sustained orchard production in which the objectivity of INM could have far reaching consequences.

In our studies, the soil microbial count in terms of bacterial and fungal count showed significant changes in response to different INM-based treatments (Table 11.7). Bacterial and fungal counts both reduced to 31 and 16×10^3 cfu g^{-1}, respectively, with T_2 incorporating MC with 75% RDF from corresponding values of 32 and 16×10^3 cfu g^{-1} with T_1 having 100% RDF as inorganic fertilizers supporting the fact that as long as good soil productivity is obtained, soil microbial health could be maintained even with inorganic fertilizers. Of course, the magnitude of such response will be of comparatively lower order when compared with organic manures or in combination with inorganic fertilizers plus microbial biofertilizers. Out of two organic manures (FYM versus vermicompost), vermicompost based treatments produced much favorable response on soil microbial counts as evident from superiority of T_7 (50 and 25 cfu g^{-1} bacterial and fungal counts, respectively) over T_3 (45 and 19×10^3 cfu g^{-1} bacterial and fungal counts, respectively). Similarly treatment T_8 (53 and 26 $\times 10^3$ cfu g^{-1} bacterial and fungal counts, respectively) was better than T_4 (44 and $20 \times$ M module as T_9 68 and 41×10^3 cfu g^{-1} bacterial and fungal counts, respectively) supporting much better soil microbial counts compared to FYM-based INM as T_5 (57 and 25×10^3 cfu g^{-1} bacterial and fungal counts, respectively) and T_6(48 and 26 × 10^3 cfu g^{-1} bacterial and fungal counts, respectively). On the contrary, inclusion of green manuring with 75% RDF plus MC (50 and 25 × 10^3 cfu g^{-1} bacterial and fungal counts, respectively, with T_{10}) and 50% RDF + MC (44 and 25 × 10^3 cfu g^{-1} soil bacterial and fungal counts, respectively, with T_{11}), although brought some favorable changes on soil microbial counts, but were more significantly of lower order when compared to other FYM or vermicompost based INM treatments (e.g., T_1, T_6, T_9, etc.).

11.7.6 SOIL MICROBIAL BIOMASS NUTRIENTS

Rhizopshere microbial environment modification through roof exudation is an important attribute that regulates only the availability of nutrient soil, but also their acquisition plants. Different INM-based treatments (Table 11.7) were accompanied with significant changes in soil microbial biomass nutrients in terms of microbial carbon (C_{mic}), microbial N (N_{mic}) and microbial P(P_{mic}). Combination of 75% RDF + MC as T_2 brought down the soil microbial biomass nutrients (146.3 mg kg^{-1} C_{mic}, 19.3 mg kg^{-1} N_{mic} and 15.2 mg kg^{-1} P_{mic}) compared to 100% RDF as T_1(152.1 mg kg^{-1} C_{mic}, 19.1 mg kg^{-1} N_{mic} and 16.1 mg kg^{-1} P_{mic}). But incorporation of organic manure either FYM (159.1 mg kg^{-1} C_{mic} 23.9 mg kg^{-1} N_{mic} and 15.9 mg kg^{-1} P_{mic} with T_3 or 164.1 mg kg^{-1} C_{mic}, 25.6 mg kg^{-1} N_{mic} and 17.0 mg kg^{-1} with T_4) or vermicompost (169.6 mg kg^{-1} C_{mic}, 29.2 mg kg^{-1} N_{mic} and 18.9 mg kg^{-1} P_{mic} with T_7 or 176.1 mg kg^{-1} C_{mic}, 29.8 mg kg^{-1} N_{mic} and 20.4 mg kg^{-1} P_{mic} with T_8) helped further in recording higher level of soil microbial biomass nutrients. But these changes were still of lower order in the absence of microbial consortium treatments. The treatments like T_5(169.7 mg kg^{-1} C_{mic}, 29.6 mg kg^{-1} N_{mic} and 17.6 mg kg^{-1} P_{mic}), T_6(169.2 mg kg^{-1} C_{mic}, 30.3 mg kg^{-1} N_{mic} and 19.3 mg kg^{-1} P_{mic}) and T_9(202.5 mg kg^{-1} C_{mic}, 49.4 mg kg^{-1} N_{mic} and 24.5 mg kg^{-1} P_{mic}) produced the best response compared to treatments with MC (e.g., T_3, T_4, T_7 and T_8). These responses supported the fact that all the three components of INM are mandatory in order to harness the best effectiveness of different INM modules.

TABLE 11.7 Changes in Total Bacterial and Fungal Count of Soil Samples (0–15 cm) in Response to Different INM-based Treatments (Pooled data)

TABLE 11.7 *(Continued)*

Treatments	Yield	SMB (cfu x 10^3 g^{-1} soil)		SMBN (mg kg^{-1})		
	(kg tree^{-1})	Bacterial count	Fungal count	C_{mic}	N_{mic}	P_{mic}
T$_1$(100% RDF)	15.6	32	16.3	152.1	19.1	16.1
T$_2$(75% RDF + MC)	8.4	31	16	146.3	19.3	15.2
T$_3$(75% RDF + 25% FYM)	9.2	45	19	159.1	23.9	15.9
T$_4$(50% RDF + 50% FYM)	9.8	44	20	164.1	25.6	17.0
T$_5$(75% RDF + 25% FYM + MC)	14.5	57	25	169.7	29.6	17.6
T$_6$(50% RDF + 50% FYM + MC)	12.8	48	26	169.2	30.3	19.3
T$_7$(75% RDF + 25% Vm)	14.4	50	25	169.6	29.2	18.9
T$_8$(50% RDF + 50% Vm)	14.6	53	26	176.1	29.8	20.4
T$_9$(75% RDF + 25% Vm + MC)	18.8	68	41	202.5	49.4	24.5
T$_{10}$(75% RDF + Gm + MC)	11.2	50	25	178.6	31.8	18.6
T$_{11}$(50% RDF + Gm + MC)	10.6	44	25	170.7	28.6	17.2
CD ($P = 0.05$)	-	3.7	2.2	2.6	1.9	3.0

– MC stands for microbial consortium developed by isolating the native microbes from the experimental soil (mixture of *Azotobacter chrococcum, Bacillus mycoides, Bacillus polymyxa, Pseudomonas flourescens* and *Trichoderma harzanium*).
– **FYM,** Vm, Gm, and RDF stand for farmyard manure, vermicompost, green manuring, and recommended doses of fertilizers, respectively.
– **SMB** and SMBN stand for soil microbial population and soil microbial biomass nutrients, respectively.
Source: Wu and Srivastava (2012).

11.8   WAY FORWARD APPROACHES

The (bio)technology beyond sustainable soil fertility faces many challenges, especially they are be addressed in relation to climate change of con-

cern are new and adequately managed generations of crops coping better with biotic and abiotic stress conditions in a climatically changing world. A large and diverse group of ecto- and endophytes colonize plant phyllo and rhizospheres and are places where plant growth promoting microorganisms interact by modifying soil fertility and yields. Soil is the largest terrestrial pool of carbon, nitrogen and sulfur, and is intimately involved in the main fluxes of these important greenhouse elements between land and atmosphere. Land use shifts and their sustainability are an important part of global change, and it is through the response of plant soil system that climate change will have to main impact on crop performance. The following challenges are further envisaged.

- Development of soil fertility-crop response databased Infocrop simulation model that takes into account the climate change.
- Development of quantitative models that simulates citrus phenological development and predicts the relationship between climate change and citrus productivity vis-à-vis soil fertility.
- Modeling changes sink capacity for carbon sequestering in response to organic manuring versus inorganic fertilization and prediction on pattern of soil carbon changes over a time lag.
- Pathogens and soil fertility maintenance under future climates.
- Alternate cropping systems and their effects on soil properties, crop yields and climate.
- Improving nutrient use efficiency with stabilizer amended fertilizers and their role in the reduction of greenhouse gases.
- Renewable energy production, nutrient availability stabilizing fertilizers and sustainability of soil fertility.
- Secondary metabolites especially of medicinal plant origin and their residual effects on soil fertility.
- Evolutionary developed resistance strategies to pathogens and their maintenance under future climates.
- Evaluating the methodologies for soil fertility appraisal.
- Changes in microbial community composition, with a dominant functional groups of microorganisms in the root zone and the consequences of changes in microbial communities for the turnover and storage of plant assimilated carbon, in the soil. Collectively known as pedometrics. Interactions of climate change and soil microbial population diversity shifts and their influence on nutrient dynamics and crop yields.

- Possible impacts of climate change on N losses associated with horticultural crops are not well understood. Dynamic stimulation models of the soil/crop/atmosphere with crop production, identify possible management alternatives to improve N use efficiency and provide insight into the impact of future climate on N-use and N-losses.

KEYWORDS

- **Carbon Sequestration**
- **Citrus Microbial Consortium**
- **Fertilizer Scheduling**
- **Integrated Nutrient Management**
- **Microbial Pool**
- **Nutrient Pool**
- **Organic Manuring**

REFERENCES

Assad, E. D., Pinto, H. S., Zullo, J., Jr., & Avila, A. M. H. (2004). Impacto Das Mudanças Climáticas no Zoneamento Agroclimático Do Café no Brasil Pesquisa Agropecuária Brasileira, 39, 1057–1064.

Bashan, Y. (1998). Inoculants of Plant Growth Promoting Bacteria for Use in Agriculture, *Biotechnol. Adv.*, 16, 729–770.

Bhargava, B. S. (2002). Leaf Analysis for Nutrient Diagnosis Recommendation and Management in Fruit Crops, *J. Indian Soc. Soil Sci.*, 50(4), 352–373.

Chance, J. E., & Rathewell, P. J. (1979). A Mathematical Model for Predicting Annual Minimum Temperature for Texas citrus, *J. Rio Grande Valley Hort. Soc.*, 32, 61–66.

Chonkhe Singh, Saxena, S. K., & Goswami, A. M. (2000). Effect of Fertilizers on Growth, Yield and Quality of Sweet Orange (*Citrus sinensis* Osbeck) cv Mosambi, *Indian J Hort.*, 57, 114–117.

Date, R. A. (2001). Advances in Inoculation Technology, A Brief Review, *Austral. J. Exp. Agri.*, 41, 321–325.

Davidson, E. A. & Janssen, E. A. (2006). Temperature Sensitivity of Soil Carbon Decomposition and Feedbacks to Climate Change, *Nature,* 7081, 165–173.

De Deyn, G. B., Cornelissen, J. H. C., & Bardgett, R. D. (2008). Plant Functional Traits and Soil Carbon Sequestration in Contrasting Biomass, *Ecol. Letters,* 11, 516–513.

Downton, W. J. S., Grant, W. J. R., & Loveys, B. R. (1987). Carbon dioxide Enrichment Increases Yield Of Valencia Orange, *Aus. J. Pl. Physiol.* 14(5), 493–501.

Drake, B. G., Gonzàlez-Meler, & Long, S. P. (1997). More Efficient Plants, a Consequence of Rising Atmosphere CO_2, *Annu. Rev. Pl. Physiol. Pl. Mol. Biol.*, 48, 609–639.

Egashira, K., Nakashima, S., & Fujiyama, M. (1990). Quantification Analysis of the Contribution of Environment, Soil and Management Factors to Crop Yields Case Study of the Yield of Rice, Potato and Orange in Nagasaki Prefecture, *Sci. Bull.*, Faculty, Agriculture, Kyushu University, 45(1–2), 9–21.

Farquhar, G. D., Caemmerer von, S., & Berry, J. A. (1980). A Biochemical Model of Photosynthetic CO_2 Assimilation in Leaves of C_3 Species. *Planta*, 149, 78–90.

Favoino, E., & Hogg, D. (2008). Compost Can Turn Agricultural Soils into a Carbon Sink, Thus Protecting Against Climate Change, *Sci. News.* www.sciencedaily.com/releases/2008/02/080225072624.htm.

Felzer, B. S., Cronin, T., Reilly, J. M., Melillo, J. M., & Wang, X. (2007). Impacts of Ozone on Trees and Crops *Computers Rendus Geosci*, 339, 784–79.

Furbank, R. T., & Hatch, M. D. (1987). Mechanism of C_4 photosynthesis, *Plant Physiol*, 85, 958–964.

Goh, K. M. (2004). Carbon Sequestration and Stabilization in Soils: Implications for Soil Productivity and Climate change, *Soil Sci. & Pl. Nutri.*, 50(4), 467–476.

Haggag, L. F., & Maksoud, M. A. (1996). Evaluation of Yield of Navel Orange Tree, Mathematical Model Procedure, *Egyptian J. Hort.*, 23(2), 197–202.

Hari Eswaran, Evert Van Den Berg, & Paul Reich (1993). Organic Carbon in Soils of the World, *Soil Sci. Soc. Am. J.*, 57, 192–194.

He, Z. L., Yao, H., Chen, G. (1997). Relationship of Crop Yield to Microbial Biomass, in Highly Weathered Soils of China, in Plant Nutrition for Sustainable Food Production and Environment, Ando, T., et al. (Eds.) Kluwer Academic Publishers, Tokyo, Japan. 745–746.

Henson, R. (2008). The Rough Guide to Climate Change (2nd ed.), Penguin Books, London, 384.

Idso, S. B., Kimball, B. A., Shaw, P. E., Widmer, W., & Vanderslice, J. T. D. (2002). The Effect of Elevated Atmospheric CO_2 on the Vitamin C Concentration of (Sour) Orange Juice, *Agri., Ecozys. & Environ.*, 90, 1–7.

Indian Horticulture Database. (2012). National Horticulture Board. Bijay Kumar, Mistry, N. C., Brajendra Singh, & Chander, P., Gandhi, (Eds.), 42–45.

Jackson, J. E., & Hamer, P. J. C. (1980). The Accuses of Year-to-Year Variation in the Average Yield of Cox's Orange Pineapple in England. *J. Hort. Sci.*, 55, 149–156.

Jackson, D. J. & Looney, N. E. (1999). Citrus, Temperature and Sub-tropical Fruit Production. (Ed.). CAB International, 2nd Edition, U.K. 1–10, 229–240.

Johnson, M. G., & Kerns, J. S. (1991). Sequestering Carbon in Soils, a Workshop to Explore the Potential of Mitigating Global Climate Change, USEPA Rep.600/3–91/031. Environ. Res. Lab., Corvallis, OR.

Jones, W. W., & Cree, C. B. (1965). Environmental Factors Related to Fruiting of Washington Navel Orange Over a 38 Year Period, *Proc. Am. Hort. Soc.*, 86, 267–271.

Kays, S. J. (1997). Postharvest Physiology of Perishable Plant Products AVI, Athens, 532.

Knorr, W., Prentice, I. C., House, J. I., & Holland, E. A. (2005). Long Term Sensitivity of Soil Carbon Turnover to Warming, *Nature* 7023, 298–301.

Kurihara, A. (1969). Fruit Growth of Satsuma Oranges Under Controlled Conditions, I. Effects of Per-Harvest Temperature on Fruit Growth Color Development and Fruit Quality of Satsuma Oranges, *Bull. No. 8*, Hort. Res. Sta., Japan, 15–30.

Lance, Frazer (2009). Climate change, will Warmer Soil be Fertile, *Environ. Hlth, Prospect* 117(2), 59–61.

Lazdunski, A. M., Ventre, I., & Sturgis, J. N. (2004). Regulatory Circuits and Communications in Gram-Negative Bacteria, *Nature Rev. Microbiol* 2, 581–592.

Levy, Y., & Syvertsen, J. P. (1981). Water Relation of Citrus in Climates with Different Evaporative Demands, *Proc. Int. Soc. Citriculture,* 2, 501–503.

Lloyd, J., & Farquhar, G. D. (2008). Effects of Rising Temperatures and [Co_2] on the Physiology of Tropical Forest Trees, *Philosophical Trans. Royal Soc. Biol. Sci.* 363, 1811–1817.

Manjunath, A., Mohan R., & Bagyaraj, D. J. (1983). Responses of Citrus to VAM Inoculation in Unsterile Soils. *Can J. Bot.,* 61, 2779–2732.

Mearns L. O. (2000). Climatic Change and Variability, in Climate Change and Global Crop Productivity, Reddy, K. R., Hodges, H. F. (Eds.), CABI Publishing, Wallingford, UK, 7–35.

Monselise, S. P. (1981). Effect of Climatic Districts, Orchard Treatment and Seal Packaging on Citrus Fruit Quality and Storage Ability, *Proc. Int. Soc. Citriculture,* 2, 705–709.

Moretti, C. L., Mattos, L. M., Calbo, A. G., & Sargent, S. A. (2010). Climate Change and Potential Impacts on Postharvest Quality of Fruit and Vegetable Crops, a review *Food Res. Int.,* 43, 1824–1832.

Moss, G. I., & Muirhead, W. A. (1971). Climatic and Tree Factors Relating to the Yield of Orange Trees, I Investigations on the Cultivars Washington Navel And Late Valencia, *Hort. Res.,* 11, 3–7.

Paustian, K, Cole, C. V., Sauerbeck, D., & Sampson, N. (1998). CO_2 mitigation by Agriculture, An overview Climate Change, 40, 135–162.

Pawlson, D. (2005). Climatology, will Soil Amplify Climate Change, *Nature* 7023, 204–205.

Pehrson, J. E. (1966). Yield Trends, *Calif. Citrog,* 51, 261–262.

Pehrson, J. (1976). Delayed Maturity in Navels, *Calif. Citrog.,* 61(6), 200.

Petit, J. R., Jouzel, J., Raynaud, D., Barkov, N. I., Barnola, J. M., & Basile, I. (1999). Climate and Atmospheric History of the Past 420, 000 Years from the Vostok Ice Core, Antarctica, *Nature,* 399, 429–436.

Radianingtyas, H., Robinson, G. K. & Bull, A. T. (2003). Characterization of a Soil Derived Bacterium Consortium Degrading Chloroaniline. *Microbiol,* 149, 3279–3287.

Raich, J. W., & Potter, C. S. (1995). Global Patterns of Carbon Dioxide Emissions from Soils, *Global Biochem, Cycle,* 9, 23–36.

Rainey, P. B. (1999). Adaptation of *Pseudomonas Fluorescens* to the Plant Rhizopshere, *Environ Microbiol,* 1, 243–257.

Reuther, W. (1973). Climate and Citrus Behavior, In the Citrus Industry, Univ. Calif., Berkeley, USA, 3, 280–337.

Robock, A. (2000). The Global Soil Moisture Data Bank. *Bull. Am. Meteor. Soc.* 81, 1281–1299.

Robock, A. (2005). Forty-Five Years of Observed Soil Moisture in Ukraine: no Summer Desiccation (yet). *Geophys. Res. Lett.* 32: L03401. doi:10.0129/2004GL021914.

Rojas, E. (1998). Floral Response of Tahiti Lime (*Citrus latifolia* Tan.) to Foliar Sprays of Hydrogen Cyanamide, *Fruits* (Paris*),* 53(1), 35–40.

Rosenzweig, C., & Hillel, D. (1993). Agriculture in a Greenhouse World Potential Consequences of Climate Change. *Nat. Geogr. Res. & Explor,* 9, 288–211.

Rosenzweig, C., Phillips, J., Goldberg, R., Carroll, J. & Hodges, T. (1995). Potential Impacts of Climate Change on Citrus and Potato Production in the US. *Agricult. Sys.,* 52(4), 455–479.

Singh, H. K. P. P., Singh, S. N., & Dhatt, A. S. (1998). Studies on Fruit Growth and Development in Kinnow, *Indian J. Hort,* 55(3), 177–182.

Srivastava, A. K. (2009). Integrated Nutrient Management, Concept and application in citrus, Citrus II. Tree and Forestry Science and Biotechnology. Tennant, P., & Beakebhia, N. (Eds.) 3 (Special Issue 1) 32–58.

Srivastava, A. K. (2011). Site Specific Potassium Management for Quality Production of Citrus. *Karnataka J. Agric. Sci.,* 24(1), 60–66.

Srivastava, A. K. (2012). Integrated Nutrient Management in Citrus, In Advances in Citrus Nutrition, A. K. Srivastava, (Ed.), Springer-Verlag, the Netherlands, 369–389.

Srivastava, A. K. (2013). Nutrient Management in Nagpur Mandarin, Frontier Developments, *Sci. J. Agric.* 2(1), 1–14.

Srivastava, A. K., & Ethel Ngullie. (2009). Integrated Nutrient Management, Theory and Practice, *Dynamic Soil, Dynamic Plant,* 3(1), 1–30.

Srivastava, A. K., Shyam Singh, & Marathe, R. A. (2002). Organic Citrus: Soil Fertility and Plant Nutrition, *J. Sustain. Agric.,* 19(3), 5–29.

Tao, J., Zhang, S., Xinmin, A., & Zhao Zhizhong. (2003). Effects of Light on Carotenoid Biosynthesis and Color Formation of Citrus Fruit Peel, *Chinese J. Appld. Ecol.* 65(11), 19–22.

Tsanava, N. G., Lominadze, Sh. D., & Kastornova, E. N. (1989). Modeling the Nitrogen Nutrition of Lemons in Western Georgia, *Agrokhimiya,* 3, 9–14.

Tubelis Antonio, & Salibe, Apparecido. (1988). Relationships between Production of Hamlin Orange Trees and the Monthly Rainfalls at the Plateau of Botucatu, *Proc. 6th Int. Citrus Con.* 1, 497–501.

Várallyay, G. (2010). The impact of Climate Change on Soils and on their Water Management. *Agron. Res.* 8(Special Issue), 385–396.

Waldrop, M. P., Zak, D. R., & Sinsa Bangh, R. L. (2004). Microbial Community Response to Nitrogen Deposition in Northern Forest Ecosystem, *Soil Biol. & Biochem.* 36, 1443–1451.

Wen, T., Xiong, Q. E., Zeng, W. G., & Liu, Y. P. (2001). Changes of Organic Acids Synthetase Activity during Fruit Development of Navel Orange (*Citrus sinensis* Osbeck), *Acta Hort. Sin.,* 28(2), 161–163.

Woolf, A. B. & Ferguson, I. B. (2000). Post Harvest Responses to High Fruit Temperatures in the Field. *Post-harvest Biol. & Technol,* 21(1), 7–20.

Wu, Q. S., & Srivastava, A. K. (2012). Rhizosphere Microbial Communities, Isolation, Characterization and Value Addition for Substrate Development, in Advances in Citrus Nutrition, Srivastava, A. K. (Ed.), Springer-Verlag, The Netherlands, 169–194.

Yelenosky, G. (1977). The Potential of Citrus to Survive Freezes. *Proc. Int. Soc. Citriculture* 1, 199–201.

CHAPTER 12

SOIL SOLARIZATION AND MOISTURE CONSERVATION PRACTICES TO COMBAT CLIMATE CHANGE

PRAMOD KUMAR[1], SOM DEV SHARMA[2],
SHAILENDRA KUMAR YADAV[3], and VINOD KUMAR[4]

[1]Regional Horticultural Research Station, Dr. Y. S. Parmar University of Horticulture and Forestry, Sharbo (Reckong Peo), District Kinnaur, Himachal Pradesh-172 107, India.

[2]Department of Fruit Science, Dr. Y. S. Parmar University of Horticulture and Forestry, Nauni, Solan, Himachal Pradesh, India.

[3]Central Soil and Water Conservation Research and Training Institute, Research Centre, Chandigarh (UT), India.

[4]Department of Food Science and Technology, Bihar Agricultural University, Sabour, Bhagalpur, Bihar, India.

CONTENTS

ABSTRACT

Climate change is a significant statistical distribution of weather patterns over periods ranging from decades to millions of years. It may be a change in average weather conditions, or in the distribution of weather around the average conditions. Temperature and CO_2 concentration are the most important components of climate change influencing agriculture. An increase of CO_2 level in atmosphere traps the solar radiation and increases the temperature near earth surface. A slight increase in temperature imposes several stress to the fruit plants mainly moisture and incomplete chilling which adversely affect the productivity and quality attribute of fruit crops. Soil solarization is a relatively new, nonchemical method of soil disinfestation, and is cost effective to methyl bromide crisis and this encompasses the benefits of the concept made possible through advances in plastic mulch technology. Solarization currently is an important and widespread practice for agri-horticultural production including fruit crops. Moreover, it is an effective tool for moisture conservation, pest control, weeds suppression, the improvement in nutrient availability and microbial activity in soil profile of fruit crops and has been mainly used commercially in regions where air temperatures are very high during the summer season. Through this article in agriculture can come to understand methods of solarization adopted in fruit cultivation and their efficiency to combat effects of climate change.

12.1   INTRODUCTION

Soil solarization is a polyethylene mulching application used to increase soil temperature that is lethal to many microorganisms including those that cause diseases in economically important agri-horticultural crops (Gill and Robert, 2010). Solarization originated during the 1970s intended for soil disinfestations by means of solar energy. Solarization was based on observations by the extension workers and farmers in the Jordan Valley, who noticed the intensive heating of the polyethylene mulched soil. However, at the time, the deleterious effect of high temperatures on soil-borne pathogens was known, but heat treatment of soils was not widely implemented due to economic considerations and practicality issues. The technique has been commercially exploited for growing high-value crops in diseased soils in geographic regions with hot temperatures and irrigated agricultural systems (Stapleton and DeVay, 1995). Soil solarization is an environment friendly, nonpesticidal and hydrothermal process that occurs in moist soil when the soil is covered by plastic film, heated by exposure to sunlight (Stapleton, 2000). Solarization is considered a preventive, organic method of killing weed seeds and soil pathogens. Solarization (also referred to as solar heating of the top soil) is a method of soil disinfestations, first described by Katan and co-workers (1976) for controlling pathogens and weeds. Solarization is often as effective as herbicides, fumigants, other potentially hazardous and expensive pest control methods. Soil solarization effects are not limited to soil biota, changes in soil chemistry involving nutrient availability, as well as other changes, such as soil salinity, soil structure, soil organic matter content (Chen et al., 2000). The physical, chemical and biological principles of soil solarization as well as commercial implementation have been well documented in the literature. The effects of solarization have been investigated for many horticultural fruit crops including apple (Sharma et al., 2011a), mango (Sharma et al., 2011b) and citrus (Sharma et al., 2011c). Polyethylene mulching was also a well-known practice that was used to increase the productivity of crops by manipulating the environment to enhance plant growth and development processes.

The effectiveness of solarization depends on how susceptible weed seeds are to temperature, and also an effective practice able to control nematodes, even though it may cause serious stress on the soil microbial community. Soil solarization is most effective against winter annual weeds that germinate under cool conditions (Elmore, 1990). Soil solarization is most effective during the summer months, less effective in cooler climates and competitively more effective in wet soils (DeVay, 1990). The higher the ambient temperature, the more quickly a kill is achieved. Solarization is reported to control parasitic weeds and certain annuals, but it is not effective against perennials (Jacobsohn, et al. 1980).

12.2 PRACTICAL UTILITY OF SOIL SOLARIZATION

Soil solarization is cost-effectiveness practice in which a transparent polyethylene sheet is the main material used and presents no dangerous or hazardous substances to surrounding plants or people (Stapleton and De

Vay, 1982). After solarization has completed, the clear polyethylene mulch can also be used as bed mulch. The thin polyethylene mulch can have the tendency to rip in certain spots. If the mulch rips, it will let out moisture and heat, which are necessary to the success of solarization. Sealing the rip with plastic tape will easily remedy any rips that occur as reported by (Sayed and Malakar, 2004). The overwhelming amount of advantages over the minimal disadvantages makes soil solarization an attractive choice for organic farmers that live in temperate climates.

12.3 PRINCIPLES OF SOIL SOLARIZATION

- Utilization of transparent polyethylene because of transmission of most of the solar radiations directly to the soil. As black polyethylene sheets absorb heat and do not conduct heat downward. The thinnest polyethylene film of 25-μ size is the most effective and economical to use. Polyethylene films contain additives that improve their properties for use in solarization. Additives include pigments, heat-retaining substances, ultraviolet stabilizers and biodegradable additives (Stevens et al., 1991). Chase et al. (1999) and Campiglia et al. (2000) observed that soil temperatures under transparent film were higher than under black mulch, while Ham et al. (1993) reported the opposite. Rieger et al. (2001) found black and clear mulches equally effective for increasing soil temperatures.
- Solarization should ideally be done during the summer months, when high air temperatures and intense solar radiations are available.
- Soil should be kept moist to increase thermal sensitivity of fungal 'resting' structures such as sclerotia, chalamydospores, etc. and to improve heat conduction. A single deep irrigation prior to treatment is likely sufficient, but additional moisture may enhance solarization process. Drip irrigation technology may be used under the film for additional water.
- Longer solarization periods will greatly improve pest organism control particularly at greater depths because physical, chemical, and biological control interactions may be involved and a variety of pests are controlled.
- A single clear plastic layer laid on the soil surface can greatly increase solar soil heating and provide good control of weeds and pathogens.

The influence of high temperatures on weeds, seeds, and pathogens is complex. It often provides a positive crop growth response in addition to weed and disease control. Deep rooted, heat tolerant weeds are usually suppressed, for example, old clover seeds germinated after solarization. Adding of compost and other soil amendments may improve control of the more resilient weedy species by increasing microbiological activity in the soil. Solarization also reduces or eliminates plant pathogens such as *Fusarium chlamydosporum*, *Gibberella fujikuroi* var. *fujikuro, Alternaria alternata, Verticillium dahlia, Rhizoctonia solani and* root rots (El-Shanawany et al., 2004; Gamliel and Katan, 1991; Pullman et al., 1981a,b).

12.4 MECHANISM OF SOLARIZATION PRACTICE

Solarization is most effective when done for 4 to 6 weeks during the hottest parts of the year. In solarization, the temperature of soil at 0–5 cm increases by about 10°C and is less in the surface layers (5 cm and 10 cm) of irrigated plots as compared to nonirrigated plots. This is likely due to the energy required to heat water as indicated in the field condensation on the lower sheeting surface. As stated previously, the thermal density decline of soil-borne organisms (plant parasitic nematodes, fungi, and some bacteria) during solarization depends on soil moisture, temperature and its exposure time (Stapleton and DeVay, 1995). Although most of the mesophilic microorganisms in soil have thermal damage thresholds, some thermophilic and thermo-tolerant microorganisms can survive temperatures achieved in solarization treatment (Stapleton and DeVay, 1995). Cloud cover, cool air temperatures, and precipitation events during the treatment period reduced solarization efficiency (Chellemi et al., 1997). In addition, other changes to the physical soil environment also occur during solarization. An increase in concentration of soluble mineral nutrients concentrations of ammonium- and nitrate-nitrogen are consistently recorded across a range of soil types after solarization, improves soil structure and increases the availability of nitrogen (N) and other essential plant nutrients, while, it does not influence soil pH, electrical conductivity and available P content of the treated soil. Soil NH_4-N levels also not affected by solarization at any depth (Katan, 1987; Stapleton and DeVay, 1995). The concentrations of soluble mineral nutrients namely, Ca, K and Mg also increase to some extent. Therefore, an increase in available mineral nutrients in soil can play a major role in the effect following solarization, including improved plant health and growth, and reduced fertilization requirements. Also, solarization also causes important biological changes in treated soils. Stapleton and DeVay (1995) reported the destruction of many mesophilic microorganisms creates a partial biological environment in which substrate and nutrients in soil are made available for recolonization.

12.5　HORTICULTURAL IMPORTANCE OF SOIL SOLARIZATION

Soil solarization is effective when applied to land before and/or during the establishment of new orchards or vineyards. Solarization been successfully used on a large scale to reduce *Verticillium* wilt symptoms in young pistachio nut orchards and has also been successfully used in vineyards (*Vitis* spp.), apple (*Malus* spp.), mango (*Mangifera* spp.), avocado, stone fruit, citrus (*Citrus* spp.) and olive orchards. In an orchard or vineyard, clear plastic is either placed around the bases of individual tree or connected to strips placed within and between the tree rows. For best results, begin solarization as soon as trees are planted. Partial shading by young trees does not prevent soil heating, nor does soil solarization appear to bother most young trees during the treatment. However, solarising certain species of trees, such as herbaceous perennials, avocado and apple, almond, mango, and citrus trees, with clear plastic may result in plant damage especially when trees are young. In addition to killing soil-borne pests, solarization of orchards can greatly reduce the amount of water needed for irrigation and increase the growth, large commercial orchards, the cost of postplant solarization should be compared to the benefits before making a treatment decision. Earlier research efforts have shown that pests that are not eradicated by solarization may recolonize roots and soil, and pathogens and nematodes may survive in roots remaining in the soil.

12.6　SOIL PREPARATION

Solarization is most effective when the plastic sheeting is placed in close contact with a smooth soil surface. Ideal soil preparation involves disking or turning the soil by hand to break up clods followed by smoothing the soil surface. Remove any large rocks, weeds or any other objects or debris that will raise or puncture the plastic.

12.7   SOIL MOISTURE

Soil moisture is a critical variable in soil solarization because it makes microorganisms and weed seeds more sensitive to heat and also transfers heat to living organisms in soil. The success of soil solarization depends on adequate moisture for maximum heat transfer; maximization of heat in soil increases with increasing soil moisture content. Soil moisture also favors cellular activities and growth of soil-borne microorganisms and weed seed germination, thereby making them more vulnerable to the lethal effects of high soil temperatures associated with soil solarization. Wet soil conducts heat better than dry soil to make soil organisms more vulnerable to heat. Soil under the plastic sheets must be saturated to at least 70% field capacity in the upper layers and moist to depths of 60 cm for soil solarization to be most effective. The interaction between temperature and soil moisture causes water cycling in soil during soil solarization. The upper soil layers (upper 5 cm) have diurnal temperature fluctuations i.e., cooling at night and heating to high temperature during sunlight hours. This diurnal fluctuation causes moisture in the upper zones in soil to move downward as a result of solar radiation during the daytime, while, at night the soil surface cools and causes an upward migration of moisture. A drip irrigation system line can be used under plastic mulch to maintain moisture in soil at desired levels.

12.8   SOIL SOLARIZATION AND WEED CONTROL

Soil solarization frequently used for suppressing weeds growth and is considered a highly effective practice to control weeds. To solarize the soil, first the soil must be in moist condition and covered with 0.125 mm thick black polyethylene or 0.1 mm thick transparent polyethylene sheet for 4 to 6 weeks. The sheet should be in a full stretched condition and the edges buried in the soil to hold the sheet in place. Due to solar radiation the soil under plastic sheet gets heated up and the temperature inside the sheet around 45–50°C. Due to heavy heat, weed seeds present in the soil under the polyethylene cover gets affected. The effect of this soil solarization mainly depends upon the exposure time of the soil and soil temperature that is to be maintained. Soil solarization is not universally effective against all weeds and may require additional weed management approaches.

12.9   SOIL NUTRIENTS AVAILABILITY

Solarization initiates changes in soil physical-chemical features that improve plant growth and development. Solarization speeds up SOM breakdown, resulting in the release of soluble nutrients such as nitrogen (NO_3, $NH4^+$), calcium (Ca^{++}), magnesium (Mg^{++}), potassium (K^+), and fulvic acid making them more available to plants (Chen et al., 2000). The concentrations of ammonium- and nitrate- nitrogen are consistently increased across a range of soil types after solarization. This has also increased availability of plant nutrients due to the relative increase in populations of rhizosphere favorable bacteria (*Bacillus* spp.) which contribute to the marked increase in growth, development and yield of fruit crop plants grown in solarized soil. The increased availability of mineral nutrients following soil solarization are particularly those tied up in the organic soil fraction primarily as a result of the death of the soil microbiota. Extractable P, K, and Ca, Mg sometimes have been found in greater amounts after soil solarization. The solarization mode of action liberates vapor, liquid N compounds and hence, increases the concentration of reduced N (NH_4-N) which then nitrify (convert to NO_3-N) after termination of the solarization to provide NO_3 for increased crop growth response.

12.10   SOIL ORGANIC AMENDMENTS

Solarization can be combined with organic amendments, such as composts, crop residues, green manures, and animal manures to increase the pesticidal effect of the combined treatments (Chellemi et al., 1997). The incorporation of these organic materials act to reduce soil borne pests by altering the composition of the resident microbiota and/or of the soil physical environment. Soil organic amendments protect soil microbial biomass and enzymatic activities from the detrimental effect of heating. Organic green manures, crop residues, and animal waste amendments such as manures (especially chicken manure) also help in the increase of the soil solarization efficacy. Organic matter additions and their subsequent decomposition increases heat generation and increase the heat carrying capacity of the soil. Volatile biotoxic compounds are released when organic matter is heated during the process of solarization. Different plant residues or manure incorporated into solarized soil may generate measurable amounts of volatiles such as ammonia, methanethiol, dimethyl sulfide, allylisothiocyanate, phenylisothiocyanate and aldehyde. Thus, organic amendments augment e biocidal activity of the soil solarization also. In addition, soil treatment with organic and inorganic ammonia based fertilizers followed by soil solarization may be effective against natural soil populations of the damping off fungus (*Pythium ultimum*), *Verticillium* spp., and root knot nematode (*Meloidogyne incognita*). Solarization appears to be an effective practice able to control soil borne pathogens, even though it may cause stress on some agriculturally beneficial soil microbial biomass. High solarization temperatures observed in high organic matter soils may kill the microbiota of soil, including nitrifying microorganisms, and favor the accumulation of NH_4-N. The lower rate of soil solarization temperatures observed in low organic matter soils that allow the survival of soil microbiota and promote aerobic conditions, with minimal liberation of nitrogenous compounds, thereby resulted in nitrification and loss of N from the soil (NO_3 is leached easily).

12.11 AGRICULTURALLY IMPORTANT AND BENEFICIAL MICRO-BIOTA

Solarization raised soil temperature is considered mild compared to steaming process in soil. Thus, soil solarization is more selective towards thermophilic and thermotolerant biota, including actinobacteria (former actinomycetes) which may survive and even flourish under soil solarization. Poor soil competitors such as many pathogens are killed by soil solarization since they tend to have specialized physiological requirements, which are more adapted to coexistence with the host plants. This results in a population shift favoring thermo-tolerant species such as fluorescent *Pseudomonads* and thermotolerant fungi. Solarization initially may reduce populations of beneficial microorganisms, but populations of beneficial, growth promoting and pathogen antagonistic bacteria and fungi quickly recolonize solarized soil. Thus, soil solarization adds a biological control component. Solar heating decreased pathogenic and ectomycorrhizal in-oculum potential and increased soil nitrate has been reported (Salerno et al., 2000). Plant pathogenic fungi weakened by high soil temperatures are more susceptible to these antagonists. Nitrogen fixing *Rhizobium* bacteria are also sensitive to high soil temperatures, but the reduction in nodulation of the roots of legumes such as peas or beans in solarized soils is also re-ported temporarily. However, in another study, Nair et al. (1990) reported an increase of root nodulation, colonization by arbuscular mycorrhizal fungi and yield of legumes.

12.11.1 SOIL MOISTURE CONSERVATION

Soil moisture conservation is essential to achieve higher yields from rain-fed agri-horticultural crops. Soil moisture is often the most unreliable and scarce resource, so the challenge is to enhance water availability for crop production. There are several ways to enhance soil moisture retention in-cluding, conservation tillage, mulching and compost use and restricting deep drainage. Mulch is defined as any material that is spread over soil acting as a protective cover. Mulch or compost is used to retain water and reduce moisture loss to the atmosphere. Advantages of mulching are the reduction of soil evaporation, suppression of weeds, insulation of the soil against extreme heat and cold, prevention of soil compaction, and control of wind and soil erosion. Mulch has a high C:N ratio (sawdust, wood chips) and may cause nitrogen deficiency on plants. Bacteria that break these materials down use nitrogen, thus depriving plants of this vital element. The best solution is to compost these materials until they start to break down. Non-chemical weed control that maintains soil quality has been identified a priority need by organic orchardists. Mulching systems, including cover crops and living mulches, offer the potential for adequate weed control along with benefits for water conservation and soil fertility

12.11.2 MOISTURE CONSERVATION MULCH MATERIALS

Mulching is an excellent horticultural technique that is beneficial to all fruit crop plants and particularly useful during dry periods. There are two main types of mulch practices used: organic and inorganic. Organic mulch-es are dried or composted plant materials that decompose over time when added to the soil, subsequently improving the soil's ability to store water, improving soil fertility, facilitating earthworms and other agriculturally beneficial microbial activity. Organic mulches include: straw (pea straw and sugarcane), cypress, yard waste (grass clippings, leaves, and pine needles), pine bark, homemade compost (farmyard manure) and grass mulch. Organic mulch materials restore beneficial soil nutrients and should be replaced periodically. Recommended depth for organic mulches is 2 to 3 inches. Too much mulch may decrease air circulation, result in waterlogging during rains, encourage root rot diseases, and provide winter shelter for hibernating insects and chewing rodents such as mice and voles. Mulch texture and depth may influence the abundance and insects and mites that reside in organic and inorganic mulches. Inorganic mulches will have a much longer life span than organic mulches, but will usually not have a very natural look or improve soil quality. Inorganic mulches are made of inert material such as rock, gravel or plastic, generally require less maintenance than organic mulch, do reduce water loss and suppress weeds, but can not contribute towards health and fertility of the soil. Stone, pebbles, gravel and rock can be used for color, texture and stability. Besides, reducing evaporation, vegetative mulches can reduce the spread of soil-borne diseases, reduce weed growth, reduce soil erosion, and provide nutrients and organic matter and aid in infiltration. Mulches improve infiltration by protecting the soil surface from the impact of raindrops and eliminating soil crusting. Mulches can however, be expensive and labor intensive to obtain, transport and apply to the soil. Mulching is usually more practical for high value crops such as vegetables and berries.

12.11.3   MULCHING RECOMMENDATIONS FOR SPECIFIC FRUIT CROPS

12.11.3.1   POME AND STONE FRUITS (APPLES, PEARS, PEACHES, NECTARINES, PLUMS AND CHERRY)

Tree fruit should be mulched to the drip line with 6 to 8 inches of straw or hay during the months of May–June. Mulch should be reapplied periodically and pulled away from trees at the end of August or early September. This will allow trees to harden off and to reduce moisture pressure around the tree.

12.11.3.2   GRAPES

Grapes are usually not mulched because they have a very deep root system. If they are mulched to control weeds, they should be treated like tree fruit.

12.11.3.3   STRAWBERRY

Mulching is generally carried out during month of November with straw or pine needles for winter protection. Mulch material is removed from the tops of plants and redistributed around plants and between rows in April to keep berries clean when they ripen, reduce fruit rot and help keep the soil cool as fruit ripen.

12.11.4   SOIL SOLARIZATION AND MULCHING

Mulching for weed control is the use of materials to cover the soil surface that block light, thereby preventing growth and germination of weeds. While mulches are commonly used in restoration operations for reseeding and soil stabilization, mulches for weed control must be applied at rates high enough to prevent light from reaching the soil surface. Mulching is most effective on controlling small seeded species. There are many types of mulches including natural ones such as straw, bark, sawdust, crop residues, and grass clippings, and artificial ones such as paper, cardboard, and plastic. While mulch applications are not commonly made to control invasive on a large scale, they are still useful. Another effective use of mulch, especially to suppress dense ground covers, combines cardboard and organic mulch. The cardboard covers and overlaps edges of the treatment area, and is then covered with organic mulch. Small drain holes in the cardboard prevent water from pooling. Left to compost in place, the cardboard will suppress vegetation underneath.

Soil solarization uses polyethylene sheeting to cover low growing, cultivated, mowed, or chopped invasive infestations and trap solar energy to heat the soil and space under the sheeting to kill and suppress invasive plants. At least 2 years of summer cover are needed to suppress most invasive plants by 90%. Other plants are killed by this method-it is not selective. Black sheeting is more effective than clear sheeting because it blocks needed sunlight, and, at an extra cost, is available with ultra violet blockers to greatly extend the useful life of sheets to more than one growing season. Summer is the most effective season, and use on wet soils increases control. After removal, the bare soil is open for reinvasion unless desirable revegetation is gained.

KEYWORDS

- **Moisture Conservation**
- **Pest Control**
- **Polythene Mulching**
- **Pome Fruits**
- **Soil Disinfestations**
- **Weed Suppression**

REFERENCES

Bainbridge, D. A. (1990). Soil Solarization for Restoration, Restoration and Management, Notes, 8(2), 96–97.

Benfield, C. B., DiTomaso, J. M., Kyzer, G. B., Orloff, S. B., Churches, K. R., Marcum, D. B., & Nader, G. A. (1999). Success of Mowing to Control Yellow Starthistle Depends on Timing and Plants Branching form, California Agriculture, 53(2), 17–21.

Campiglia, E., Temperini, O., Mancinelli, R., & Saccardo, F. (2000). Effects of Soil Solarization on the Weed Control of Vegetable Crops and on the Cauliflower and Fennel Production in the Open Field, in Eighth International Symposium on Timing Field Production of Vegetable Crops (Stofella, P. J., Cantliffe, D. J., & Damato, G. Eds.) Acta Horticulture, 533, 249–255.

Chase, C. A., Sinclair, T. R., Chellemi, D. O., Olson, S. M., Gilreath, J. P., & Locascio, S. J. (1999). Heat-Retentive Films for Increasing Temperatures during Solarization in a Humid, Cloudy Environment, Hort Science, 34, 1095–1089.

Chellemi, D. O., Olson, S. M., Mitchell, D. J., Secker, I., & McSorley, R. (1997). Adaptation of Soil Solarization to the Integrated Management of Soilborne Pests of Tomato under Humid Conditions. Phytopathology, 87, 250–258.

Chellemi, D. O., Olson, S. M., Mitchell, D. J., Secker, I., McSorley, R. (1997). Adaptation of Soil Solarization to the Integrated Management of Soilborne Pests of Tomato under Humid Conditions. Phytopathology, 87, 250–258.

Chen, Y., Katan, J., Gamliel, A., Aviad, T., & Schnitzer, M. (2000). Involvement of soluble organic

DeVay, J. E. (1990). Historical Review and Principles of Soil Solarization, in DeVay, J. E., Stapleton, J. J., & Elmore, C. L. (eds.), Soil Solarization. United Nations, Rome.

El-Shanawany, A. A., El-Ghamery, A. A., El-Sheikh, H. H., & Bashandy, A. A. (2004). Soil Solarization and the Composition of Soil Fungal Community in Upper Egypt, Ass. Univ. Bulletin of Environmental Research, 7(1), 137–151.

Elmore, C. L. (1990). Use of Solarization for Weed Control, In DeVay, J. E., Stapleton, J. J., & Elmore, C. L. (eds.), Soil Solarization, United Nations, Rome.

Gill, H. A. K., & Robert, M. C. S. (2010). Effect of Integrating Soil Solarization and Organic Mulching on the Soil Surface Insect Community Florida Entomologist, 93(2), 308–309.

Grinstein, A., & Hetzroni, A. (1991). The Technology of Soil Solarization, In Katan, J., & DeVay, J. E. (eds.) Soil Solarization, CRC Publications, Boca Raton, 159–170.

Ham, J. M., Kluitenberg, G. J., & Lamont, W. J. (1993). Optical Properties of Plastic Mulches Affect the Field Temperature Regime, Journal of the American Society for Horticultural Science, 118, 188–193.

Hanson, E. (1996). Tools and Techniques, Chapter 3 in Invasive plants, Randall, J. M., & Marinelli, M. eds. Handbook, 149. Brooklyn Botanical Garden, Inc., Brooklyn, New York. 111pp.

Jacobson, R., Greenberger, A., Katan, J., Levi, M., &. Alon, H. (1980). Control of Egyptian Broomrape (Orobanche aegyptiaca) and other Weeds by means of Solar Heating of the Soil by Polyethylene Mulching, Weed Science, 28, 312–316.

Katan, J. (1987). Soil Solarization, In Chet, I. (eds.), Innovative Approaches to Plant Disease Control. Wiley, New York, 77–105.

Katan, J., Greenberger, A., Alon, H., & Greenstein, A. (1976). Solar Heating by Polyethylene Mulching for the Control of Diseases Caused by Soil-Borne Pathogens, Phytopathology, 66, 683–688.

Katan, J., Grinstein, A., Greenberger, A., Yarden, O., & DeVay, J. E. (1987). First decade (1976–1986) of Soil Solarization (Solar Heating), A Chronological Bibliography, Phytoparasitica. 15, 229–255.

Nair, S. K., Peethambaran, C. K., Geetha, D., Nayar, K., & Wilson, K. I. (1990). Effect of Soil Solarization on Nodulation, Infection by Mycorrizal Fungi and Yield of Cowpea. Plant Soil, 125, 153–154.

Pickart, A. J., & Sawyer, J. O. (1998). Ecology and Restoration of Northern California Coastal Dunes, California Native Plant Society, Sacramento, CA. 152pp.

Porras, M., Barraua, C., & Romeroa, F. (2007). Effects of Soil Solarization and Trichoderma on Strawberry Production, Crop Protection, 26(5), 782–787.

Pullman, G. S., De Vay, J. E., & Garber, R. H. (1981a). Soil Solarization and Thermal Death, Alogarithmic Relationship between Time and Temperature for Four Soil-Borne Plant Pathogens, Phytopathology, 71(9), 959–964.

Pullman, G. S., DeVay, J. E., Elmore, C. L., & Hart, W. H. (1984). Soil Solarization, A Nonchemical Method for Controlling Diseases and Pests. UC Cooperative Extension, Berkeley, C. A., 8p.

Pullman, G. S., Devay, J. E., Garber, R. H., & Weinhold, A. R. (1981b). Soil Solarization, Effects on Verticillium wilt of Cotton and Soil-borne Populations of dahliae, V., Pythium spp., Rhizoctonia solani and Thielaviopsis basicola. Phytopathology, 71, 954–959.

Rieger, M., Krewer, G., & Lewis, P. (2001). Solarization and Chemical Alternatives to Methyl Bromide for Preplant Soil Treatment of Strawberries, Hort Technology, 11, 258–264.

Salerno, M. I., Lori, G. A., Giménez, D. O., Giménez, J. E., & Beltrano, J. (2000). Use of Soil Solarization to Improve Growth of Eucalyptus Forest Nursery Seedlings, in Argentina, New Forests, 20(3), 235–248.

Sayed, Abu, & Malakar, P. K. (2004). Soil Solarization for Healthy Seedlings and Increased Crop Production, Wheat Research Centre, Bangladesh Agricultural Research Institute, Web Cornell University.

Sharma, S. D., Kumar, P., & Bhardwaj, S. K. (2011b). Screening of AM Fungi and Azotobacter Chroococcum under Natural, Solarization, Chemical Sterilization and Moisture Conservation Practices for Commercial Mango Nursery Production, in North-West Himalayas, Scientia Horticulturae, 128, 506–514.

Sharma, S. D., Kumar, P., Bhardwaj, S. K., Chandel, A. (2011c). Symbiotic Effectiveness of Arbuscular Mycorrhizal Technology and Azotobacterization for Citrus Nursery Management under Soil Disinfestation and Moisture Conservation Mulch Practices, Scientia Horticulturae (In Press).

Sharma, S. D., Kumar, P., Bhardwaj, S. K., Yadav, S. K. (2011a). Screening and Selecting Novel AM Fungi and Azotobacter Strain for Inoculating Apple Under Soil Solarization and Chemical Disinfestations with Mulch Practices for Sustainable Nursery Management, Scientia Horticulturae, 130(1), 164–174.

Stapleton, J. I., Quick, J., & DeVay, I. E. (1985). Soil solarization, Effects on Soil Properties, Crop Fertilization and Plant Growth, Soil Biology and Biochemistry, 17, 369–373.

Stapleton, J. J. (1990). Thermal Inactivation of Crop Pests and Pathogens and Other Soil Changes Caused by Solarization, In DeVay, J. E., Stapleton, J. J., & Elmore, C. L. (eds.), Soil Solarization. United Nations, Rome.

Stapleton, J. J. (2000). Soil Solarization in Various Agricultural Production Systems, Crop Protection. 19, 837–841.

Stapleton, J. J., & De Vay, J. E. (1982). Effect of Soil Solarization on Population of Selected Soil-Borne Microorganisms and Growth of Deciduous Fruit Tree Seedling, Phytopathology, 72, 233–226.

Stapleton, J. J., & DeVay, J. E. (1995). Soil Solarization, A Natural Mechanism of Integrated Pest Management, In Reuveni, R. (Ed.), Novel Approaches to Integrated Pest Management. Lewis Publishers, Boca Raton, 309–322.

Stevens, C., Khan, V. A., Brown, J., Hochmuth, G., Splittstoesser, W., & Granberry, D. (1991). Plastic Chemistry and Technology as Related to Plasticulture and Solar Heating of Soil, in Soil Solarization, (Katan, J., & DeVay, J. E., eds.) CRC Press, Boca Raton, Florida. 141–158.

CHAPTER 13

BIOCHAR TECHNOLOGY FOR SUSTAINABLE HORTICULTURE

ANSHUMAN KOHLI[1], YANENDRA KUMAR SINGH,
MANOJ KUMAR DWIVEDI, and SHASHI BHUSHAN KUMAR

Department of Soil Science and Agricultural Chemistry, Bihar Agricultural University, Sabour – 813210 (Bihar), India;
[1]Email: anshuman_kohli@hotmail.com

CONTENTS

ABSTRACT

The floodplain soils of Bihar comprising of alluvial deposits of various rivers in their lower course have shaped the agricultural destiny of the state. Although soil fertility is not generally a severe concern in these soils because of regular deposition of fertile alluvium, these soils which are in fact just sedimentary deposits, pose serious soil physical problems due to concentration of very similar textural separates in a season's debris. Besides there is little or no profile development in the characteristic Inceptisols and Entisols in the state. A large chunk of these problems are related to the physical quality and can be mitigated by improving the structure and stability of the floodplain soils. Adding stabilized carbon such as biochars to the soils and cultivating horticultural crops for an economic advantage is a proposition for these soils. Various advantages of biochar addition in soils including increasing cation exchange capacity, increasing water holding capacity, offering hiding space for useful bacteria and fungi, modifying the soil hydrothermal regime, affecting the dynamics of mineral nutrients in soils, etc., are enumerated followed by a brief discussion of biochar technology in horticulture. Biochar can be a substrate of growing ornamental plants and also a substrate for protected cultivation in green houses or poly houses. Biochar is a suitable material for making the earth ball for saplings to be transplanted and also a potential inoculant carrier for biofertilizers. Moreover, it is overall a good soil conditioner for ameliorating soil physical constraints encountered in floodplain soils. All these advantages can potentially lead to sustainably high horticultural crop yields from biochar interventions. However, there is still a need for strategic research efforts, to allow elucidation of mechanisms, differentiated by environmental and management factors and to include studies over longer time frames.

13.1   INTRODUCTION

Floodplains are the lands comprising of alluvial deposits of soils in the vicinity of the streams or rivers. Although inherently fertile and usually productive, these soils have their own specific problems. These include, among others, frequent seasonal floods and accompanying soil erosion and siltation, water logging and scarcity at different times of the year, textural problems associated with high sand or silt contents, and, above all, little or no profile development as characteristic of Inceptisols and Entisols. Floodplain soils have buried soil horizons due to deposition of debris during the receding floods. Pedogenesis is hardly evident with nondescript horizonation and poor structure. Thus floodplain soils suffer from moderate to severe physical problems, which are accentuated by their vulnerability to flood induced erosion and sedimentation. Still, the course of rivers exhibits sufficient variability in the extent of these problems. So do we need to identify the critical factors that make some areas more vulnerable than the others? In other words, can we identify any critical component that accounts for these differences in vulnerability?

A large chunk of these problems are related to the physical quality and can be mitigated by improving the structure and stability of the floodplain soils. An approach towards this end being considered in this proposition is adding stabilized carbon such as biochars to the soils and cultivating horticultural crops for an economic advantage. One of the important goals of producing biochar and burying it in soils is to sequester atmospheric carbon perpetually in the soils. The practice of smoldering agricultural wastes in pits or trenches for increasing soil productivity several centuries ago by these people, still visible as *Terra Pretta de Indio* or Indian Black Earths, has found renewed interest in recent years in view of the anthropogenic modification of the global carbon cycle. Large amounts of CO_2 are cycling between atmosphere and plants on an annual basis and most of the worlds organic C is already stored in soils. Increasing atmospheric CO_2 is an important global issue and long-term storage of carbon in soil is considered an important option to mitigate the increasing level of CO_2 in the atmosphere (Lal, 2009). A similar material called biochar can be produced artificially when biomass, such as wood, manure or leaves is heated in a closed container with little or no available air. Biochar application in soils is a carbon-negative technology and can remove CO_2 on Giga ton scales to combat climate change because biochar halts the decay process and captures CO_2 in a virtually permanent carbon stock preventing its rerelease into the atmosphere. Biochar yields several potential cobenefits such as renewable bioenergy, improved agricultural productivity specifically in low fertility and degraded soils, reduced losses of nutrients and agricultural chemicals in runoff and improved water holding capacity of soils (Woolf et al., 2010). Hence, in this proposition, we consider various options of using biochars in cultivating horticultural crops that have pertinence in managing the flood plains of Bihar.

13.2 BIOCHAR TECHNOLOGY

Biochar is charcoal, but it's special. At its best, biochar is made from sustainably sourced materials. This makes it cleaner and greener than regular charcoal. Biochar is a carbonaceous material produced by thermo-chemical pyrolysis of organic materials. Pyrolysis metamorphoses the subjected materials to produce an inert product known as biochar along with liberation of pyrolytic vapors comprising of volatile gaseous compounds. Biochar, in physical terms, is similar to coal, so also in terms of the source materials and the conditions of metamorphosis. Metamorphosis of buried organic materials under intense temperature and pressure leads to step wise formation of various types of coals such as peat, lignite, bituminous and anthracite, in this order. Further metamorphosis would lead to formation of the purest form of carbon, that is, graphite. Similar metamorphosis occurs under synthetic pyrolysis leading to the production of biochars. Both coals as well as biochars are metamorphosed organic materials with far greater carbon content in comparison to the original organic materials. In other words, pyrolysis leads to preferential release of elements other than carbon from the feedstock leading to a relatively more carbonized end product. When carbonization of an organic feedstock occurs under natural conditions over a time scale of millions of years, coal is formed. But pyrolysis of lingo-cellulosic materials under synthetic conditions leads to production of charcoal. Another important misconception with pyrolysis of biomass is that this is similar to burning of biomass. But the fact is that while on one hand burning is oxidation of the biomass, pyrolysis of biomass on the other hand results in the production of a highly reduced carbon form charcoal which if used for soil improvement is known as biochar. Carbon in charcoal is so reduced that it is used as a reductant in metallurgy.

To begin with, an overview of the expected advantages of biochar applications to soils is proposed. There are several outstanding arguments regarding the potential of biochars, like increasing cation exchange capacity, increasing water holding capacity, offering hiding space for useful bacteria and fungi, modifying the soil hydrothermal regime, affecting the dynamics of mineral nutrients in soils, etc. Biomass carbon in biochar is super-stable and will remain in soil for hundreds of years. Biochar promotes a microbial population explosion and these microbes sequester more carbon as their living and dead bodies, which causes a doubling or more of plant growth, which sequesters even more carbon into living biomass. Biochar loosens, lightens, opens, and aerates soil so it requires less machinery and energy for tillage. Biochar plus microbes holds nutrients in soil, to reduce carbon and nitrogen emissions and leaching, thus reduce or eliminate synthetic, energy intensive fertilizers. Making biochar by gasification produces gas and liquid biofuels to replace fossil fuels for energy and, as a bonus; biochar creates nutrient-dense soil to grow nutrient-dense food to reverse the growing epidemic of chronic and degenerative diseases, and their rising health care costs. The influence of biochar on soil properties is likely to vary significantly between different biochars because biochar properties are governed by the biomass source and the pyrolysis.

Conditions such as temperature and activation treatment (Chan et al., 2007, 2008; Chan and Xu, 2009; Gaskin et al., 2008; Novak et al., 2009; Nguyen et al., 2010). Singh et al. (2001) reported that Biochar pH varied significantly among feedstocks, ranging from 6.93 to 10.26. Higher pyrolysis temperature increased the pH of most biochars. The total C content of the biochars varied widely, with average values ranging from 165 to 836 g/kg. The eucalyptus wood and leaf biochars had the highest total C content and cow manure biochars the lowest. All these effects need to be explained and quantified so that we know exactly the likely effects of soil application of various biochars before being advocated on a larger scale.

13.3   APPLICATIONS OF BIOCHAR IN HORTICULTURE

In horticulture, biochars have potential applications over a wide range of crops, soils, crop stages and motives. A few of them are being projected in this proposition based on researches conducted in different parts of the world.

13.3.1   BIOCHAR—A SUBSTRATE FOR ORNAMENTAL PLANTS

Biochar made from wood has been reported to have positive influences on ornamental features of potted plants for high ornamental value. This is as a result of greater stem diameter, Leaf area, flowering and rooting with amelioration of the growing medium with biochar.

13.3.2 BIOCHAR APPLICATIONS IN PROTECTED CULTIVATION (GREENHOUSE TECHNOLOGY)

Cultivation of high value crops under poly house or greenhouse conditions requires ideal conditions for optimum production. Even a slight stress can cause drastic economic loss. So biochar application to ameliorate the growing medium under protected cultivation for crops like tomato, straw-berries, cutflowers and other offseason vegetables presents a superior medium for growth of plants right from germination to the bearing stage. Biochar amendment not only results in better soil physical conditions to reduce abiotic stresses of mechanical impedance and water scarcity, but also helps ward off biotic stresses such as plant diseases under green house conditions with its better potential to harbor beneficial soil microorgan-isms and biocontrol agents.

13.3.3 BIOCHAR—A SUITABLE MATERIAL FOR FRUIT NURSERIES

Biochar has excellent water retention and nutrient supplying capacities. It promotes aggregation of soil particles but still remains soft and porous. Thus it is a material perfectly suitable for ameliorating the soil to be used as earth ball while transporting fruit saplings from nurseries to new or-chards.

13.3.4 BIOCHAR—A POTENTIAL INOCULANTS CARRIER FOR BIOFERTILIZERS

Biochar has been reported to be a good potential carrier material for bio-fertilizers. As such it has been shown to harbor Rhizobium and Pseudo-monas in its pores and can be used in the biofertilizer manufacturing with good shelf life. Glodowska et al. (2013) demonstrated that the survival of bacteria depends on the type of biochar and that the biochars with neutral pH, wwater-holdingcapacity between 70–75% and porosity between 65–75% create a suitable habitat for bacteria.

13.3.5 BIOCHAR—A SOIL CONDITIONER FOR FRUIT TREES

Biochar application in the soils of fruit orchards can potentially improve the soil physical condition leading to an overall increase in growth. Fruit trees with biochar application in the basins would need less frequent irrigation and be flourishing with activities of soil microorganisms. The efficiency of fertilizer application is also expected to increase in biochar-amended soils due to reduced leaching and runoff losses of nutrients. Biochar applications in degraded lands being used for fruit protection like aonla, guava, wild pomegranate and other minor fruits would also benefit the overall system. Not only would the crop growth and yields be improved, the soils would improve in their physical condition and fertility. Torres (2011) has reported that nutrient-poor feedstocks such as cobs and maize stover are most effective as organic amendments when pyrolyzed. Even modest benefits from biochar application can have a very significant impact on per hectare profitability of crops, particularly high-value crops such as vegetables, producing economic benefits both for farmers and the associated food chain. These benefits, if realized, could have a significant knock-on impact on job creation and economic activity across the region; Light and sandy soils that require high water input and high fertilizer input are likely to show greatest benefit from biochar application, which links directly with the high-value crops that are grown on these soils (Collison et al., 2011).

13.4 THE WAY AHEAD

Increased crop yield and quality is a commonly reported benefit of adding biochar to soils. However, experimental results are variable and dependent on the experimental set-up, soil properties and conditions, while causative mechanisms are yet to be fully elucidated. There is a need for strategic research efforts, to allow elucidation of mechanisms, differentiated by environmental and management factors and to include studies over longer time frames.

KEYWORDS

- **Biochar**
- **Climate Change**
- **Ecofriendly approach**
- **Mitigation**
- **Soil Health Management**
- **Sustianable Horticulture**

REFERENCES

Chan, K. Y., & Xu, Z. (2009). Biochar Nutrient Properties and their Enhancement. In: 'Biochar for Environmental Management, Science and Technology' (Lehmann, J., Joseph, S., Eds.) 67–84. (Earthscan London).

Chan, K. Y., & Xu, Z. (2009). Biochar Nutrient Properties and their Enhancement. In: 'Biochar for Environmental Management Science and Technology.' (Eds Lehmann, J., Joseph, S.) 67–84. (Earthscan: London).

Chan, K. Y., Van Zwieten, L., Meszaros, I. A., Downie, A., & Joseph, S. (2007). Agronomic Values of Greenwaste Biochar as a Soil Amendment. Australian Journal of Soil Research, 45, 629–634. doi: 10.1071/SR07109.

Chan, K. Y., Van Zwieten, L., Meszaros, I. A., Downie, A., & Joseph, S. (2008). Using Poultry Litter Biochars as Soil Amendments, Australian Journal of Soil Research, 46, 437–444. doi: 10.1071/SR08036.

Collison Martin, Collison Lynn, Sakrabani Ruben, Tofield Bruce, & Wallage Zoe (2011). Biochar and Carbon Sequestration, A Regional Perspective, A Report Prepared for East of England Development Agency (EEDA). http://www.uea.ac.uk/polopoly_fs/1.118134!LCIC%20EEDA%20BIOCHAR%20REVIEW%2020–04–09.pdf

Gaskin, J. W., Steiner, C., Harris, K., Das, K. C., & Bibens, B. (2008). Effect of Low Temperature Pyrolysis Conditions on Biochars for Agricultural Use. Transactions of the ASABE 51, 2061–2069.

Lal, R. (2009). Challenges and Opportunities in Soil Organic Matter Research, *European Journal of Soil Science*, 60, 158–169. doi: 10.1111/j.1365–2389.2008.01114.x.

Nguyen, B. T., Lehmann, J., Hockaday, W. C., Joseph, S., & Masiello, C. A. (2010). Temperature Sensitivity of Black Carbon Decomposition and Oxidation. Environmental Science and Technology 44, 3324–3331. doi: 10.1021/es903016y.

Novak, J. M., Lima, I., Xing, B., Gaskin, J. W., Steiner, C., Das, K. C., Ahmedna, M., Rehrah, D., Watts, D. W., Busscher, W. J., & Schomberg, H. (2009). Characterization of Designer Biochar Produced at Different Temperatures and their effects on a Loamy Sand. Annals of Environmental Science, 3, 195–206.

Singh, B., Singh, B. P., &. Cowie, A. L. (2010). Characterization and Evaluation of Biochars for their Application as a Soil Amendment, *Australian Journal of Soil Research*, 48, 516–525.

Torres, Dorisel (2011). Biochar Production with Cook Stoves and Use as a Soil Conditioner, in Western Kenya. Thesis (M. S. of Soil & Crop Sciences), Cornell University 292 p + ix preliminary pages.

Woolf, D., Amonette, J. E., Street-Perrott, F. A1, Lehmann, J., & Joseph, S. (2010). Sustainable Biochar to Mitigate Global Climate Change *Nature Communications,* 56, 1–9. doi: 10.1038/ ncomms1053.

CHAPTER 14

MYCORRHIZAL FUNGI IN SUSTAINABLE HORTICULTURAL PRODUCTION UNDER CHANGING CLIMATE SITUATIONS

S. K. SINGH[1,3], V. B. PATEL[2], A. K. SINGH[1], and M. K. VERMA[1]

[1]Division of Fruits and Horticultural Technology, Indian Agricultural Research Institute, New Delhi 110012, India.

[2]Department of Horticulture, Bihar Agricultural University, Bhagalpur 813210, Bihar, India.

[3]E-mail: sanjaydr2@gmail.com

CONTENTS

ABSTRACT

Soil is the host of several symbionts and free-living beneficial microbes apart from different flora and fauna. Mycorrhizae, arbuscular mycorrhizal fungi (AMF) represent symbiotic association between plant roots and certain soil fungi of order Glomales. Such an association plays a key role in nutrient management in varied ecosystems and also protects the host plants against environmental and biotic stresses. About 80% of the plants are colonized by mycorrhizae, hence their proper use can boost the production in depleted or degraded soils and assist in mitigating the harmful effects of different biotic and abiotic stresses. AMF biotechnology is usable for crops right from transplant stage in plug plant production, as is the case with horticultural/ forestry nurseries. Mycorrhizal symbiosis is deciphered and they are known to affect plant growth and health, as bio-fertilizers and bio-protectors. Maximum benefits can be obtained from inoculation with efficient AM fungal strain and careful selection of compatible host/ fungus/ substrate combinations. Interactions between AM fungi and other bio-agents have been found beneficial. With molecular and biotechnological gains, AMF science has become more advanced not only based on taxonomic classification and their beneficial effects on plant but different gene(s) responsible for successful harnessing of such an association in rhizosphere and their role in climate change situations are the future approaches.

14.1 INTRODUCTION

In the present day of depleting environmental conditions, shrinking natural resources, climate change, etc. warrant effective technologies to be developed for mitigating the ill effects of them for safe horticulture production. Soil beneficial microflora have a definite role to play making crop production ecofriendly and sustainable. A study conducted at the North Carolina State University shows that important and common soil microscopic organisms, including (AMF), play a role in sequestering carbon below ground, trapping it from escaping into the atmosphere as greenhouse gases. AMF help hold this carbon in the ground by decelerating decomposition of soil organic matter, which prevents the carbon in the decomposing material. A contrary study shows that different microbes may aggravate the carbon release due to increased microbial activity under elevated CO_2 levels, which need to be elucidated.

14.2 ROLE OF AMF

Harnessing AMF, in perennial and annual horticultural systems give us with alternative for successful agriculture in marginal and submarginal areas. The symbiotic association between perennial plants and microorganisms play an important role in soil fertilization, improving growth, mineral nutrition, abiotic and biotic stresses tolerance, phytoremidiation, etc. Arbuscular mycorrhizal fungi are the common symbiotic association, which produce fungal structure (vesicles and arbuscules) in cortical root cells. This symbiotic association is found in most of the plants of arctic, temperate and tropical regions, including aquatic and desert environments. Mycorrhiza is the mutualistic symbiosis (nonpathogenic association) between soil-borne fungi with the roots of higher plants. The endomycorrhizae are characterized by inter and intracellular mycelial growth in the root cortex, forming specific fungal structures, referred to as vesicles and arbuscles. This characteristic growth gives the endomycorrhizae the alternate name; vesicular arbuscular mycorrhiza (VAM), presently called AMF since all the mycorrhizae do not form vesicles. About 80% of all terrestrial plant species form this type of symbiosis (Smith and Read, 1997). The arbuscular mycorrhizal fungi belong to taxonomic order called Glomales, which comprises of six genera (Fig. 14.1).

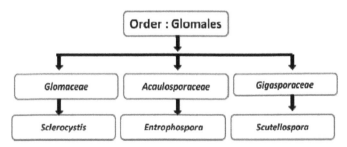

FIGURE 14.1 The current taxonomy of AMF with examples of different genera in soil (adapted from Dodd, 2000).

14.3   AMF IN BIOTIC AND ABIOTIC STRESS MANAGEMENT

Coville (1921) for the first time reported intracellular infection in litchi roots with mycorrhizal fungi when planting material was imported to USA from China. Mycorrhizal associations are also found to ameliorate and increase tolerance to adverse soil conditions, influence response to severe climatic conditions and have compatibility with different hosts; they increase plant productivity and are important for the present day agriculture under different ecosystems. The different roles of AMF in horticulture and environment are presented hereunder.

14.3.1   SOIL FERTILITY

Three main components are involved in AM association: (1) the soil, (2) the fungus, and (3) the plant. The fungal component involves the fungal structure within the cell of the root and the extraradical mycelium in the soil. The last component may be quite extensive between fungus, plant and soil. The increased efficiency of mycorrhizal roots versus nonmycorrhizal roots is caused by the active uptake and transport of different nutrients by mycorrhizae (Table 14.1).

TABLE 14.1   Effect of AMF on Absorption of Different Plant Nutrients

Nutrient	Reference(s)
P	Harley and Smith, (1983), Al-Karaki and Al-Radad (1997), Chandreshekara et al. (1995)
N	Ázcón et al. (1996), Subramanian and Charest (1999)
K	Liu et al. (2002)
Mg	Liu et al. (2002)
Cu	Gildon and Tinker (1983), Li et al. (1991)
Zn	Faber et al. (1990), Gildon and Tinker (1983), Chen et al. (2003), Jamal et al. (2002)
Ca	Liu et al. (2002)
Fe	Caris et al. (1998)
Cd	Guo et al. (1996) Gonzalez et al. (2002)
Ni	Jamal et al. (2002), Guo et al. (1996)
Ur	Rufykiri et al. (2002)

Arbuscular mycorrhizal fungi have been shown to improve productivity in low fertility soils and are particularly important for increasing the uptake of slow diffusing ions such as PO_4^{3-}, immobile nutrients such as P, Zn and Cu. Under drought conditions, the uptake of highly mobile nutrients such as NO_3^- can also be enhanced by mycorrhizal associations. Improved P nutrition has been shown to increase in infertile and P fixing soils of the tropics (Dodd, 2000). Some of the AMF in enhancing yield and produce quality are presented in Table 14.2.

TABLE 14.2 AMF and Yield Increase in Horticultural Crops

Crop	AMF +/– bioagent	Yield/Quality	Reference
Papaya 'Maradol'	*Glomus* sp.	Growth, production, and fruit quality	Vazquez-Hernandez et al. (2011)
Strawberry	*Glomus intraradices*	Quality	Morales et al. (2010)
Grape	*Glomus* sp.	Quality	Schreiner (2010)
Plantain (*Musa* AAB cv. Horn)	*Glomus manihotis* *Scutellospora heterogama*	Nutrient uptake	Gonzalez and Cuenca (2006)
Sweet passion fruit (*Passiflora alata*)	*Gigaspora albida* and *Scutellospora heterograma*	Production	Silva et al. (2008)
Sweet orange cv. Mosambi	Mixed AMF strains of IARI), *Azospirillum*, and micronutrients sprays	Yield and higher mixed strains of IARI), *Azospirillum*, and micronutrients sprays	Patel et al. (2009)
Kinnow mandarin	*Glomus deserticola*	Yield and quality	Usha et al. (2012)
Banana and plantain (*Musa* spp.)	Mixed AMF	Growth, nutrient uptake and control of root damage by nematodes	Jefwa et al. (2010)
Apple	AMF + *Azotobacter chroococcum*	Yield, P and Zn uptake	Sharma et al. (2005)
Mango	*Azotobacter, Azospirillum,* mixed strain of AMF and phosphate solubilizing bacteria	Growth, fruit yield and quality	Patel et al. (2005)

14.3.2 WATER STRESS

Under drought stress conditions, the extension of root surface by the extraradical mycelia of AM fungi enhances the acquisition of water and nutrients. It has been demonstrated that AMF colonized plants in field are better adapted and attributed this response to reduced water stress. Mycorrhizae improves water uptake by exploring the soil volume, improving plant nutrition, and/or regulating stomata through hormone synthesis. It has been shown that phosphorus levels, tissue dry weight and transpiration of AM seedlings were greater than noninoculated citrus plants. It is attributed that improved establishment of citrus to drought conditions was mainly due to improved P uptake and reduced plant stress. On the contrary, it has been observed that low transpiration rate and higher water use efficiency in mycorrhizal *ber* seedlings due to increased stomatal resistance provided by the AM colonization under arid and semiarid conditions. Mycorrhizal root colonization may indirectly influence the stomatal behavior of host leaves and higher in mycorrhizal than nonmycorrhizal plants. The stomatal conductance and leaf water potential are also altered (Auge et al., 1986, 1987, 2001). This effect was attributed to more efficient extraction of soil moisture by mycorrhizal root systems in drought conditions. Furthermore, it has been observed that there is enhanced nitrate reductase activity in roots and leaves of AM infected plants due to improved P nutrition. Some of the role of AMF symbiosis in fruit crops is mentioned hereunder (Table 14.3).

TABLE 14.3 AMF in Mitigation of Low Moisture and Drought Conditions

Crop	AMF + bioagent	Beneficial effect	Reference
Grape 'Cabernet Sauvignon'	Mixed AMF	Drought stress	Schreiner et al. (2007)
Grape 'Pinot Noir'	Mixed AMF	Root spread, higher Mg and Zn uptake and moisture stress	Schreiner (2005)
Carrizo citrange	*Glomus intraradices*	Drought and transplant stress	Johnson and Hummel (1985)
Sour orange and Carrizo citrange	-do-	Drought and enhanced growth	Dutra et al. (1996)

TABLE 14.3 *(Continued)*

Crop	AMF + bioagent	Beneficial effect	Reference
Tangerine	*G. mosseae* and *G. geosporum*	Drought and enhanced growth	Wu et al. (2007)
Ber	*Glomus* sp.	Net photosynthesis and transpiration	Mathur and Vyas (1995 & 1996)

14.3.3 DROUGHT AND SALINITY STRESS

Plant colonized by AM fungi can tolerate and recover more rapidly from soil water deficits than plants without AM fungi plants colonized by AM fungi can tolerate and recover more rapidly from soil water deficit than plants without AM fungi (Sanchez-Diaz and Honrubia, 1994). Mycorrhizal symbiosis can protect host plants against detrimental effects caused by drought stress. Alleviation of salt stress by arbuscular mycorrhizal *Glomus* species has been shown (Ruiz-Lozano et al., 1996). Several mechanisms have been proposed to explain the protection of AMF symbiosis, such as changes in plant hormones, increased leaf gas exchange and photosynthetic rate; direct hyphal water uptake from the soil and transfer to the host plant, enhanced activity of enzymes involved in antioxidant defense, nitrate assimilation, enhanced water uptake through improved hydraulic conductivity and increasing leaf conductance and photosynthetic activity osmotic adjustment and changes in cell-wall elasticity.

Often mycorrhizal improvement of drought tolerance occurs *via* improved acquisition of phosphorus, nitrogen and other growth promoting nutrients by AMF plants. The influence of AMF on water uptake and transport may be a secondary consequence of enhanced host phosphorus nutrition. Mycorrhiza can also reduce the impact of environmental stresses such as salinity. The introduction of AMF to sites with saline soils may improve early plant tolerance and growth and increased protection from salt stress. High level of proline is known to afford protection to various enzyme systems against dehydration. AMF inoculation has been found to increase the sugar levels in the plants. The higher sugar accumulation favors the plant in maintaining the osmotic balance and preventing dehydration of tissues thereby, helping the plants to grow normal even under stress conditions. Reducing, nonreducing and total sugars are enhanced in mycorrhizal plants compared to control.

14.3.4   REMEDIATION OF SOIL POLLUTANTS

Mycorrhizal fungal strain tolerant to heavy metals has provided evidence for their rapid adaptation to contaminated soils (Bhalerao, 2013; Khan, 2007). Several heavy metal tolerant AMF strains have been isolated, that is, cadmium-tolerant *Glomus mosseae* isolates. Copper (Cu) was absorbed and accumulated in the extraradical mycelium of AMF isolates, as observed in a study with *Glomus* spp. thus, protecting the host from damage. Arbuscular mycorrhizal fungi tolerant to aluminum have also been reported. *Glomus caledonicum* is the promising mycorrhizal fungus for bioremediation of heavy metal contaminated soil, while it can uptake and translocate of uranium towards the roots. Mycorrhizae are also found to ameliorate toxicity of the trace metals in polluted soils. Since heavy metal uptake and tolerance depend on both plant and soil factors, including soil microbes, interactions between plant root and their symbionts such as AM fungi can play an important role in successful survival and growth of plants in contaminated soils. Mycorrhizal associations increase the absorptive surface area of the plant due to extrametrical fungal hyphae exploring rhizospheres beyond the root-hair zone, which in turn enhances water and mineral uptake. AM fungi can further serve as a filtration barrier against transfer of heavy metals to plant shoots. The protection and enhanced capability of uptake of minerals result in greater biomass production, a prerequisite for successful remediation.

14.3.5   RESTORATION OF DEGRADED AREAS

The soils of degraded sites are low in available nutrients and lack the N_2-fixing bacteria and mycorrhizal fungi usually associated with root rhizosphere. As such, land restoration in semiarid areas faces a number of constraints related nutrient and water shortage. Mycorrhizae enhance the ability of the plant to coop with water stress situations, nutrient deficiency and drought, mycorrhizal inoculation with suitable fungi has been proposed as a promising tool for improving restoration success in semiarid degraded areas. Therefore, by adding beneficial microorganisms in the rhizosphere, the use of AMF-infected plants could reduce the amount of inorganic fertilizers needed for production.

14.3.6 MANAGEMENT OF SOIL-BORNE PATHOGENS

AMF are known to act against soil-borne diseases leading to reductions in incidence and/or severity of root rot or wilt caused by fungi such as *Rhizoctonia, Fusarium, Verticillium, Phytophthora, Pythium* and *Aphanomyces*. The phenomenon of AMF protecting plants from root pathogens is known from studies involving root-infecting pathogens, for example, *Phytophthora parasitica* or *Fusarium* sp., root-invading nematodes and horticultural species. *Glomus mosseae* induced local and systemic resistance to *P. parasitica* and was effective in reducing symptoms produced by the pathogen. Mycorrhizal plants exhibit increased tolerance towards fungal root diseases causing wilts and rots owing to better cuticulized with more lignified xylem element, allowing increased flow of nutrients, greater mechanical strength and reduced impact of harmful pathogens.

Colonization of plant roots by AMF has been shown to lead to reduced damage by nematodes (Azcon-Aguilar and Barea, 1996). Plant parasitic nematodes are among the most common pest constraint to banana production (Gowen et al., 2005). The benefits of AMF in nematode management in banana has been demonstrated by Elsen et al. (2004), who showed that AMF could suppress *Radopholus similis* densities by almost 50% in pots. Fernández et al. (2003) also demonstrated that colonization by *Glomus* spp., particularly *G. intraradices, G. manihotis* and *G. mosseae* could significantly reduce nematode damage caused by *R. similis* and *Meloidogyne incognita* on banana in pots. In Cameroon, inoculation of a *Glomus* spp. on plantain in pots significantly suppressed *R. similis* (Fogain and Njifenjou, 2002).

Phenols and enzymes such as polyphenol oxidase are important components of plant defense mechanism against the diseases. Dual inoculation with AMF and *Rhizobium* markedly increased the level of leaf chlorophyll, total soluble sugars, total phenols and free amino acids content in roots compared with those inoculated only with *Rhizobium*. Mycorrhizal formation is positively correlated with relative turgidity, activities of peroxidase and polyphenol oxidase (catechol oxidase), available phosphorus and total phenolic contents of the plants. Accumulation of phenols in AM plants along with peroxidase, peroxidase (PRO) and polyphenol oxidase (PPO) and, higher accumulation of phenols impart disease resistance. Some of the successful application of AMF in disease management is shown in Table 14.4.

TABLE 14.4   Successful Events of AMF in Disease and Pest Control

Crop	AMF + bioagent	Disease causing agent	Reference
Strawberry	*Glomus* sp. + *Bacillus* sp.	*Verticillium dahliae*	Tahmatsidou et al. (2006)
Olive	*Glomus intraradices*	*Verticillium dahliae*	Kapulnik et al. (2010)
Papaya	*Glomus mosseae, G. manihotis*	*Meloidogyne incognita*	Jaizme-Vegas et al. (2006)
Apple	*G. mosseae* and *G. versiforme*	*Pratylenchus penetrans*	Forge et al. (2001)
Quince	*Glomus intraradices*	*Pratylenchus penetrans*	Calvet et al. (1995)
Plum rootstock	*Glomus mosseae*	*Pratylenchus vulnus*	Camprubi et al. (1993)

14.3.7   CONTROLLING PHYTOPHAGOUS INSECTS

Interaction with herbivorous insects is also altered in mycorrhizal plants, as the symbiosis has an impact on the growth and/or survival of those insects. In general, AM reduce the incidence of generalist chewing insects, while sap feeding or specialist insects show increases in performance on mycorrhizal plants. As a result, insects are affected by the enhanced defense capacity of mycorrhizal plants, for example, reduced the performance of potato aphids in tomato due to *G. mosseae*.

14.3.8   BIO-HARDENING OF TISSUE CULTURE RAISED PLANTLETS AND NURSERY PLUG PLANTS

Micropropagation has got several successful applications, though suffers from low survival rates and poor growth, while shifting plantlets to field conditions (*ex vitro*), which hinders its commercial application. Inoculation of AMF to the roots of micropropagated plantlets plays a beneficial role on their posttransplanting performance (Rai, 2005, Kapoor et al., 2008). It has been reported that bio-hardening of micropropagated plantlets with AMF improves plant performance and plays a significant role in ensuring better health of plantlets. Several plants of floricultural, horticultural and forest tree species have been found effective. Acclimatization phase is the beginning of autotrophic existence of the plant, with the initiation of physiological processes necessary for survival. During acclimatization, the plantlets must increase absorption of water and minerals as well as the photosynthetic rate. AMF are well known to increase the vigor of plants by increasing absorption of water and mineral nutrients, especially phosphorus. Moreover, AMF can protect host plants from root pathogens and mitigate the effects of extreme variations in temperature, pH and water stress. Successful AMF inoculation at the beginning of acclimatization period or even during in vitro conditions and in vivo propagation has been demonstrated (Tables 14.5 and 14.6).

TABLE 14.5 AMF as Bio-Hardening Agent for Tissue Culture Raised Plants

Crop	AMF + bioagent	Beneficial effect	Reference
Strawberry cvs. Senga Sengana and Jonsok	Mixed AMF	Low fertilizer dependent	Vestberg et al. (2000)
Lemon (*Citrus limon* (L.) Burm. 'Zagara Bianca'	*Glomus mosseae* and mixed strains	Improved survival, and root development	Quatrini et al. (2003)
Grape rootstocks SO4, Paulsen 1103	*Glomus etunicatum* and *Scutellospora heterogama*	Improved survival, growth and nutritional status	Anzanello et al. (2011)
Grape (*Vitis vinifera* L.)	*Glomus mosseae*	Improved survival, improved physiological and nutritional status, higher relative water content and photosynthetic rate, higher N, P, Mg and Fe	Krishna et al. (2005, 2006)

TABLE 14.5 *(Continued)*

Grape rootstocks	Mixed AMF	Improved survival, transplantation shock, improved physiological and nutritional status, higher relative water content and photosynthetic rate	Ramajayam et al. (2013)
Grape embryo rescued hybrid plants	Mixed AMF	Transplantation shock, improved physiological and nutritional status	Singh et al. (2011)
Strawberry	Mixed AMF	Survival, growth, photosynthetic rate, drought tolerance	Borkowska (2002)
Apple, peach and plum rootstocks	*Glomus* sp.	Growth increase and survival.	Sbrana et al. (1994)
Apple and plum rootstocks	*Glomus mosseae*, *Glomus intraradices* and *Glomus viscosum*	Growth increase, survival and P uptake	Fotuna et al. (1996)
Pomegranate cv. G137	*Glomus* sp.	Improved survival, transplantation shock, improved physiological and nutritional status, higher relative water content and photosynthetic rate	Singh et al. (2012)

TABLE 14.6 AMF and Nursery Plants Hardening

Crop	AMF +/– bioagent	Effect	Reference
Passion fruit	*Glomus clarum*	Higher flavodnoids to counter moisture stress	Soares et al. (2005)
Citrus volkameriana Tan. & Pasq	Mixed AMF	Photosynthetic rate, relative growth rate of the stem, leaf area and total dry matter	Alarcon et al. (2003)

TABLE 14.6 *(Continued)*

Crop	AMF +/– bioagent	Effect	Reference
Papaya (*Carica papaya* L.) and pineapple [*Ananas comosus*]	*Glomus mosseae*	P uptake, field survival, growth	Rodriguez et al. (2011)
Mandarin orange	Mixed AMF	Growth and nutrition management in low fertility soil	Panja and Chaudhuri (2004)
Mango	*G. fasciculatum* + *Azotobacter chroococcum*	Increased height, diameter, leaf area and total root length, microbial consortium of the rhizosphere soil and leaf N, P, K and Zn	Sharma et al. (2011)
Prunus rootstock	*Glomus mosseae, G. intraradices* and *G. etunicatum*	Higher growth	Calvet et al. (2004)

14.4 AMF AND CLIMATE CHANGE PERSPECTIVES

The main areas, in which the benefits of introducing inoculant AMF into a plant growth system will accrue, are those in which they are lacking indigenous inoculum of AMF. Many manipulation experiments focusing on climate change impacts on ecosystem functioning were conducted under controlled conditions, without considering the mycorrhizal fungi living in symbiosis with the plants. Under such situation, indirect AMF effects on root exudation, root longevity and decomposition, soil aggregate stability, and nutrient acquisition are probably crucial as well. In this context, recent research that highlighted the central role of mycorrhizal fungi may play regarding feedbacks on global change. Arbuscular mycorrhizal fungi also are sensitive to climate change. Elevated CO_2 concentrations can indirectly affect AMF through increased C allocation from the host plant to the fungus a temperature-induced increase in fungal growth was associated with increased specific root length. Photosynthates were more rapidly transferred to and respired by AMF when exposed to elevated temperatures. These effects are to be studied in different ecosystems.

14.5   CONCLUSIONS

Despite a concerted effort on different aspects of AMF association, the issues which mostly concerns present day agriculture such as drought and disease tolerance and enhanced crop growth and yield are to be understood well. The AMF diversity existing in the country need to be collected, identified, characterized and conserved for future use and the management of indigenous AMF needs focus of their less understood aspects. There is need to determine the basic characteristics, such as host-fungus specificity interaction and competitiveness between AMF and, effectiveness of the AMF under possible environmental stresses, well-watered, drought stress, salinity, N and P deficiency conditions. Furthermore, role of AMF under climate change situation needs to be studied in details for sustainable production in such situations.

ACKNOWLEDGEMENTS

The authors are grateful to the Director, NBAIM, Mau (UP), for the financial assistance provided under AMAAS network project of the ICAR.

KEYWORDS

- Arbuscular Mycorrhiza Fungi
- Biotic and Abiotic Stress Mamangement
- Climate Change
- Fruit Production

REFERENCES

Alarcón, A., & Ferrera-Cerrato, R. (2003). Aplicación de fósforo e inoculación de hongos Micorrízicos arbusculares en el Crecimiento y estado Nutricional de *Citrus volkameriana* Tan & Pasq. *Terra,* 21, 91–99.

Al-Karaki, G. N., & Al-Radad, A. (1997). Effects of Arbuscular Fungi and Drought Stress on Growth and Nutrient Uptake of Two Wheat Genotypes Differing in their Drought Resistance, *Mycorrhiza*, 7, 83–88.

Allen, M. F. (1991). The ecology of mycorrhizae. Cambridge University Press, Cambridge.

Anzanello, Rafael, Paulo, de Souza, Vitor Dutra, & Casamali, Bruno (2011). Use of Arbuscular Mycorrhizal (AMF) Fungi in Micropropagated Grape Rootstocks, *Bragantia* 70(2), 409–415.

Augé, R. M., Schekel, K. A., & Wample, R. L. (1986). Greater Leaf Conductance of Well-Watered VA mycorrhizal Rose Plants is not related to Phosphorus Nutrition, *New Phytol*, 103, 107–116.

Augé, R. M., Schekel, K. A., & Wanple, R. L. (1987). Rose Leaf Elasticity Changes in Response to Mycorrhizal Colonization and Drought Acclimation, *Physiol. Plant*, 70, 175–182.

Augé, R. M., Stodola, Ann J. W., Tims, J. E., & Saxton, A. M. (2001). Moisture Retention Properties of a Mycorrhizal Soil, *Plant Soil*, 230, 87–97.

Azcón, R., Gomes, M., & Tobart, R. (1996). Physiological and Nutritional Responses by *Lactuca sativa* L. to Nitrogen Sources and Mycorrhizal Fungi under Drought Stress Conditions, *Biol. Fert. Soils*, 22, 156–161.

Azcon-Aguillar, C., & Barea, J. A. (1996). Arbuscular Mycorrhizas and Biological Control of Soil-Borne Plant Pathogens, An Overview of the Mechanisms Involved, *Mycorrhiza*, 6, 457–464.

Barea, J. M., & Azcón-Aguilar, C. (1982). Production of Plant Growth Regulating Substances by Vesicular-Arbuscular Mycorrhizal, *Glomus mosseae*, *Appl. Env. Microbiol*, 43, 810–813.

Bhalerao, S. A. (2013). Arbuscular Mycorrhizal Fungi, a Potential Biotechnological Tool for Phytoremediation of Heavy Metal Contaminated Soils, *Int. J. Sci. Nature*, 4, 1–15.

Borkowska, Bożenna (2002). Growth and Photosynthetic Activity of Micropropagated Strawberry Plants Inoculated with Endomycorrhizal Fungi (AMF) and Growing Under Drought Stress, Acta Physiologiae Plant, 24(4), 365–370.

Branzanti, B., Gianinazzi-Pearson, V., & Gianinazzi, S. (1992). Influence of Phosphate Fertilization on the Growth and Nutrient Status of Micropropagated Apple Infected with Endomycorrhizal Fungi During the Weaning Stage, *Agronomie*, 12, 841–845.

Brundrett, M., Beegher, N., Dell, B., Groove, T., & Malajczuk, N. (1996). *Working with Mycorrhizas in Forestry and Agriculture*, ACIAR Monograph, 32, 374p.

Calvet, Cinta, Estaúna, Victoria, Camprubı, Amèlia, Hernández-Dorregob, Adriana, Pinochetc, Jorge Marıa and Moreno, & Angeles (2004). Aptitude for Mycorrhizal Root Colonization in *Prunus* Rootstocks, *Scientia Horticulturae*, 100, 39–49.

Calvet, Cinta, Pinochet, Jorge, Camprubí, Amella & Fernández, Carolina (1995). Increased Tolerance to the Root-Lesion Nematode *Pratylenchus Vulnus* in mycorrhizal micropropagated BA-29 Quince Rootstock, *Mycorrhiza*, 5(4), 253–258.

Camprubi, Pinochet, J., Calvet, C. & Estaun, V. (1993). Effects of the root-lesion nematode *Pratylenchus vulnus* and the vesicular-arbuscular mycorrhizal fungus Glomus mosseae on the growth of three plum rootstocks. Plant and Soil 153(2), 223–229.

Caris, C., Hoerdt, W., Hwkins, H. J., Roenheld, V., & George, E. (1998). Studies on the Iron Transport by Arbuscular Mycorrhizal Hyphae from Soil to Peanut and Sorghum Plants, *Mycorrhiza* 8, 35–39.

Chandreshekara, C. P., Patil, V. C., & Sreenivasa, M. N. (1995). VA-mycorrhiza Mediated P Effect on Growth and Yield of Sunflower (*Helianthus annus* L.) at Different P Levels, *Plant Soil*, 176, 325–328.

Chen, B. D., Li, X. L., Tao, H. Q., Christie, P., & Wong, M. H. (2003). The Role of Arbuscular Mycorrhiza in Zinc Uptake by Red Clover Growing in Calcareous Soil Spiked with Various Quantities of Zinc, *Chemosphere*, 50(6), 839–846.

Coville, F. V. (1921). The Lychee (*Litchi chinensis*), A Mycorrhizal Plant. In: Groff, G. W. (Ed.), *The Lychee and Lungan*," New York, 151–152.

Cumming, Jr., & Ning, J. (2003). Arbuscular Mycorrhizal Fungi Enhance Aluminum Resistance of Broomsedge (*Andrpongon virginicus* L). *J. Exp. Bot.*, 54(386), 1477–1459.

da Silva, T. F. B., da Silva, S., de Oliveira, A. B., Caroliny Emilia, R., dos Santos, A. C., & Paiva, L. M. (2010). Influence of the Density of Arbuscular Mycchorizic Fungus in the Production of Sweet Passion Fruit (*Passiflora alata* Curtis), *Revista Caatinga*, 22(4), 1–6.

Dodd, J. C. (2000). The Role of Arbuscular Mycorrhizal Fungi in Agro and Natural Ecosystems, *Outlook on Agriculture*, 29(1), 55–62.

Dutra, P. V., Abad, M., Almel, V, Agustí, M. (1996). Auxin Interaction with the Vesicular-Arbuscular Mycorrhizal Fungus *Glomus intraradices* Schenck and Smith Improves Vegetative Growth of Two Citrus Rootstocks, *Scientia Horticulturae*, 66, 77–83.

Elsen, A., Orajay, J., De Waele, D. (2004). Expression of Nematode Resistance in *in Vitro* roots of Three Musa Genotypes in Response to *Radopholus Similis*. Nematology Monographs and Perspectives, 2, 303–314.

Estrada-Luna, A. A., & Davies, F. T. (2003). Arbuscular Mycorrhizal Fungi Influence Water Relations, Gas Exchange, Abscisic Acid and Growth of Micropropagated Chile Ancho Pepper (*Capsicum Annum*) Plantlets during Acclimatization and Post-Acclimatization, *J. Plant Physiol*, 160, 1073–1083.

Faber, B. A., Zasoski, R. J., Burau, R. G., & Uriu, K. (1990). Zinc Uptake by Corn as Affected by Vesicular-Arbuscular Mycorrhizae, *Plant Soil*, 121–130.

Fernández, E., Mena, J., González, J., & Márquez, M. E. (2003). Biological Control of Nematodes in Banana, 193–200. In Turner, D. W., & Rosales, F. E. (eds.), *Banana Root System, Towards a Better Understanding for its Productive Management*. INIBAP, Montpellier.

Fogain, R., & Njifenjou, S. (2002). Effect of Mycorrhizal *Glomus* sp. on Growth of Plantain and on the Development of *Radopholus similis* under Controlled conditions, *Afr. Plant Prot.* 8, 1–4.

Forge, Thomas, Muehlchen, Andrea, Hackenberg, Clemens, Neilsen, & Vrain, Thierry (2001). Effects of Preplant Inoculation of Apple (*Malus domestica* Borkh.) with Arbuscular Mycorrhizal Fungi on Population Growth of the Root-Lesion Nematode, *Pratylenchus penetrans*. Plant and Soil, 236(2), 185–196.

Fortuna, P., Citernesi, A. S., Morini, S., Vitagliano, C., & Giovannetti, M. (1996). Influence of Arbuscular Mycorrhizae and Phosphate Fertilization on Shoot Apical Growth of Micropropagated Apple and Plum Rootstocks, *Tree Physiol.*, 16, 757–763.

Gange, A. C., Brown, V. K., & Farmer, L. M. (1990). A Test of Mycorrhizal Benefit in an Early Successional Plant Community, *New Phytol.*, 115, 85–91.

George, E., Gorgus, E., Schmeisser, A. & Marschner, H. (1996). A Method to Measure Nutrient Uptake from Soil by Mycorrhizal Hyphae. In *Mycorrhizas in Integrated System from Genes to Plant Development*, Azcon-Aguilar, & Barea, J. M. (eds). Luxembourg. European Community.

George, E., Romheld, V. & Marschner, H. (1994). Contribution of Mycorrhizal Fungi to Micronutrient Uptake by Plants, In *Biochemistry Of Metal Micronutrients In The Rhizosphere*, Monthey, J. A., Crowley, D. E., & Luster, D. G. (eds.), CRC Press, Boca Raton FL, 93–109.

Gerderman, J. W. (1975). Vesicular-Arbuscular Mycorrhizae, In Torey, J. C., & Clarkson, D. P (eds). *The Development and Function of Roots*. Academic Press, NY, USA, 575–591.

Gildon, A., & Tinker, P. B. (1983). Interactions of Vesicular-Arbuscular Mycorrhizal Infection and Heavy Metals in Plants. I., The Effects Of Heavy Metals on the Development of Vesicular-Arbuscular Mycorrhizae, *New Phytol.*, 95, 247–261.

Goicoechea, N., Doleza, K., Antolin, M. C., Strand, M., & Sanchez-Diaz, M. (1995). Influence of Mycorrhizae and *Rhizobium* on Cytokinin Content in Drought Stressed alfa lafa. *J. Exp. Bot.*, 46, 1543–1549.

Gonzalez, M., & Cuenca, G. (2008). Response of Plantain Plants (Musa AAB cv. Horn) to Inoculation with Indigenous and Introduced Arbuscular Mycorrhizal Fungi (AMF) under field Conditions, *Rev. Fac. Agron.*, 25(3), 470–495.

Gonzalez-Chavez, C., D'Haen, J., Vangronsveld, J., & Dodd, J. C. (2002). Copper Soprtion and Accumulation by the Extraradical Mycelium of Different *Glomus* spp. (arbuscular mycorrhizal fungi) Isolated from the Same Polluted Soil, *Plant Soil*, 240(2), 287–297.

Gowen, S. C., Quénéhervé, P., & Fogain, R. (2005). Nematode Parasites of Bananas and Plantains, 611–643, In Luc, M., Sikora, R. A., &. Bridge, J. (eds.), *Plant Parasitic Nematodes in ubtropical and Tropical Agriculture* (2nd Edn). CAB International, Wallingford.

Graham, J. H., & Syversen, J. P. (1984). Influence of Vesicular Arbuscular Mycorrhizal on the Hydraulic Conductivity of Roots of Two Citrus Rootstocks, *New Phytol*, 97, 277–284.

Guo, Y., George, E., & Marschner, H. (1996). Contribution of an Arbuscular Mycorrhizal Fungus to uptake of Cadmium and Nickel in Bean by Maize Plants, *Plant Soil*, 184, 195–205.

Habate, M., & Manjunath, A. (1991). Categories of Vesicular-Arbuscular Mycorrhizal Dependency of Host Species, *Mycorrhiza*, 1, 3–12.

Hardie, K. (1985). The Effect of Removal Of Extra-Radical Hyphae On Water Uptake By Vesicular-Arbuscular Mycorrhizal Plants, *New Phytol.*, 101, 677–684.

Harley, J. L., & Smith, S. E. (1983). Mycorrhizal Symbiosis, University of Michigan, Academic Press, 483.

Jacobsen, I., Abbott, L. K., & Robson, A. (1992). External Hyphae of Vesicular-Arbuscular Mycorrhizal Fungi Associated with *Trofoluim subterraneum* L. I. Spread of Hyphae and Phosphorus Inflow into roots, *New Phytol.*, 120, 371–380.

Jaizme-vega, M. C., Rodríguez, A. S., & Barroso, L. A. (2006). Effect of the Combined Inoculation of Arbuscular Mycorrhizal Fungi and Plant-Growth Promoting Rhizobacteria on Papaya (*Carica papaya* L.) Infected with the Root-Knot Nematode, *Meloidogyne incognita*, *Fruits*, 61, 1–12.

Jamal, A., Ayub, N., Usman, M., & Khan, A. G. (2002). Arbuscular Mycorrhizal Fungi Enhance Zinc and Nickel Uptake from Contaminated Soil by Soybean and Lentil. *Int. J. Phytoremed.* 4(3), 203–221.

Jeffries, P. (1987). Use of Mycorrhiza in Agriculture, *Crit. Rev. Biotechnol*, 5, 319–357.

Johnson, C. R., & Hummel, R. L. (1985). Influence of Mycorrhizae and Drought Stress on Growth of *Poncirus* x *Citrus* seedlings, *HortSci.*, 20, 754–755.

Kapoor Rupam, Sharma Deepika, & Bhatnagar, A. K. (2008). Arbuscular Mycorrhizae in Micropropagation Systems and their Potential Application, *Scientia Hort.*, 116, 227–239.

Kapulnik, Y., Tsror, L., Zipori, I., Hazanovsky, M., Wininger, S., Dag, A. (2010). Effect of MF Application on Growth, Productivity and Susceptibility to Verticillium Wilt of Olives Grown Under desert conditions. *Symbiosis*, 52, 103–111.

Khan, A. G. (2007). Mycorrhizoremediation an Enhanced form of Phytoremediation, *J. Zheji-ang Univ. Sci. B*, 7(7), 503–514.

Krishna, H., Singh, S. K., Minakshi Patel, V. B., Khawale, R. N., Deshmukh, P. S., & Jindal, P. C. (2006). Arbuscular-mycorrhizal Fungi Alleviate Transplantation Shock in Micropropagated Grapevine (*Vitis vinifera* L). *J. Hort. Sci. Biotech*, 81, 259–263.

Krishna, H., Singh, S. K., Sharma, R. R., Khawale, R. N., Minakshi, & Patel, V. B. (2005). Biochemical Changes in vitro Raised Grape Plantlets due to Mycorrization During Hardening. *Scientia Horticulturae*, 106, 554–567.

Levy, Y., & Krikun, J. (1980). Effect of Vesicular-Arbuscular Mycorrhiza on *Citrus Jambhiri* Water relation, *New Phytol.*, 85(1), 25–31.

Li, X. L., George, E., & Marschner, H. (1991). Extension of the Phosphorus Depletion Zone in VA-Mycorrhizal White Clover in a Calcareous Soil, *Plant Soil*, 136, 41–48.

Liao, J. P., Lin, X. G., Cao, Z. H., Shi, Y. Q., & Wong, M. H. (2003). Interactions between Arbuscular Mycorrhizae and Heavy Metals under Sand Culture Experiment, *Chemosphere*, 50(6), 847–853.

Liu, A., Hamel, C., Elmi, A., Costa, C., Ma, B., & Smith, D. L. (2002). Concentrations of K, Ca and Mg in Maize Colonized by Arbuscular Mycorrhizal Fungi under Field Conditions, *Canadian J. Soil Sci.*, 82(3), 271–278.

Marin, M., Ybarra, M., Fe, A., & Garcia-Ferriz, L. (2002). Effect of Arbuscular Mycorrhizal Fungi and Pesticides on *Cyanara cardunculatus* growth, *Agric. Fd Sci. Finland*, 11(3), 245–251.

Mathur, N., & Vyas, A. (1995). Influence of VA mycorrhizae on net Photosynthesis and Transpiration on *Ziziphus mauritiana*. *J. Plant Physiol.*, 147, 328–330.

Mathur, N., & Vyas, A. (1996). Physiological changes in *Ziziphus mauritiana* by different VAM fungi. *Indian Forester*, 501–505.

Mathur, N., & Vyas, A. (1995). *In vitro* Production of *Glomus deserticola* in Association with *Ziziphus nummularia*, *Plant Cell Rep.*, 14, 735–737.

Menge, J. A., Davies, R. E., Johnson, E. L. V., & Zentmeyer, G. A. (1978). Mycorrhizal Fungi Increase Growth and Reduce Transplant Injury in Avocado, *Calif. Agric.*, 32, 6–10.

Morales, Castellanos, V., Villegas, J., Vierleingig, H., & Navarro, C. R. (2010). Nitrogen Availability drives the effect of *Glomus intraradices* on the Growth of Strawberry (Fragaria *ananassa Duch) plants, *J. Sci. Food Agric.*, 90(11), 1774–82.

Mosse, B., Stribley, D. P., & Le Tacon, F. (1981). Ecology of Mycorrhizae and Mycorrhizal Fungi. *Adv. Microbiol. Ecol.*, 2, 137–210.

Panja, B. N., & Chaudhuri, S. (2004). Exploitation of Soil Arbuscular Mycorrhizal Potential for AM-Dependent Mandarin Orange Plants by Pre-Cropping with Mycotrophic Crops, Applied Soil Ecol., 26(3), 249–255.

Panwar, J. A. & Vyas, A. (2000). Changes in Peroxidase and Polyphenol Oxidase Activities in Roots of *Tamarix aphylla* by different AM fungi, *J. Ecol-Physiol.*, 3, 103–106.

Patel, V. B., Singh, S. K., Asrey, R., Nain, L., Singh, A. K., & Singh, L. (2009). Microbial and Inorganic Fertilizers Application Influenced Vegetative Growth, Yield, Leaf Nutrient Status and Soil Microbial Biomass in Sweet Orange cv. Mosambi. *Indian J. Hort.*, 66(2), 163–168.

Patel, V. B., Singh, S. K., Asrey, Ram, & Sharma, Y. K. (2005). Response of Organic Manures and Biofertilizer on Growth, Fruit Yield and Quality of Amrapali Mango Under High-Density Orcharding, *Karnataka J. Hort. J.*, 3, 51–56.

Pennington, J. C. (1986). Feasibility of using Mycorrhizal Fungi for Enhancement of Plant Establishment on Degraded Material Disposal Sites, a Literature Review, Miscellaneous Paper D-86-3, US Army Engineer Waterways Experiment Station, Vicksburg, Mississippi, USA.

Powell, C. L. l., & Bagyaraj, D. J. (1984). Mycorrhiza, V. A. (eds). CRC Press, Inc.

Quatrini, P., Gentile, M., Carimi, F., Pasquale, F. & Puglia, A. M. (2003). Effect of Native Arbuscular Mycorrhizal Fungi and *Glomus mosseae* on Acclimatization and Development of Micropropagated *Citrus limon* (L) Burm. *J. Hort. Sci. Biotech*, 78(1), 39–45.

Rai, M. K. (2001). Current Advances in Mycorrhization in Micropropagation, *In vitro Cell. Dev. Biol. Plant,* 37, 158–167.

Ramajyam, D., Singh, S. K., Patel, V. B., & Alizadeh, M. (2013). Mycorrhization Alleviates Salt Stress in Grape Rootstocks During in vitro acclimatization, *Indian J. Hort.,* 70(1), 26–32.

Rodríguez, Romero, Suea, Ana Azcón, Rosarioa, & Jaizme, Vega, Del Carmen, María (2011). Early Mycorrhization of two Tropical Crops, Papaya (*Carica Papaya* L.) and Pineapple [*Ananas comosus* (L.) Merr.], Reduces the Necessity of P fertilization during the Nursery Stage, *Fruits,* 66(1), 3–10.

Rufykiri, G., Thiry, Y., Wang, L., Delvaux, B., & Declerck, S. (2002). Uranium uptake and Translocation by the Arbuscular Fungus, *Glomus intraradices*, under Root-Organ Culture Conditions, *New Phytol.,* 156(2), 275–281.

Ruiz-Lozano, J. M., & Azcon, R. (1996). Mycorrhizal Colonization and Drought Stress as Factors Affecting Nitrate Reductase Activity in Lettuce Plants, *Agric. Ecozys. Env.,* 60, 175–181.

Ruiz-Lozano, J. M., Azcon, R., & Gomez, M. (1996). Alleviation of Salt Stress by Arbuscular Mycorrhizal *Glomus* Species in *Lactuca sativa* plants, *Physiol. Plant,* 98, 767–772.

Ruiz-Lozano, J. M., Azcón, R., & Plama, J. M. (1996b). Superoxide Dismutase Activity in Arbuscular Mycorrhizal *Lactuca sativa* plants subjected to drought stress, *New Phytol.,* 134, 327–333.

Safir, G. R. (1987). Ecophysiology of VA Mycorrhizal plants. (eds.) CRC Press, Inc.

Salamanca, C. P., Harrea, M. A., & Barea, J. M. (1992). Mycorrhizal Inoculation of Micropropagated woody legumes used in revegetation programs for decertified Mediterranean Ecosystems, *Agronomie,* 12: 869–872.

Sanchez-Diaz, M., & Honrubia, M. (1994). Water Relations and Alleviation of Drought Stress in Mycorrhizal Plants. In: *Impact of Arbuscular Mycorrhizas on Sustainable Agriculture and Natural Ecosystems*, S., Gianni Nazi, & Schuepp, H. (eds.), 167–178.

Sbrana Cristiana, Giovannetti, Manuela, & Vitagliano. (1994). The Effect of Mycorrhizal Infection on Survival and Growth Renewal of Micropropagated Fruit Rootstocks, *Mycorrhiza,* 5, 153–156.

Schreiner, R. P. (2005). Mycorrhizas in Mineral Acquisition of Minerals, Proc. Soil Environment Vine Mineral Nutrition, 49–60.

Schreiner, R. P. (2010). Foliar Application of Phosphorus has Minimal Impact on 'Pinot noir' Growth, Mycorrhizal Colonization, or Fruit Quality, *HortScience,* 45, 815–821.

Schreiner, R. P., Mihara, K. L., McDaniel, H., & Bethlenfalvay, G. J. (1997). Mycorrhizal Functioning Influence Plant and Soil Functions and Interactions, *Plant Soil,* 188, 199–209.

Shamshiri, M. H., Usha, K., & Singh, Bhupinder. (2012). Growth and Nutrient Uptake Responses of Kinnow to vesicular arbuscular mycorrhizae. *ISRN Agron.,* 2012, Article ID 535846, 7 pp. doi:10.5402/2012/535846.

Sharma, Som Dev, Kumar Pramod & Bhardwaj, Satish Kumar (2011). Screening of AM fungi and Azotobacter Chroococcum under Natural, Solarization, Chemical Sterilization and Moisture Conservation Practices for Commercial Mango Nursery Production In North-West Himalayas, *Scientia Horticulturae*, 128(4), 506–514.

Singh, N. V., Singh, S. K., & Singh, A. K. (2011). Standardization of Embryo Rescue Technique and Bio-hardening of Grape Hybrids (*Vitis vinifera* L.) using Arbuscular Mycorrhizal Fungi (AMF) under Sub-Tropical Conditions, *Vitis,* 50(3), 115–118.

Singh, N. V., Singh, Sanjay K., Singh, Anand K., Meshram, Suroshe, Sachin S., Mishra Dwijesh. C. (2012). Arbuscular Mycorrhizal Fungi (AMF) Induced Hardening of Micropropagated Pomegranate (*Punica granatum* L.) plantlets, *Scientia Horticulturae*, 136, 122–127.

Singh, S. K., Minakshi, G., Khawale, R. N., Patel, V. B., Krishna, H., & Saxena, A. K. (2004). Mycorrhization as an Aid for Bioharderning of in vitro raised Grape (*Vitis vinifera* L.) Plantlets, In: *Acta Hort.*, 662, 289–295.

Singh, S. K., Patel, V. B., Saxena, A. K., Khawale, R. N., Krishna, Hare, & Minakshi (2004). Mycorrhizae Induced Hardening of Tissue Culture Raised Grape (*Vitis vinifera* L.) Plantlets, *Indian J. Hort.*, 61(1), 23–29.

Smith, S. E., & Read, D. J. (1997). *Mycorrhizal Symbiosis*, Academic Press, Inc. San Diego, California.

Soares, Ana Cristina Fermino, Martins, M. A., Mathias, Lêda, & Freitas, Marta Simone Mendonça (2005). Arbuscular Mycorrhizal Fungi and the Occurrence of Flavonoids in Roots of Passion Fruit Seedlings, *Sci. Agric.*, 62(4), 331–336.

Srivastava, D. J., Kapoor, R. J., Srivastava, S. K., & Mukerjee, K. G. (1996). Vesicular Arbuscular Mycorrhiza an overview, In Mukerjee, K. G. (Ed.). Concepts in Mycorrhizal Research, Kluwer Acad. Pub., Dordrecht, 1–39.

Subramanian, K. S., & Charest, C. (1999). Acquisition of N by External Hyphae of an Arbuscular Mycorrhizal Fungus and its Impact on Physiological Responses in Maize Under Drought-Stressed and well Watered Conditions, *Mycorrhiza,* 9, 69–75.

Tagu, D, & Martin, F. (1996). Molecular Analysis of Cell Wall Proteins Expressed During the Early Steps of Ectomycorrhiza Development, *New Phytol.*, 133, 73–85.

Tahmatsidou, Vasiliki, O'Sullivan, John, Cassells, Alan, C., Voyiatzis, Demetrios, & Paroussi, Georgia (2006). Comparison of AMF and PGPR Inoculants for the Suppression of *Verticillium* wilt of strawberry (*Fragaria ananassa* cv. Selva). *Applied Soil Ecol.,* 32, 316–324.

Thanuja, T. V., Hedge, R. V., & Sreenivasa, M. N. (2002). Induction of Rooting and Root Growth in Black Pepper Cuttings (*Piper Nigrum* L.) with Inoculation of Arbuscular Mycorrhizae, *Scientia Hort.,* 92(3–4), 339–346.

Troeh, Z. I., & Loynachan, T. E. (2003). Endomycorrhizal Fungal Survival in Continuous Corn, Soybean and Fallow, *Agron. J.*, 95(1), 224–230.

Van der Heijden, M. G. A., Boller, T., Wiemken, A., & Sanders, J. A. (1998). Different Arbuscular Mycorrhizal Fungal Species are Potential Determinants of Plant Community Structure. *Ecology,* 79(6), 2082–2091.

Varma, A. (1998). Functions and Application of Arbuscular Mycorrhizal Fungi in Arid and Semi-arid Soils, In *Mycorrhiza, Structure, Function, Molecular Biology and Biotechnology.* A. Varma, & Hock, B. (eds). 2nd edition. Library of Congress Cataloging-in-Publication Data.

Vázquez-Hernández, M. V., Arévalo-Galarza, L., Jaen-Contrerasa, D., Escamilla-García, J. L., Mora-Aguilera, A., Hernández-Castro, E., Cibrián-Tovara, J., & Téliz-Ortiz, D. (2011). Effect of *Glomus mosseae* and *Entrophospora colombiana* on Plant Growth, Production, and Fruit Quality of 'Maradol' Papaya (*Carica papaya* L.). *Scientia Hort.,* 128, 255–260.

Whisenant, S. G. (1999). *Repairing Damaged Wildlands.* Cambridge University Press.

Wu, Q. S., Zou, Y. N., Xia, R. X., & Wang, M. Y. (2007). Five Glomus Species Affect Water Relations of Citrus Tangerine During Drought Stress. *Botanical Studies,* 48, 147–154.

CHAPTER 15

IMPACT OF CLIMATE CHANGE ON PLANT PATHOGENS

RITESH KUMAR[1], SUDARSHAN MAURYA,
JAIPAL SINGH CHOUDHARY, and SHIVENDRA KUMAR

ICAR Research Complex for Eastern Region, Research Centre, Plandu,
Ranchi-834010, India;
[1]E-mail: ritesh.nstt@gmail.com

CONTENTS

15.1 INTRODUCTION

Climate change is one of the burning and complex problems facing mankind today. The overriding complexity of the problem is attributed to its deeper global consequence on a vast range of issues impacting the very survival of life on Earth. Climate change is defined as any long-term significant change in the "average weather" that a given region experiences or in other words it is the shift in the average statistics of weather for long-term at a specific time for a specific region. Average weather may include temperature, precipitation and wind patterns. It involves changes in the variability or average state of the atmosphere over durations ranging from decades to millions of years. These changes can be caused by dynamic process on earth, external forces including variations in sunlight intensity and by human activities. Climate change in the usage of the Intergovernmental Panel on Climate Change (IPCC) refers to 'a change in the state of the climate that can be identified (e.g., using statistical tests) by changes in the mean and/or the variability of its properties that persists for an extended period, typically for decades or longer. It refers to any change in climate over time, whether due to natural variability or as a result of human activity' (IPCC, 2007).

Since climate is a direct input into the agricultural production process, the agricultural sector has been a natural focus for research. Vulnerability to climate change may be greater in the developing world, as in India, where agriculture typically plays a larger economic role and where poverty and agriculture are both salient. Climate change is likely to reduce agricultural yields significantly, and that this damage could be severe unless adaptation to higher temperatures is rapid and complete. The most imminent climatic changes in recent times is the increase in the atmospheric temperatures due to increased levels of greenhouse gases such as carbon dioxide (CO_2), methane (CH_4), ozone (O_3), nitrous oxide (N_2O) and chlorofluoro carbons (CFCs). Because of the increasing concentrations of those radiative or greenhouse gases, there is much concern about future changes in our climate and direct or indirect effect on agriculture (Garg et al., 2001; IPCC, 2001; Krupa, 2003; Aggarwal, 2003; Mall et al., 2006). In recent years, with the growing recognition of the possibility of global climate change, an increasing emphasis on world food security in general and its regional impacts in particular have come to forefront of the scientific community. Crop growth, development, water use and yield under normal conditions are largely determined by weather during the growing season. Even with minor deviations from the normal weather, the efficiency of extremely applied inputs and food production is seriously impaired.

Climate Change is projected to have significant impacts on conditions affecting agriculture, including temperature, precipitation and glacial run off. It affects agriculture in more ways than one. It can affect crop yield as well as the types of crops that can be grown in certain areas, by impacting agricultural inputs such as water for irrigation, amounts of solar radiation that affect plant growth, as well as the prevalence of pests. Effect of climate change on agriculture or more precisely on insect pests and diseases of agricultural crops is multidimensional. Changes in temperature, moisture and atmospheric gases can fuel growth and generation rates of plants, fungi and insects, altering the interactions between pests, their natural enemies and their hosts. Changes in land cover, such as deforestation or desertification, can make remaining plants and animals increasingly vulnerable to pests and diseases.

15.2 CHALLENGES, OPERATING ENVIRONMENT, OPPORTUNITIES AND TARGETS

Pests and diseases have historically affected food production either directly through losses in food crops and animal production, or indirectly through lost profits from insufficient cash crop yields. While new pests and diseases have regularly emerged throughout history, climate change is now throwing any number of unknowns into the equation. Today, these losses are being exacerbated by the changing climate and its increasing volatility, threatening food security and rural livelihoods throughout the world. Developing countries with a high dependence on agriculture are the most in danger to today's changing patterns of pests and disease. Hundreds of millions of smallholder farmers depend exclusively on agriculture and for their survival. As rural farmers struggle to produce food, poor people in nearby urban areas are left to contend with less availability in addition to higher food prices. National economies will also suffer as new pests and diseases either reduce agricultural products' access to International markets or incur higher costs associated with inspection, treatment and compliance.

Plant pests, which include insects, pathogens and weeds, continue to be one of the biggest constraints to food and agricultural production. Despite of continued release of high yielding disease and pest resistant varieties and introduction of new pesticides 12–15% yield losses were observed which is caused by pest and diseases. In spite of that they not only reduce the crop yield but they also contaminate the agricultural produce by secreting toxins, which creates several health hazards. Climate change and plant disease development rising concern of agricultural producers and scientists how is climate change going to affect agriculture as a whole Plant pathology is one of the agricultural disciplines that is directly affected by the climate change. Garret et al. (2006) established the connection in the climate change and plant pathogens by empirical and modeling experiments. According to them the first direct impact of climate change on plant disease would be the balance of the encounter rate between pathogen and host by changing rates of the two species.

15.3   EFFECT OF ELEVATED TEMPERATURE ON PLANT PATHOGENS

Temperature has potential impacts on plant disease affecting the host and pathogen. Rise in temperatures caused by increasing green house gases is likely to affect crops differently from region to region. For example, moderate warming (increase of 1 to 3°C in mean temperature) is expected to benefit crop yields in temperate regions, while in lower latitudes especially seasonally dry tropics, even moderate temperature increases (1 to 2°C) are likely to have negative impacts for major cereal crops. Warming of more than 3°C is expected to have negative effect on production in all regions (IPCC, 2011). The Third Assessment Report of the IPCC, 2001 concluded that climate change would hit the poorest countries severely in terms of reducing the agricultural products. The Report claimed that crop yield would be reduced in most tropical and subtropical regions due to decreased water availability, and new or changed insect/pest incidence. Fungi that cause plant disease grow best in moderate temperature ranges. Temperate climate zones that include seasons with cold average temperatures are likely to experience longer periods of temperatures suitable for pathogen growth and reproduction if climates warm. It has been predicted that the expansion of geographic area of *Phytophthora* in response to increased temperature, which allow for overwintering of this oomycetes in new areas (Bergot et al., 2004). Research has shown that host plants such as wheat and oats become more susceptible to rust diseases with increased temperature; but some forage species become more resistant to fungi with increased temperature (Coakley et al., 1999). For example earlier onset of warm temperatures could result in an earlier threat from tomato late blight (caused by *Phytophthora infestans*) with the potential for more severe epidemics and increases in the number of fungicide applications needed for control. Rising temperature would increase fertilizer requirement for the same production targets and result in higher GHG emissions, ammonia volatilization and cost of crop production (http://en.wikipedia.org/wiki/Climate_change_and_agriculture).

15.4   EFFECT OF MOISTURE CHANGE ON PLANT PATHOGENS AND DISEASE

Moisture also affects both host and pathogens in many ways. Increase in moisture level facilitates infection in several cases such as in apple scab, late blight, and several vegetable root pathogens. Other pathogens like the powdery mildew species tend to prosper in conditions with lower (but not low) moisture. Several climate change models predict frequent and extreme precipitation that could result in more and longer periods with favorable pathogen environments. Some climate change models envisage higher atmospheric water vapor concentrations with increased temperature, which would favor pathogen and disease development.

15.5   EFFECT OF ELEVATED CO_2 ON PLANT PATHOGENS AND DISEASE

Elevation in CO_2 level is a major impact of climate change and it can affect the host and pathogen in multifold ways. Researchers have shown that with increase in CO_2 level, higher growth rates of leaves and stems observed for plants which may result in denser canopies with higher humidity that favor pathogens. Another ill impact of increased CO_2 level is lower plant decomposition rates, which could increase the crop residue on which disease organisms can overwinter, consequential higher inoculum levels at the beginning of the growing season, and earlier and faster disease epidemics. Increase in CO_2 level also result in greater fungal spore production and may induce physiological changes to the host plant that can increase host resistance to pathogens (Coakley et al., 1999) In Eastern Plateau and Hill region also realizes the alteration of environments, some agricultural crops like wise tuber and bulb crops will certainly fetched higher yield than today's some reports also indicated that the size of tuber increased due to elevated temperature and CO_2 Moreover, the exponential trends of transportation and human activities may acts synergistically with temperature (Anderson et al., 2004; Epstein, 2001). Moreover, the stripe rusts of wheat spreading which is closely associated with changes in rainfall patterns. Spore germination of *Puccinia* increases with increasing temperature, as well as root rot pathogens also reproduces disease inoculums quickly. Due to rising temperature it has realizes the incidence of powdery mildew appears much earlier where they occurs very late (observed data). It has been realizes that the elevated CO_2 increases the activities of biotrophic and nectrotrophic pathogens and the resultants the disease severity increased drastically as well as it would influences the phytopathogenic infection of crop plants. In rice, enhanced susceptibility to *Magnaporthe oryzae* under elevated CO_2 was attributed to lower leaf silicon content (Kobayashi et al., 2006). In soyabean elevated CO_2 increased incidence of *Septoria* brown spot but reduced downy mildew (Garrett et al., 2006; Eastburn et al., 2010).

15.6 IMPACT OF CLIMATE CHANGE ON DISEASE MANAGEMENT PRACTICES

Physiological changes in host plants may result in higher disease resistance under climate change scenarios, but due to more rapid disease cycles there is a greater chance of pathogens evolving to overcome host plant resistance. With increased CO_2, moisture, and temperature, fungicide and bactericide efficacy may change. Systemic fungicides could be affected negatively by physiological changes that slow uptake rates, such as smaller stomatal opening or thicker epicuticular waxes in crop plants grown under higher temperatures. These same fungicides could be affected positively by increased plant metabolic rates that could increase fungicide uptake. The more frequent rainfall events predicted by climate change models could result in farmers finding it difficult to keep residues of contact fungicides on plants, triggering more frequent applications. Biological control of pathogens will also be difficult in climate change scenarios as microbial population is greatly influenced by changes in temperature and moisture regimes As a result in some cases antagonistic organisms may out-compete pathogens while in others pathogens may be favored.

15.7 CONCLUSIONS

Climate change is not an isolated issue, but it is addition to several problems already faced in agriculture. There are various scenarios on the consequences of climate change. If the annual number of days of active vegetation (with temperature above 5°C) within the 100year period will be increased which also influenced increased the rate of infection of foliar diseases, which also increased the load of pesticides in protection of foliar diseases earlier than in the past or to increase the number of pesticide treatments. In the era of climate change, diagnostics of pests and diseases and capacity building of farmers, extension and even research personnel for adaptation to changed pest scenario under future climates assumes significance, wherein Integrated Crop management in Private Public Partnership mode could be very effective. Today's challenge is to 'produce more from less.' Farmers' decisions are of vital importance for good yields of crops. Forecasted weather products and area-wide weather networks are becoming more prevalent. Accurate information concerning possible yield losses due to occurrence of a pest is needed by growers or plant protection specialists to decide on cost-effective control measures. Thus, future research and education in Crop Protection in every part of India including the Eastern Plateau and Hill region, does need to address the issue of future climates in disease-pest management.

KEYWORDS

- **Climate Change**
- **Elevated CO$_2$**
- **Elevated Temperature**
- **Environment**
- **Moisture Change**
- **Plant Pathogen**

REFERENCES

Aggarwal, P. K. (2003). 'Impact of Climate Change on Indian agriculture,' *J. Plant Biol.*, 30(2), 189–198.

Anderson, P. K., Cunningham, A. A., Patel, N. G., Morales, F. J., Epstein, P. R., & Daszak, P. (2004). Emerging Infectious Diseases of Plants, Pathogen Pollution, Climate Change and Agrotechnology Drivers, *Ecol. Evol.*, 19, 535–44.

Bergot, M., Cloppet, E., Perarnaud, V., Deque, M., Marcais, B., & Desprez-Loustau, M. l. (2004). Simulation of Potential Range Expansion of Oak Disease Caused by *Phytophthora cinnamomi* under climate change. *Glob. Change Biol.*, 10, 1539–1552.

Climate Change and Agriculture (http://en.wikipedia.org/wiki/Climate_change_and_agriculture).

Eastburn, D. M., Degennaro, M. M., DeLucia, E. H., Dermody, O., McElrone, A. J. (2010) Elevated Atmospheric Carbon Dioxide and Ozone Alter Soybean Diseases at Surface, Glob Ch Biol, 16, 320–330.

Epstein, P. R. (2001). Climate Change and Emerging Infectious Diseases, Microbes Infect, 3, 747–54.

FAO. (1981). Food Loss Prevention in Perishable Crops. FAO Agric. Serv. Bull. 43, UN Food & Agric. Org., Rome, Italy, 72 p.

FAO. (2012). UN Food & Agric. Org., Rome, Italy.

Garg, A., Shukla, P. R., Bhattacharya, S., & Dadhwal, V. K. (2001). 'Sub-region (district) and Sector Level SO_2 and NOx emissions for India, Assessment of Inventories and Mitigation flexibility, *Atmos. Environ.*, 35, 703–713.

Garrett, K. A., Dendy, S. P., Frank, E. E., Rouse, M. N., & Travers, S. E. (2006). Climate Change Effects on Plant Disease, Genomes to Ecosystems, *Ann Rev Phytopathol.*, 44, 489–509.

IPCC (Intergovernmental Panel for Climate Change) (2001), Climate Change (2001). The Scientific Basis, Contribution of Working Group I to the Third Assessment Report of the Intergovernmental Panel on Climate Change [Houghton, J. T., Ding, Y., Griggs, D. J., Noguer, M., van der Linden, P. J., Dai, X., Maskell, K., & Johnson, C. A. (eds.)], Cambridge University Press, Cambridge, UK, 881pp.

IPCC. (2007). Fourth Assessment Report: Climate Change, Downloaded from http://www.ipcc. ch on 27 Dec 2011 at 2216 hrs. IST.

Kobayashi, T., Ishiguro, K., Nakajima, T., Kim, H. Y., Okada, M., & Kobayashi, K. (2006). Effects of Elevated Atmospheric CO_2 Concentration on the Infection of Rice Blast and Sheath Blight, *Phytopathol*, 96, 425–431.

Krupa, S. (2003). Atmosphere and Agriculture in the New Millennium. *Environ. Poll.*, 126, 293–300.

Mall, R. K., Singh, R., Gupta, A., Srinivasan G., & Rathore, L. S. (2006). Impact of Climate Change on Indian agriculture, A Review *Climatic Change* 78, 445–478.

CHAPTER 16

QUALITY OF FRUITS IN THE CHANGING CLIMATE

MD. ABU NAYYER[1], MD. WASIM SIDDIQUI[2],
and KALYAN BARMAN[1]

[1]Department of Horticulture (Fruit and Fruit Technology), Bihar Agricultural University, Sabour, Bhagalpur, Bihar (813210) India

[2]Department of Food Science and Technology, Bihar Agricultural University, Sabour, Bhagalpur, Bihar (813210) India; E-mail: wasim_serene@yahoo.com

CONTENTS

16.1 INTRODUCTION

Climate change is perceived to be the greatest threat to the food security and mankind in twenty-first century. Over the past few decades, increase in average air temperature on earth and its associated effects on climate and crops have became a concern worldwide, particularly after the 4th Assessment Report of Intergovernmental Panel on Climate Change (IPCC, 2007). Since preindustrial era to the year 2009, the carbon dioxide (CO_2) concentration in the atmosphere has increased from 280 ppm to 384 ppm coupled with an increase in mean temperature of 0.76°C. Further, according to the studies carried out by IPCC, it is predicted that by the end of this century increase in average air temperature in Asian countries could range between 1.8 to 6.0°C and CO_2 concentration may reach up to 700 ppm or more (IPCC, 2007). South Asian countries are predicted to have least increase in temperature, in the range of 1.8 to 5.0°C except for the Himalayas (IPCC, 2007). On the other hand, changes in rainfall pattern have also been predicted in the range of 5 to 20% during the winter season and 40 to 15% during the summer season (IPCC, 2007). As a consequence of rising atmospheric temperature, there will be frequent occurrence of drought, flood and heat waves.

Temperature increase and the effects of greenhouse gases are among the most important issues associated with climate change. The major greenhouse gasses (GHGs) responsible for climate change are carbon dioxide (CO_2), methane (CH_4) and nitrous oxide (N_2O). Collectively, these three gasses are responsible for 99% of global warming. Moreover, perflourocarbons (PFCs), hydroflourocarbons (HFCs), sulfur hexafluoride (SF_6) and ozone (O_3) also make a small contribution to global warming. Among the different GHGs, carbon dioxide is the most important GHG and is alone responsible for 77% of global warming. The global warming potential of different GHGs is measured in terms of carbon dioxide equivalence (CO_{2eq}), which is the warming effect exerted by one molecule of CO_2 over a given period of time, usually 20 or 100 years. Methane and nitrous oxide have a global warming potential over a 100 year period are about 23 CO_{2eq} and 298 CO_{2eq}, respectively.

16.2   WEATHER VARIABLES AND ITS EFFECT ON FRUIT QUALITY

16.2.1   TEMPERATURE

Fruit crops are sensitive to temperature and each plant species has its own characteristic response to temperature. They have specific temperature requirements for growth, higher yield and quality. Increase in atmospheric temperature affects photosynthesis directly, causing alterations in sugars, organic acids, firmness, flavonoids contents and antioxidant capacity of fruits. High temperature reduces the net carbon gain in crops, particularly in C_3 plants, by increasing photorespiration.

Higher temperature on fruit surface caused by prolonged exposure to sunlight hastens ripening and other associated events. Grapes exposed to direct sunlight ripened faster than those ripened in shaded areas within the canopy. This delay in ripening is associated with reduced cell wall enzyme activity (cellulose and polygalacturonase) due to high temperature. However, this delay did not occur via a direct effect on the enzymes associated with cell wall degradation. Fruit firmness is also affected by high temperature conditions during growth. Changes in cell wall composition, cell number and cell turgor properties were postulated as being associated with the observed phenomenon.

Temperate fruit crops require sufficient accumulated chilling or vernalization to develop fruitful buds and break winter dormancy in the spring. Inadequate chilling due to enhanced greenhouse warming may result in prolonged dormancy, leading to reduced fruit quality and yield. Flavor of the fruit is also affected by temperature. Flavor is dependent on the balance between organic acids (citric, malic, tartaric acid, etc.) and sugars (sucrose, fructose, and glucose). Fruits like apple exposed to direct sunlight develop higher sugar content compared to those fruits grown on shaded sides of the canopy (Brooks and Fisher, 1926). In case of grapes, high temperature during growing season increases sugars and decreases tartaric acid content (Kliewer and Lider, 1970). It is reported that an increases in 10°C temperature lead to 50% reduction in tartaric acid content. Moisture content of 'Hass' avocados were found to be increased when grown under higher temperatures (45±2°C), than comparatively lower temperatures (30±2°C) (Woolf et al., 1999). The concentration of specific fatty acids also increased (e.g., palmitic acid by 30%) at higher temperatures.

Functional component in fruit crops can also be altered due to exposure of fruits to high temperatures during the growing season. It has been observed that 'Kent' strawberries grown in warmer nights (18–22°C) and warmer days (25°C) contain higher antioxidant capacity than fruits grown under cooler (12°C) days (Wang and Zheng, 2001). This increase in antioxidant capacity was due to increase in flavonoid content due to high temperature condition. McKeon et al. (2006) also addressed the effects of temperature and verified that higher temperature tended to reduce vitamin content in fruit.

High temperature and/or direct sunlight also influenced the mineral accumulation in fruit crops. For example, 'Hass' avocado fruits exposed to direct sunlight during growing period contain higher calcium (100%), magnesium (51%) and potassium (60%) contents when compared to fruits grown under shaded conditions (Woolf et al., 1999).

Fruit coloration may be severely affected because the biosynthesis of coloring pigments is strongly influenced by the temperature. The nutritional value and antioxidant potential of fruit may be affected due to decrease in skin pigments. The deep red color development in apple is a result of low temperatures during the night in autumn, just before harvesting. If the temperatures are not low enough, most apples fail to turn into their specific red shades. Besides, higher day and night temperatures have a direct influence in strawberry fruit color development. Berries grown under those conditions are redder and darker (Galletta and Bringhurst, 1990).

16.2.2 CARBON DIOXIDE

The atmosphere of earth consists chiefly of 78.1% nitrogen and 20.9% oxygen, with carbon dioxide (0.031%) and argon (0.93%) comprising next most abundant gases. Among these, nitrogen and oxygen are not considered to play an important role in global warming as these gases are virtually transparent to terrestrial radiation. While, carbon dioxide, water vapor and minute amounts of other gases (methane, nitrous oxide and nitrous oxide) that absorb the earth's infrared radiation, trapping heat are responsible for global warming. Since a significant part of all the energy emanated from earth occurs in the form of infrared radiation, increase in CO_2 concentration mean more energy will be retained in the atmosphere, contributing to global warming (Lloyd and Farquhar, 2008). As a consequence of climate change, the CO_2 concentrations in the atmosphere have increased approximately 35% from preindustrial times to 2005 (IPCC, 2007).

There is growing evidence suggesting that many crops (notably C_3 crops) may respond positively to increased atmospheric carbon dioxide in the absence of other stressful conditions (Long et al., 2004). On the contrary, positive impact of elevated CO_2 can be equalized by other effects of climate change, such as elevated temperatures, higher troposphere ozone concentrations and altered patterns of precipitation (Easterling et al., 2007). Changes in CO_2 concentration in the earth's atmosphere can alter the growth and physiological behavior of plants. Many research findings during the last decade concluded that increased atmospheric CO_2 alters net photosynthesis, biomass production, sugars and organic acids content, stomatal conductance, firmness, seed yield, water and nutrient use efficiency and plant water potential.

Fruit quality and aroma volatile composition in field-grown strawberries (*Fragaria ananassa* Duch.) is also affected by elevated carbon dioxide (Wang and Bunce, 2004). Increased CO_2 concentration (ambient + 300 and ambient + 600 μmol mol^{-1} CO_2) than ambient resulted in high fruit dry matter, fructose, glucose and total sugars content and low citric and malic acid content. The volatile compounds such as ethyl hexanoate, ethyl butanoate, methyl hexanoate, methyl butanonate, hexyl acetate, hexyl hexanoate, furaneol, linalool and methyl octanoate are also enhanced due to high CO_2 exposure. Thus, total amount of these compounds were higher in berries grown in CO_2 enriched conditions than those grown in ambient conditions. The highest CO_2 enrichment (600 μmol mol^{-1}) condition yielded fruit with the highest levels of these aroma compounds.

A recent study by Sun et al. (2012) demonstrated that elevated CO_2 could alleviate the negative effect of high temperature on fruit yield of strawberry (*Fragaria ananassa* Duch. cv. Toyonoka) at different levels of nitrogen. Higher carbon dioxide concentration increased dry matter–content, fructose, glucose, total sugars and sweetness index per dry matter, but decreased fruit nitrogen content, total antioxidant capacity and all antioxidant compounds per dry matter in strawberry fruit. They concluded that elevated CO_2 improved the production of strawberry (including yield and quality) at low temperature, but decreased it at high temperature.

Elevated atmospheric carbon dioxide during growing season in grape is reported to affect the quality of wines (Bindi et al., 2001). Higher CO_2 concentration has significant effect on fruit dry weight, with an increase ranging from 40–45% in 550 mmol CO_2/mol treatments and from 45–50% in 700 mmol CO_2/mol treatments. During middle of ripening season, rising CO_2 levels up to a maximum increases tartaric acid (8%) and total sugars (14%) contents. However, when the grapes reached the maturity stage, the CO_2 effect on both quality parameters almost completely disappeared. Kimball et al. (2007) reported that sour orange (*Citrus aurantium*) plants grown for 17 years in elevated CO_2(300 ppm above ambient), respond by increasing fruit production by 70% instead of acclimation. It also increased biomass accumulation (70% extra) due to higher wood growth.

Altered carbon availability to mango fruit affected both dry mass and water mass of peel, pulp and stone (Léchaudel et al., 2002). Changing carbon availability to mango fruit influenced both the dry mass and the water mass of its three main compartments: skin, pulp and stone (Léchaudel et al., 2002). Since dry matter accumulation is affected by the availability of assimilate supply, changes in its structural component, including cell walls, and its nonstructural one, consisting of soluble sugars, acids, minerals and starch, have been investigated according to their sensitivities to leaf-to-fruit ratio treatments. The influence of dilution during fruit growth on quality traits such as flesh taste or shelf life has been considered by expressing concentrations of the main biochemical and mineral compounds per unit of fresh mass (Léchaudel et al., 2005).

16.2.3 RAINFALL

Global warming has affected ocean wind current intensity and directions. Slowly and steadily annual rainfall is moving towards deficit with erratic monthly trends. Distribution of rainfall is disturbed and more rains are received in July–August and sometimes extending in to September–October. It disturbs entire crop cycle in the year. Higher rains/cloud bursts are causing floods in some parts and in other parts deficit rains are causing drought. There is also a shift in rainfall pattern.

Due to change in climate and its impact on both occurrence and distribution pattern of rainfall has deterring effect on fruit quality especially at growth and developmental stages. Onset of unusual rains during flowering of mango caused negative effects and higher incidence of diseases and pests. The occurrence of anthracnose and powdery mildew pathogens has been reported to infect fruits quiescently at this stage and become visible during postharvest conditions. Rains during fruiting periods may blacken fruits (in mango) or prevent desirable fruit coloration (in guava), making the produce less appealing for the consumers. Increase in humidity can initiate un-seasonal flowering. Various quality traits such as fruit coloration, spottiness, fruit texture and taste can be altered by change in temperature, humidity and rainfall. Sudden rains after a long dry spell result the fruit cracking in several fruits (litchi, pomegranate, apple, etc.) affecting postharvest quality, storability, and consumer acceptability. Due to deficit in rainfall water stress occurs which alter the fruit size, according to the quantity of water shortage and the period when stress occur. Reduced water availability (40% of the daily evapotranspiration) has been reported to increase pulp dry matter content with development of fruit (Diczbalis et al., 1995). Late water stress accentuate decline in fruit calcium concentration (Simmons et al., 1995) and strongly affect fructose concentration (Léchaudel et al., 2005).

16.2.4 OZONE

In the troposphere, ozone is formed as a result of series of photochemical reactions involving carbon monoxide (CO), methane (CH_4) and other hydrocarbons in the presence of nitrogen species (NO + NO_2) (Schlesinger, 1991). During the period of higher temperature and solar irradiation in summer seasons, ozone is formed (Mauzerall and Wang, 2001). However, its production also occurs naturally in other seasons, reaching the peak in the spring (Singh et al., 1978). Ozone enters in the plant tissue through stomata, causing direct cellular damage especially in the palisade cells (Mauzerall and Wang, 2001). The occurrence of damage by ozone is as a consequence of changes in membrane permeability and may or may not result in visible injury, reduced growth and, ultimately, reduced yield (Krupa and Manning, 1988).

Strawberry fruits stored for three days under cold storage (2°C) in a ozone-enriched atmosphere (0.35 µl/l) showed 3 times increase in vitamin C content than berries stored at the same temperature under normal atmosphere as well as a 40% reduction in emissions of volatile esters in ozonized fruits (Perez et al., 1999). In another study, quality of persimmon fruits (cv. Fuyu) harvested at two different harvest dates was evaluated after ozone exposure. Fruits were exposed to 0.15 mol/mol (V/V) of ozone for 30 days at 15°C and 90% RH. The authors reported that ozone exposure was capable to maintain firmness of second harvested fruits, which were naturally softer that first harvested fruits, over commercial limits even after 30 days at 15°C plus shelf life. Fruits treated with ozone showed the highest values of weight loss and maximum electrolyte leakage. However, ozone exposure had no significant effect on color, ethanol, soluble solids, and pH (Salvador et al., 2006).

KEYWORDS

- **Climate Change Effects**
- **Fruit Quality**
- **Nutrition**
- **Postharvest**
- **Shelf Life**
- **Weather Variables**

REFERENCES

Bindi, M., Fibbi, L., & Miglietta, F. (2001). Free air CO_2 Enrichment (FACE) of grapevine (Vitis vinifera L.) II. Growth and Quality of Grape and Wine in Response to Elevated CO_2 concentrations. *Euro. J. Agron.*, 14, 145–155.

Brooks, C., & Fisher, D. F. (1926). Some High Temperature Effects in Apples, Contrasts in the Two Sides of an Apple. *J. Agric. Res.*, 23, 1–16.

Diczbalis, Y., Hofman, P., Landrigan, M., Kulkarni, V., & Smith, L. (1995). Mango Irrigation Management for Fruit Yield, Maturity and Quality. In: Proceedings of Mango (2000) marketing seminar and production workshop. *Brisbane Australia*, 85–90.

Easterling, W. E., Aggarwal, P. K., Batima, P., Brander, L. M., Erda, L., & Howden, S. M. (2007). Food, Fibre and Forest Products. In. Parry, M. L., Canziani, O. F., Palutikof, J. P., van der Linden, P. J., & Hanson, C. E. (Eds.), Climate change (2007) Impacts, Adaptation and Vulnerability. Contribution of Working Group II to the fourth assessment report of the intergovernmental panel on climate change 273–313. Cambridge University Press.

Galletta, G. J., & Bringhurst, R. S. (1990). Strawberry management. In Galletta, G. J., & Bringhurst, R. S. (Eds.), Small Fruit Crop Management, (83–156). Prentice-Hall: Englewood Cliffs.

IPCC. (2007). Climate Change. In Solomon, S., Qin, D., Manning, M., Chen, Z., Marquis, M., Averyt,. K. B., Tignor, M., Miller, H. L. (Eds.). The Physical Science Basis. Contribution of Working Group I to the Fourth Assessment Report of the Intergovernmental Panel on Climate Change (996p). Cambridge, United Kingdom: Cambridge University Press.

Kimball, B. A., Idso, S. B., Johnson, S., & Rillig, M. T. (2007). Seventeen Years of Carbon Dioxide enrichment of sour orange trees, Final results. *Global Change Biol.*, 13, 2171–2183.

Kliewer, M. W., & Lider, L. A. (1970). Effects of Day Temperature and Light Intensity on Growth and Composition of Vitis vinifera L., fruits. *J. Am. Soc. Hortic. Sci.,* 95, 766–769.

Krupa, S. V., & Manning, W. J. (1988). Atmospheric ozone, Formation and Effects on Vegetation. *Environ. Pollu.*, 50, 101–137.

Léchaudel, M., & Joas, J. (2007). An overview of preharvest factors influencing mango fruit growth, quality and postharvest behavior. *Braz. J. Plant Physiol.*, 19(4), 287–298.

Long, S. P., Ainsworth, E. A., Leakey, A. D. B., & Ort, D. R. (2006). Food for thought: Lower-than-expected crop yield stimulation with rising CO_2 conditions. *Science*, 312, 1918–1921.

Mauzerall, D. L. & Wang, X. (2001). Protecting agricultural crops from the effects of tropospheric ozone exposure: Reconciling science and standard setting in the United States, Europe, and Asia. *Ann. Rev. Ener., Environ.*, 26, 237–268.

McKeon, A. W., Warland, J., & McDonald, M. R. (2006). Long-term Climate and Weather Patterns in Reaction to Crop Yield, A Mini Review. *Canadian J. Bot.*, 84, 1031–1036.

Perez, A. G., Sanz, C., Rios, J. J., Olias, R., & Olias, J. M. (1999). Effects of Ozone Treatment on Postharvest Strawberry Quality, *J. Agric. Food Chem.*, 47(4), 1652–1656.

Salvador, A., Abad, I., Arnal, L., & Martinez-Javegam, J. M. (2006). Effect of Ozone on Postharvest Quality of Persimmon. *J. Food Sci.*, 71(6), 443–446.

Schlesinger, W. H. (1991). Biogeochemistry, An analysis of Global Change, New York, Academic Press, 443p.

Simmons, S. L., Hofman, P. J., & Hetherington, S. E. (1995). The Effects of Water Stress on Mango Fruit Quality, In Proceedings of Mango (2000) Marketing Seminar and Production Workshop. *Brisbane Australia*, 191–197.

Wang, S. Y., & Zheng, W. (2001). Effect of Plant Growth Temperature on Antioxidant Capacity in Strawberry, *J. Agric. Food Chem.*, 49, 4977–4982.

Woolf, A. B., Ferguson, I. B., Requejo-Tapia, L. C., Boyd, L., Laing, W. A., & White, A. (1999). 'Impact of Sun Exposure on Harvest Quality of 'Hass' avocado fruit, *Revista Chaingo Serie Horticultura*, 5, 352–358.

CHAPTER 17

HOMA THERAPY AN EFFECTIVE TOOL IN MITIGATING SOIL, WATER AND ENVIRONMENTAL CRISES

R. K. PATHAK[1] and E. ULRICH BERK[2]

[1]Homa Teacher, Five Fold Path Mission and ex-Director, Central Institute for Subtropical Horticulture, Lucknow, E-mail: pathakrkripal@yahoo.co.in

[2]President, Deutsche Gesellschaft fur Homa Therapie, Haldenhof, Germany; E-mail: dght@homatherapie

CONTENTS

17.1   INTRODUCTION

Indiscriminate use of agro-chemicals over 5–6 decades had adversely affected soil fertility, crop productivity, produce quality, water and the environment. This is now causing global warming and climatic disasters, which are the major challenges to the humanity. It is speculated that 40 percent diseases of crops, human being, and animals are associated with pollution and presence of toxic chemicals in the food chain. The adverse effects of intensive agriculture have compelled to think for a sustainable system of agriculture. As a result, numbers of alternative organic farming systems such as Biodynamic Farming, Natural Farming, Nateuco Farming, Rishi Krishi, Panchagavya Farming, Jaivik Krishi, and Homa Organic Farming emerged in recent pasts from different parts of the country (Pathak and Ram, 2009). If practiced in earnest way, all these systems are capable of assuring yield equal to those obtained with the use of agrochemicals. In order to assure sustainable agriculture, we tried to integrate few techniques from organic farming systems, which can be implemented by the common farmers and can strengthen rhizosphere and biosphere simultaneously. The same is now promoted as "Homa Jaivik Krishi".

It is pertinent to record that Homa Jaivik Krishi is based on systematic and synergistic harnessing of solar energy, which is available in plenty in India. It is important to record that most of the countries are trying to harness solar energy to meet their energy requirement, but as on today, it is very expensive. The concept, which we are trying for sustainable agriculture through Homa Jaivik Krish will be a giant step in addressing food and social security of the country (Pathak, 2011; Pathak and Ram, 2013). Solar energy is an infinite nonpolluting, renewable source of energy. Some facts regarding solar energy are enumerated below.

- Sun is giant power house;
- Infinite, nonpolluting renewable source of energy;
- Solar energy equivalent to almost 75,000 trillion kwh hits earth every day;
- A meager 0.1 percent of this staggering figure is sufficient to meet world's energy needs;
- Visible range is not only light we see but also is energy used in photosynthesis.
- Government of India is giving major emphasis for harnessing solar energy; but as on today, techniques are very expensive;

- If not harnessed, it is radiated back in the atmosphere;
- The same we are trying to harness through "Homa Jaivik Krishi" techniques.

17.2 STEPS FOR HARNESSING SOLAR ENERGY: THROUGH HOMA JAIVIK KRISHI APPROACH

- Promoting Farming System Approach;
- Adopting use of agriculture calendar for various farm activities;
- Encouraging massive plantation for habitat development in the organic village;
- Integration of legumes in the cropping system;
- Integration of indigenous cows with hump and use of her products in farming;
- Enhancing biological quality of soil through increased humus to harness & retain the energy;
- Integration of Homa Organic Farming, most potential source to harness solar energy and an effective tool to heal the ailing environment.

Homa Organic Farming has been attempted in large number of crops and it has shown exciting result with respect to quantum production, its quality (Paranjpe, 2005). Agnihotra practices also positively influence the taste, texture, color, disease resistance in large number of crops. Harvesting losses are reduced and shelf life is much better than crop grown by any other agricultural methods. Agnihotra ash is wonderful tonic and used to store seed, treat seed/seedlings, enhance soil fertility, plant vigour and for pestmanagement. This is also placed in water bodies for improvement of water quality and its availability (2014). Agnihotra ash is an important component in preparation of a special Bio-enhancer known as Biosol used as foliar spray on plants and on soil (Basakar, 2012; Weir, 2009). Application of biosol improves crop productivity, its quality and soil fertility which has been reported in crops like cabbage, tomato and soybean. As per experience from different continents the following benefits have been reported on wide range of crops through Homa Jaivik Krishi (Pathak, 2009, 2012; Pathak and Ram, 2013).

- Rejuvenation of all kinds of soil;

- Increase in yield and quality improvement in cereats, pulses, vegetables, fruits, nuts, plantation crops, medicinal crops, forest tress including pastures;
- Homa Therapy controls and eradicates weeds;
- It enhances soil biological activities including earthworms in rhizosphere and honey-bee activity in biosphere.

Homa Therapy with Agnihotra as its basic tool has wide-reaching beneficial effects on soil, water resources and on ailing environment. It is pertinent to mention that environmental pollution is creating all these crises. In fact, the foods, which we consume, water which we drink and the environment in which we live, all are polluted and problem is becoming alarming advocated as early as in 1977(Paranjpe, 1977). These calls for introspection and change in the mindset of every individual and organizations associated with these basic issues. Interestingly, Homa Organic Farming an ancient technology provides an alternative in resolving these crises (Paranjpe, 2005). As on today, it is being implemented in more than 71 countries in all continents by thousands of people particularly under the guidance of volunteers of Five Fold Path Mission and few other organizations.

Agnihotra is a gift to humanity from ancient- most Vedic Sciences of bio energy, medicine, agriculture and climate engineering. Agnihotra is the basic fire in HOMA THERAPY. It is the process of purification of the atmosphere through the agency of fire, prepared in the copper pyramid, tuned to the biorhythm of sunrise/sunset. Basically it is science of pyramidology, biorhythm of nature, burning of organic substances, and sonic power of chanting special mantras and its electro magnetic effects which is extended to a larger area by establishing resonance point. By practice of Agnihotra, one can notice that tension of mind disappears and one begins to experience peace. The mind is reshaped so nicely, so delicately, so effortlessly by simply sitting in the Agnihotra atmosphere (Paranjpe, 2005).

Tremendous amounts of energy are gathered around the Agnihotra copper pyramid just at Agnihotra time. A magnetic type field is created, one, which neutralizes negative energies and reinforces positive energies (Narang, 2007). Therefore, a positive energy pattern is created by one who does perform regular Agnihotra.

There are two basic energy systems in the physical world: heat and sound. In performing Agnihotra, both these two energies, namely, the heat from fire and sound from chanting of Vedic Mantras, are blended to acieve the desired physical, psychological and spiritual benefits, (Narang, 2007).

When Agnihotra is performed, Agnihotra smoke gathers practices of harmful

In fact radiation from the atmosphere and on very subtle level neutralizes their radioactive effect. Nothing is destroyed, merely changed from one form to another. While healing and purifying the atmosphere through the instrument of Agnihotra the uttering of mantras spread subtle energies in the surrounding atmosphere as the oblations are offered. The emerging red, yellow, blue colors of the flame resulting from the burning of the oblations also have unifying link with Agnihotra mantra. Thus, all beneficial qualities of chromatography too are present in the Agnihotra mantra (Potdar, 1999).

A Western Scientist from Poland has said that Agnihotra is a wonder weapon to counter pollution. Agnihotra heals the atmosphere. Modric (2004) was of view that in Agnihotra, it is the role of special vessel made of copper, its ziggurat shape, a form related to the horn antennas used in high frequency transmission. The ash, which is available after Agnihotra, is disinfectant, anticoagulant, and tissue contracting was well established. He agreed that ash had pesiticidal and fungicidal properties. He was of the firm view that he was dealing with a complex problem that could potentially affect the whole environment, countering the toxins of modern technology developed over the last century by the industrial revolution, and that the process might have enormous implications for our existence. He claimed that Agnihotra ceremony was energetically quite complex, involving at least three energetic aspects, or field phenomenon, having to do with fire and the ash, with radiation of an undefined nature, and with ESP, or psychism. He said that there is possibility of electromagnetic radiation during the ceremony. He suggested that systematic research is needed to provide scientific explanations for exciting impact, which has been claimed in agriculture, animals and human health.

When Agnihotra fire is burnt there is not just energy from the fire, but subtle energies are generated and thrust into the atmosphere by the fire. The quality of materials burnt is an important prerequisite to its full healing effect. Much healing energy emanates from the Agnihotra pyramid (Potdar, 1999). Salient features of impact of this healing fire associated with soil, water and environment have been dealt.

17.3 SOIL

Soil is the most important production factor for crops and at the same time also most influenced by the farmer. In recent decades, there has been heavy pressure on land because of urbanization and for number of other nonfarm uses; diversion takes place not only from wastelands but also from ecologically important areas, such as forests or pasture land, lakes and ponds and even land suitable for agriculture production.

According to broad estimates, about 120 million ha of land in India is suffering from one or other reasons. Imbalanced fertilizer application, unscientific ways of water use and land preparation have affected vast stretches of once fertile land (NAAS, 2010). Soils are very diverse and complex systems full of life. The soil itself can be viewed as a living organism, because it is a habitat for plants, animals and microorganisms, which are all interlinked with each other. With increased doses of agro chemicals the top few inches of the soil on Earth are likely to be totally destroyed so that nothing will grow. The answer to our ills lies just 6 inches beneath the surface of Earth. Due to indiscriminate use of agro-chemicals, the soil, water, atmosphere, subsoil all are polluted by metallic, nonmetallic and gaseous toxicants of different types. The soil in large areas of forest is nearly dead. Hence there is urgent need to device techniques, which can help in maintenance of soil fertility.

Homa Therapy Farming is the way out. Changes in soil composition due to pollution become drastic and the results become disastrous. Soil becomes unable to sustain plants and life. The way out is to inject nutrients and fragrance into the atmosphere by widespread practice of Homa activities (Paranjpe, 2005).

17.3.1 SOME FACTS REGARDING SOIL

A teaspoon of active soil is the habitat of millions of soil organisms. Some are of animal origin, some are of plant origin. The organisms vary greatly in size. Some are visible to the naked eye, such as earthworms, mites, spring-tails or termites. Most of them, however, or so small that they can only be seen with a microscope, thus they are called microorganisms. The most important microorganisms are bacteria, fungus and protozoa. Microorganisms are the key elements to the quality and fertility of soils, but for humans, they do their work invisibly. The greater the variety of species and higher their number, the natural fertility of soil will be better (Rao and Chhonkar, 1998). This same has been summarized.

- Soil is dynamic and living system and therefore in a continuous process of transformation.
- It grows fast and vanishes very fast by erosion if some one does not respect her.
- Without soil organisms, soil is dead.
- Not all microbes are hostile.
- Most soil microbes are very important helpers of the farmers.

17.3.1.1 LESSON TO REMEMBER

"The soil humus losses on planet might become ecologically dangerous if not arrested now, because humus is the major accumulator of solar energy at the earth surface at present and the guardian of soil productivity, guaranteeing ecological stability of the biosphere." Rozanov (1990) from Moscow cautioned.

- Humus retains nutrients and moisture and releases slowly as per need of plants;
- It captures humidity from dew and subtle energy from planets, retains it;
- It also transmutate nutrients as plants need it;
- If the soil is deficient in humus content, subtle energy coming from planets are radiated back in atmosphere;
- So all human efforts need to be made to enhance soil humus content;

Unfortunately so for we have abused the soil with cocktail of chemicals, and many other human activities, therefore she has started in showing her discomfort through incidence of number of pest, diseases, weeds and rampant soil erosion. If not arrested now, survival of humanity at planet earth will face crises in every sphere of life (Howard, 2004).

17.3.2   CURRENT STATUS OF INDIAN SOIL

Organic carbon content in most of soils is less than (>0.5%);
   • Large part of India's soils are deficient in three or more critical nutrients;
   • Soils of Punjab, Haryana, UP, Bihar which produce 50 percent of grain and feed 40 percent population are seeing multiple nutrient deficiencies;
   • N-deficient in Punjab, Haryana, UP, Rajasthan, Gujarat, Maharashtra, parts of Bihar, Jharkhand, MP, AP and Tamilnadu.

17.3.3   MEASURES TO IMPROVE SOIL FERTILITY

We had been trying to conceive a technique, which can show a way for sustainable agriculture as this crucial juncture of pollution all around. Now, we are of the view that soil fertility can be enhanced by adopting following measures (Pathak, 2011, Pathak and Ram, 2013 Rupela, 2014).
   • Adoption of farming system approach;
   • Practice of crop rotation, mixed cropping, multistoried cropping,
   • Inclusion of legumes in the system as green manure, inter/cover crops and aurogreen crops.
   • Regular addition of organic manure and other measures such as mulching, use of enriched compost and regular use of enriched compost and frequent use Bio enhancers.
   • Integration with Homa Organic Farming and frequent use Agnihotra ash and Biosol special bio enhancer has good impact in improving soil fertility as evident from the following reports.
   Ulrich, (2009) a strong promoter of Homa Organic Farming, suggested that the soils needs to be rejuvenated first by Homa Therapy. In the rejuvenated soil different types of microorganisms, starting from the level of viruses, bacteria, fungi, algae, thrive. Thus, a healthy micro-flora and micro-fauna are created. This helps in creation of micro environment, which is comparatively less toxic to the growing plants. He further explained

that, soil in Homa atmosphere holds moisture better than any soil. It is due to the ghee and the feedback of Homa on atmosphere. When a nutritional rain comes, the nutrients and moisture are sustained as a unit in the soil. This makes for better quality vegetation. Placement of Agnihotra ash on top of soil before it is tilled or cultivated it does a lot of help to nourish the plant and make the plant happy. Agnihotra and Agnihotra ash, when put on soil, help stabilize the amount of nitrogen and potassium present.

Reports on Homa Impact show that:

- Acidic soil was brought back to normal-Soil with a pH of 4.4 was brought back to normal by organic farming methods like composting and mulching and Homa Therapy including the use of Agnihotra ash (Poland). The soil that was not suitable for crop production, now large number of fruits, vegetable, medical herbs are grown and the place has become paradise (Bizberg, 2009 –www. Homatherapypoland.org).
- Some preliminary observation at one of the KVK in Unnao district with ameluration of sodic soil has shown interesting results.
- High sodicity in soil could be brought back to normal by Homa atmosphere and by adding some Agnihotra ash along with vermi compost showed remarkable impact in minimization of soil pH in first year it self as indicated in the Table 17.1.

TABLE 17.1 Impact of Agnihotra and Use of Its Ash in Sodic Land Ameluration Based on Soil pH Content

Sl. No.	Soil treatment	Soil pH
1	With agro chemicas	9.86
2	With vermi compost	9.06
3	Vermi compost+ Agnihotra ash	7.67

- Wheat was taken as crop;
- In control, recommended dose of N,P,K was given
- In vermi compost, 5 tones of vermi compost per ha was applied;
- In Agnihotra treatment 5 tones vermi compost+ one kg ash/ha was applied;

Results showed that:

- Field treated both with ash and vermi compost showed lower soil pH;

- Potash and phosphorus content were also found abundant quantaties in ash treated plot.
- As there are large areas in India of previously fertile land, which now are lying barren because of high sodicity/salinity, Homa Therapy, could give a solution to a big problem.
- The soil improves through Homa Organic Farming Technology even in difficult climatic conditions like acid rain.
- At Tapovan Homa Farm of about seven acres in Ratnapimpri, Taluka: Parola, District: Jalgaon, Maharashtra, India, where Agnihotra and other Homas are practiced since 2001. The soil was very poor. Always there is a water problem in that area as wells dry up in summer. After one year of practicing 24 h fire, there was a total change and the soil received much nutrition (www.homa1com; www.homa-therapyindia.com).
- In the surrounding area, it looks like all deserts and only these HOMA acres are lush, shining green and full of biodiversity. The crops yields are highest per acre compared to neighbouring farms. Fruits and vegetables have exquisite taste and aroma; they are resistant to disease. The soil has been nourished by the effect of the 24 h HOMA fire in the atmosphere, Anon (2014).
- Soil in resonance atmosphere holds moisture better than any soil, due largely to the feedback of Homa on the atmosphere. When the nutritional rain comes due to the practice of Homa healing fires of Homa Organic Farming Technology, (Agnihotra and Om Tryambakam Homa), the nutrients and moisture are sustained as a unit in the soil. This makes for better quality vegetation. Agnihotra Homa and Agnihotra ash, when put in the soil, help stabilize the amount of nitrogen and potassium present (Berk, 2009).
- In Homa atmosphere the metabolic process of plants is speeded up. It is the ghee used in Agnihotra process that is the catalytic factor and on a more subtle level the mantras interacting with the combined effect of the burnt ghee and rice. This combination enters the soil after returning from the solar range. It enters the plant by, one might say, attaching itself to minerals and water absorbed by the root system of the plants. The ghee acts as a catalyst, creating a chemical reaction with the plant, aiding in enzyme and vitamin production and encouraging and increasing the cyclic rate. In other words plants mature faster, taste better and are better just by performance of Homa in the

garden." It results in enhanced aeration of soil and its moisture holding capacity (Paranjpe, 2005).

- It has been observed that when "*Agnihotra* Ash" is added to the normal soil it increases the water soluble phosphate content of the soil (Kartz, 2007; Lai, 2007) and nutrients are absorbed readily by the root hairs of the plants. Solubility of Phosphate as influenced by incorporation of Agnihotra ash is evident from the Table 17.2.

TABLE 17.2   Solubility of Phosphate Influenced by Incorporation of Agnihotra Ash

Soil Used	Phosphate % per gram of Soil		Phosphate % per gram Ash	
	Non-Homa	Homa	Non-Homa	Homa
No Soil, Only Ash			3.40–8.90	
Weld Loam	0.42–1.72		21.00–86.00	
Red Feather Loamy Sand	0.23–1.15		11.50–57.50	

Report from Braunschweig University by Dr. Sylvia Kratz.

- Absorption of macronutrients like nitrogen, phosphorus, potash because of small cells and active transport also improves in Homa atmosphere (Paranjpe, 2005).
- In Homa Jaivik Krishi, in addition to regular performance of Agnihotra and use of its ash and Biosol, the other associated activities such as habitat development, inclusion of legumes in the system, crop rotation, mulching recycling of organic waste through use of bio enhancers, all help in enhancement of soil humus all these thus helps in continuous soil fertility improvement (Ghosh, 2002).

17.4   WATER RESOURCES

Water is an important natural resource made available in plenty by the Mother Nature. It is one of the critical inputs, which is essentially required for survival of humidity. Agriculture uses nearly 80 percent of the water resources in the country. According to IWMI study, food crops alone use 74 percent of total consumptive water use. As for supply is concerned, the ground water component is declining. Progressively larger blocks are being declared 'dark zones.' The situation is more worrisome because of the low water use efficiency. As in case of land, quality of water resources is also deteriorating. Salinity and alkanity is continuously increasing and damaging agricultural land, especially in areas irrigated by ground water sources, while water logging is a major problem in the areas irrigated by surface irrigation particularly through canals.

In recent years, water, is becoming one of the major constraints for the welfare of human beings. In future, water shortages are likely to be wide-spread. Our current concerns are over the long-term sustainability of water infrastructure, and its ability to sustain future food, water requirements to human being and other activities. The public water infrastructure is show-ing signs of crumbling, and is impacting the land, and water resource qual-ity base of agriculture, affecting crop productivity. Indiscriminate over ex-ploitation of ground water resources is shrinking the total resource-base itself, as wells fail because of fall in water tables beyond acceptable limit.

17.4.1 SOME FACTS CURRENT WATER RESOURCES

Shree Vasant, strong promoter of Homa technology stated in an article written in Nov 2002 that 'If Homa is practiced on a large scale, the at-mosphere is healed and the water resources get purified, leading to better absorption of sun's rays.' I believe this is a clue as to how Homa brings water resources back into healthy conditions, which is evident from the statement of Paranjpe (2005), which, I quote.

Water resources on earth are finding it difficult to absorb energy from the sun (due to pollution). This will result in depletion of marine life and imbalance in nature. Water pollution now takes its toll in rich as well as poorer countries. No water is relatively safe to drink now.'

HOMA Therapy and Agnihotra ash improve the quality of water and make it potable. Agnihotra ash is extremely medicinal and has a powerful purifying effect on water.

It has been reported that addition of Agnihotra ash to a fish tank would enhance the process of purification of the water by retarding the process of multiplication of harmful bacteria and algae in the water freshening through by absorbing odors. The observations clearly indicated that Agni-hotra ash does, indeed affect the quality of water and the growth of algae in the medium. The ecosystem wherein we live, Agnihotra ash treatment of polluted and acidified lakes and streams can be initiated (Pathak and Ulrich, 2014).

17.4.2 REPORT INDICATING THE DE-POLLUTING EFFECT OF AGNIHOTRA ASH ON WATER

The following report indicates the de-polluting effect of Agnihotra ash on water and on biotic life when industrial effluent was present in the water.

- A Bio assay test was carried out to study the effect of Agnihotra ash on biotic life. Different concentrations of effluents were prepared. Ash dose was given and fish were kept under observation for 48 h. The effluent was collected from textile process industry. Biotic life was possible in concentrations of 15% effluent with 0.5 gm/L ash and 20% effluent with 2 gm/L ash. Whereas biotic life was absent in 20% effluent with 0.5 gm/L ash and 15% effluent with no ash (Frits and Ringma, 2009).
- The Chemical Oxygen Demand (COD) of the effluent before and after Agnihotra ash treatment was compared.
- The study revealed concentration of COD after Agnihotra ash treatment is reduced which indicates Agnihotra ash helps to purify the water.

Ground water quality is controlled by geological factors (arsenic, iron, fluoride, etc.), overdraft, fertilizers and pesticides use, and saline water intrusion in the coastal regions. The current status of water can be viewed from the following statement.

- High levels fluorides have been reported in several pockets of 200 districts in West Bengal, Bihar, Chhattisgarh, Assam and in certain pockets of Uttar Pradesh.
- High concentrations of iron in ground water have been reported in Assam, West Bengal, Orissa, Chhattisgarh and Karnataka.
- Nitrate pollution has been reported in intensively irrigated and high agricultural productivity regions, and in urban areas due to improper and inadequate sewage disposal.
- Punjab and Haryana consumes N fertilizers more than 100 kg/ha and have the highest nitrate pollution in their waters.
- Nitrate pollution (above the WHO permissible limit) also occurs at several locations in Gujarat, Maharashtra and Orissa.

- The trend in N fertilizer consumptions in MP, AP, West Bengal, Karnataka, points to nitrate pollution becoming a major threat to their sustainable water resource development.
- High nitrate concentrations (more than 45 mg/L) have been found in many districts of AP, Bihar, Delhi, Haryana, Punjab, Himachal Pradesh, Karnataka, Kerala, MP, Maharashtra, Orissa, Tamil Nadu, Rajasthan, etc.
- About two-lakh square kilometers has been estimated to be affected by saline water of EC in excess of 4 dS/m. In several places in Haryana and Rajasthan the EC values greater than 10 dS/m make water nonpotable.
- Coastal salinity, caused by excessive withdrawal of fresh ground water, has been observed in Tamil Nadu, Saurashtra, and in an 8–10 km wide belt near coast of Subarnrekha, Salandi, and Brahmani Rivers.
- Several reports show that by performing Agnihotra and putting Agnihotra ash in the well, the water quality improves considerably. Non-potable water become good drinking water in one case the pH came down from 9.5 to 7.2, and the salinity from 1150 ppm to 720 ppm (report from Australia).
- On one farm in Austria officials closed one well as the water was not even good enough for the animals to drink. With Homa and putting Agnihotra ash to the well after only two months another inspection was done and they found out now it was good quality drinking water (Ringma, 2009).
- A simple experiment shows that if you put Agnihotra ash in a container with putrid water within a few days the water becomes clear again.
- One experiment done in a Polish institute for Environmental Biology, showed that by adding Agnihotra ash to water with some decomposing plant matter the beneficial microorganisms grew much better than in the control environment.

Dr. Masuru Emoto-JAPAN did pioneer work on water quality and dedicated his life to research on water and human health; he was of the view that human body consists of 75% and brain it self has 69% of water, which is crucial on health. This fact realization created idea how can water affect out wellbeing? Can we heal ourselves through water? Yes! Water is a vital healing agent. He developed a technology by quick-freezing of water and through advance photography it giges wonderful results of crystal formation depending upon its purity. He was of the view that it was typical case of subtle energy (www.rh4.info) as enumerated below.

- Emoto-photos show difference in water patterns after you utter mantras;
- Same effect has been observed with adding Agnihotra ash in the water source;
- Water is not just H_2O;
- In order to fully understand how Agnihotra and Homa Therapy work, we probably have to go one step deeper to level of subtle energies generated during Agnihotra;
- He took water from rivers of Japan, Germany and London, in all the rivers water observed to be polluted and the showed hazy photographs;
- Wonderful crystal formation was noticed from water collected from springs from japan;
- He got similar photographs from the polluted water by placing Agnihotra ash;
- He was of the view that regular performance of Agnihotra and placement of Ash in water bodies may be helpful in resolving water crises in the country.

17.5   ENVIRONMENT

According to the ancient Science of VRUKSHA AYURVEDA (VRUKSHA means TREE), which is now presented as Homa Organic Farming Technology, atmosphere is the biggest single factor which affects plant kingdom, soil and water quality (Rameshwar, et al., 2009). The principles of life must be restated now on the Earth. We withdraw nutrients from the environment. They must be somehow replaced.

17.5.1   RADIOACTIVITY

Dangers of radiation are now a reality, but we also know the steps to put into place to avoid destruction. The ancient Vedic Knowledge states that radioactivity can be neutralized by Agnihotra and Agnihotra ash. Scientific studies and individual experiences have shown that Agnihotra neutralizes radiation and its effets. As planet earth now faces new challenges as a reslt of Fukushima, performance of Agnihotra is even more vital and crucial to practice then even before (Anon, 2014).

This statement sounds quite extraordinary or even incredible as there is no technique known to modern science to reduce radioactivity. You cannot change the half life of radioactive elements. But experience shows that with Homa methods this is possible:

We have the report of one Homa Farm in Austria on which there was no increased radioactivity neither in the milk nor in vegetables after Chernobyl – although on all surrounding farms they had this problem (Karin, 2009). This gets further support with recent experiment done done at the Academy of Sciences in Kiev where highly radioactive rice was tested. After soaking this rice in Agnihotra ash water, the radioactivity was totally neutralized.

Homa is the means. When Agnihotra is performed there is a turbulence of electricity's and ethers created by the combination of Mantras and fire that extends all the way to the solar range. This turbulence leads to quick upheaval of the nutrient structure in the area (Parajnpe, 2005). Ghee is the vehicle and Mantras are the power. Homa must be done now on large scale. The Earth has been robbed of its element by pollution. Homa can replace these things, which are vital to all life. The Earth is wasting away. Homa is the way to save it (for more information visit website: http:// www.homa1.com).

The purpose of Agnihotra is not to burn the substances that are used in the form of oblations, rather it is vaporize them. Thus they diffuse into surrounding air and transform the air quality (Narang, 2007).

So now, let Science investigate if, in effect, Agnihotra is an efficient way to make "Atmosphere More Nutritious and Fragrant" and thus improve soil quality and water resources. When cow dung is burnt along with rice and ghee by chanting mantras the disinfectant effect goes to the atmosphere and the atmosphere is disinfected and purified. The impact is radiated to a larger area by establishing Resonance Point. It is claimed that up to 60 hectares polluted area can be covered with same efforts (Johnson, 2008). The statement explains that:

- Experiments done with Agnihotra showed that indoor microbial pollution is greatly reduced.

- Regular practice of Agnihotra controls pathogenic bacteria in an area where resonance has been established.
- The concentration of negative ions is an important indicator of atmospheric pollution: The more negative ions in the air, the less pollution.
- Normally smoke particles are charged positive. This can be easily tested if we blow cigarette smoke towards an instrument measuring the electric charge of the air. It will show that the concentration of negative ions is getting less.
- But if you perform Agnihotra and place the same device above the pyramid, the smoke of the Agnihotra fire shows a higher content of negative ions.
- Agnihotra is thus purifying the air in an area around the pyramid.

Polluted planet is in a denatured state, that is, nature is thrown out of balance due to rampant pollution through human activities. This renders nature unable to perform properly and this is a very serious situation in relation to nature's capacity to provide food. Already we are seeing food shortage on a planetary scale. Homa Organic farming needs to be implemented world wide to bring nature back into balance so that we have quality food for all and peaceful natural surroundings to enjoy quality of life. This can be achieved by the help for the environment.

- The principles of life must be restated now on the planet earth.
- Everyone must be made aware that HOMA is necessary to survive.
- We withdraw nutrients from the environment.
- They must be some how replaced. HOMA is the means.
- When Agnihotra is performed there is a turbulence of electricity's and ethers created by the combination of mantras and fire that extends all the way to the solar range. This turbulence leads to a quick upheaval of the nutrient structure in the area.
- More people should begin to perform Agnihotra under a tree. This makes the tree happy and the tree dances and sings. Also the birds are much attracted to Agnihotra. It is healing to them.

HOMA trees will soon begin to grow. They will be a species in themselves. They are now generating in intense HOMA atmosphere. Places like Amazon region of Peru is where these have been observed.

Agnihotra, thus appears to be a promising scientific,cost effective, eco-friendly method to counter the ever increasing deadly pollution of the environment and purify and enrich the environment with healthy ingredients (Narang, 2007).

Besides the above impact, there are other numbers of evidences where Agnihotrie's claim that they have been saved from man made disasters and natural calamaties. Just to quote few instances in Bhoplal, India, where poisonous MIC gas leaked from Union Carbide pesticide factory is the site for biggest industrial accoident in the history. In the same area of Bhopal, there were many families who practice Agnihotra. They immediately performed Om Tryambakam Homa and were saved. Like wise report from Chile wo practice Agnihotra states that while earthquake were destroying homes and families, where Agnihotra and Homas were performed, people were saved (Anon., 2014).

Agnnihotra is gift to humanity. Thousands of people from every wake of life, from every race, religion and creed, perform Agnihotra around world. We can all focus on health of our great mother earth.

It is pertinent to mention here that the aforesaid experiences on soil, water and environment of merely Homa Organic Farming. In Homa Jaivik Krishi, there are other components like habitat management through plantation; rain water conservation, inclusion of legumes in the system, practice of green manure, auro green crops, crop rotation, use of enriched compost, bio enhancers, enhancement of humus content in the soil, all these factors will be helpful in addition to Homa Organic farming. Thus, there is need to initiate systematic research on different aspects, develop scientific data, understand the complete package and try in few specific areas and after getting fully satisfied it needs assertive promotion.

KEYWORDS

- **Biodynamic Farming**
- **Biosol**
- **Nateuco Farming**
- **Panchgavaya**
- **Resilient**
- **Rishi Krishi**

REFERENCES

Anon. (2014). Fire and Sound Healing for Fukushima, Satsang 41(6), 3 and 13.

Anon. (2010). State of Indian Agriculture, The Indo-Gangetic Plain, National Academy of Agricultural Sciences, New Delhi, India.

Berk, U. (2009). Agnihotra and Homa Therapy Scientific Perspective, Proceedings in Brainstorming Conference on Bringing Homa organic Farming into the main stream of Indian Agriculture, organized in collaboration with Planning Commission and Five Fold Path Mission, 46–52.

Biswas, T. D., & Naraanasamy, G. (1998). Soil Organic Matter and Organic Residue Management for Sustainable Productivity, ISSS Bulletin No.19 Indian Society of Soil Science, New Delhi.

Elsbieta, G. (1985). Observations with water, Extract from Satsang 12(10,11) October, 1985.

Emoto, M., Aranda, & Paranjpe, V. V. (2010). Holy Fires-Holy Crystals in Fire & Water-The Magic Power of Creation.www.rh4.info.

Ghosh, S. (2004). Principles of Organic Farming in the Tropics, 53–68.

Heschl, K. (2009). Experience with Agnihotra and Radioactivity, 44 in Proceedings in Brainstorming Conference on Bringing Homa Organic Farming into the main stream of Indian Agriculture, organized in collaboration with Planning Commission and Five Fold Path Mission.

Howard, A. (2004). An Agricultural Testament, Other India Press, Mapusa, 403507, Goa, India.

Johnson, B. (2008). Homa Organic Farming (Training Manual), Published by Five Fold Path Mission, Homa Therapy Go Shala, P.O. Mandaleshwar, Maheshwar, Dist. Khargaon, M. P.

Johnson, B., & Heschl, K. (2009). Homa Organic Farming, 84-9 in Proceedings in Brainstorming Conference on Bringing Homa organic Farming into the main stream of Indian Agriculture, Organized in Collaboration with Planning Commission and Five Fold Path Mission.

Kratz, S., & Schnug, E. Landbauforschung Volkenrode 3/2007(57). Homa Farming a Vedic fire for agriculture: Influence of Agnihotra ash on water solubility of soil. E-mail: sylvia.kratz@fal.de.

Lai, T. M. Agnihotra cited on.11.07.2007. Ash and Water Soluble Phosphates (on line). See http://www.agnihotra.org/science.htm.

Modric, Mato. (2004). In Purified with fire 243–254 in Secrets of the Soil. Tomkins, P., & Bird, C., ed.; Rupa & Co., Ansari Road, New Delhi, 110002.

Narang, I. (2007). The Science of Agnihotra, Maharishi Dayanand Charitable Trust Delhi, 110092.

Paranjpe, V.V. (1977). Light Towards Divine Path. Published in U.S.A. from The Agnihotra Press, Inc, Maryland,21133.

Paranjpe, V. V. (2005). Homa Therapy Our Last Chance, Five Fold Path Mission, Five Fold Path Publications, "Shree Niwas" 40 Ashok Nagar, Dhule, 424001, Maharashtra.

Pathak, R. K. (2009) Expectations from Homa Organic Farming, Proceedings in Brainstorming Conference on Bringing Homa organic Farming into the main stream of Indian Agriculture, organized in collaboration with Planning Commission and Five Fold Path Mission, 64–68.

Pathak, R. K. (2011). Homa Jaivik Krishi. A Ray of Hope for Sustainable Horticulture MARDI's Initiatives, 15–16, Souvenir and Abstract, National Symposium cum Brainstorming Workshop on Organic Agriculture, April 19–20, 2011.

Pathak, R. K., & Ram, R. A. (2013). Homa Jaivik Krishi. A Ray of Hope for Sustainable Horti-culture, 39–46, Horticulture for Economic Prosperity and Nutritional Security in twenty-first century, (ed.) Hazarika, T. K., & Nautiyal, B. P. Westville Publishing House, New Delhi.

Pathak, R.K. & Ulrich, B. (2014). Agnihotra : An effective tool in resolving atmospheric pol-lution, mitigating soil and water crises. pp. 1–10. Lead talk delivered in National Seminar on Role of Organic Farming in Climate Resilient and Sustainable Agriculture. Organized by Navsari Agricultural University, Navsari-396450, Gujarat.

Potdar, M. M. (1999). Agnihotra for Equilibrium of Nature and Enhancement of Human Life, Institute for Studies in Vedic Sciences, Shivpuri, Akkakkot, 413216, Maharashtra.

Rameshwar, R., Poonam, P., & Atul, D. (2009). Research on Homa Organic Farming. Proceed-ings in Brainstorming Conference on Bringing Homa organic Farming into the main stream of Indian Agriculture, organized in collaboration with Planning Commission and Five Fold Path Mission, 53–55.

Rao, D. L. N., & Chhonkar, P. K. Organic Matter in Relation to Soil Biological Quality and Sustainability. Bulletin of the Indian of Soil Science, 19, 80–89.

Rozanov, B. G. (1990). In Trans. 14th Int. Cong. Soil Sci., Kyoto, Japan, 53.

Rupela, O. P. (2014). Aurogreen – an Innovative Method of Green-manuring-Personal com-munication.

Ulrich, Berk. (2009). Agnihotra and Homa Theray –Scientific Persepective: Proceedings in Brainstorming Conference on Bringing Homa organic Farming into the main stream of Indian Agriculture, organized in collaboration with Planning Commission and Five Fold Path Mis-sion, pp. 46–52.

CHAPTER 18

AWARENESS ABOUT CLIMATE CHANGE: PERCEPTION AND ACTION

ADITYA[1] and B. KUMAR[2]

[1]Department of Extension Education, Bihar Agriculture College, Sabour-813 210

[2]DRI-cum-Dean PGS, Bihar Agricultural University, Sabour-813 210
E-mail: inc.aditya@gmail.com

CONTENTS

ABSTRACT

Climate change has emerged as one of the most devastating environmental threats. It has become increasingly essential to assess the awareness regarding climate change in the general population for framing the mitigation activities. Keeping this in view, an online survey was conducted to know responses of educators, researchers, and students on their awareness level, probable actions, and suggestions in matters related to climate change. The awareness level was quite important for policy planning at their level, which could have impact on the later generations. The survey was conducted through Google Drive online analysis and survey tool consisting of both open-ended and closed-ended questions. Subjects had poor appreciation of facts such as effect of climate change on farmers but established a consistency on the responses that individuals at their own level can mostly mitigate climate change.

18.1   INTRODUCTION

Climate change has emerged as one of the most important issues in global discussion forums in the present times. Global climate change impacts on human and natural systems are since long predicted to be severe. As evidence of climate change and its impact continues to be gathered, it has become quite clear that many of the causes of climate change are anthropogenic in nature through lifestyles, consumption and choices that pollute and exploit resources in an unsustainable manner. It is also predicted that climate change will have detrimental effects upon agriculture, fisheries, wildlife and may even result in collapsing ecosystems.

Increasing evidence have been provided from national public opinion surveys throughout the world that seriousness and immediacy of climate change have deteriorated during the years (Jowit, 2010; Kaufman, 2010; McCright and Dunlap, 2011; Pew Center for People and the Press, 2008, 2009; Saad, 2009; Satzman et al., 2010; Weber and Stern, 2011;). The probable causes attributed to this trend may be the lack of a mega campaign to promote "climate change issues" by the political parties (Hoggan, 2009; McCright and Dunlap, 2011), media publicity which plays a crucial role in generating interests and public opinion, published researches of scientists in popular newspapers and journals among the few.

18.2   OBJECTIVE

The objective of the paper is to understand the awareness on climate change, their perception and planned course of action from a sample of internet-literate community comprising researchers/academicians, students and professionals. The outcome would enable the policy makers and the intellectuals to create further awareness of the situation for mitigating the grave nature of climate change.

18.3   MATERIALS AND METHODS

In this limelight, a questionnaire was framed through Google Drive online survey tool consisting of both open and close-ended questions to determine the awareness level of the respondents, action planned and suggestions in matters related to climate change issues. The invitation to participate in the online survey was strictly sent to adult individuals above 18 years of age. Responses were evaluated by the Google drive online testing tool using proportions and percentage as the main statistical analysis used in the software. Researchers/Academicians, professionals and students comprised the variety of the respondents.

18.4   RESULTS

The respondent comprised of Internet literate individuals. A total of 129 respondents participated in the survey in which the responses of eight respondents were screened out due to inappropriate submission. Thus, a total of 121 entries were finally analyzed. Out of these, 35 (28.92%) respondents were having qualification of degree or equivalent, while 55 (45.45%) had a post graduate degree and the remaining 31 (25.61%) had a doctorate degree and above. Academicians/Researchers constituted 66 (54.54%) in number, students 38 (31.40%) and professionals 17 (14.04%).

A good number 107 (88.42%) had heard about the IPCC (International Panel on Climate Change) and the Kyoto Protocol while 14 (11.57%) were ignorant of it. The main source of information on climate change was attributed to the internet 30 (36.30%) followed by television 26 (31.46%), specialized publications/academic journals 22 (26.62%), college/university 20 (24.20%), government agencies 8 (9.68%), environmental groups like worldwide fund for nature, Greenpeace, etc. and radio had equal respondents 6 (7.26%) and family/friends as the source of information contributed to only 3 (3.63%) of the respondents.

The responses on seriousness of the issue were opined from the respondents. The issue of air pollution was viewed as a serious problem by 104 (85.95%) of respondents while 17 (14.04%) were not worried regarding it. Pollution of river and seas was viewed as a serious problem by 72 (59.50%), not worried by 11 (9.09%) and will improve in the coming years by 38 (31.40%). Flooding was a serious problem for 60 (49.58%), not worried 33 (27.27%) and will improve in the coming years by 28 (23.14%) respondents. Litter was deemed as a serious problem by 38 (31.40%), not worried 26 (21.48%) and will improve in the coming years by 36 (29.75%). Traffic/congestion was felt as a serious problem by 72 (59.50%), not worried by 18 (14.87%) and will improve in the coming years by 31 (25.61%) respondents. Genetically Modified foods was viewed as a serious problem by 48 (39.66%), not worried regarding their future use 52 (42.97%) and will improve in the coming years by 21 (17.35%) respondents. The hole in the ozone layer was viewed as a serious problem by 88 (72.72%), not worried by 15 (12.39%) and will improve in the coming years by 18 (14.87%) respondents. Using up earth's resources was classified as a serious problem by 82 (67.76%), not worried by 12 (9.91%) and will improve in the coming years by 27 (22.31%) of the respondents. The extinction of species was termed as a serious problem by 102 (84.29%), not worried by 8 (6.61%) and will improve in the coming years by 11 (9.09%) respondents. The radioactive and electronic waste was viewed as a serious problem by 80 (66.11%), not worried by 11 (9.09%) while 30 (24.79%) felt that the scenario will improve in the coming years. Overpopulation was termed as a serious problem by 92 (76.03%), not worried by 13 (10.74%) and will improve in the coming years by 16 (13.22%) respondents.

As per 13 (15.73%) respondents belonged to a member of any environmental organizations (Word Wide Fund for Nature, GreenPeace, Groups and Forums, etc.) while 108 (89.25%) did not belong to any such groups and forums.

According to majority; 70 (84.70%) respondents, the main responsibility of tackling climate change should rest with the individuals themselves followed by International organization 18 (14.87%), local/state government and business or industry had equal number 12 (14.52%), national government by 6 (7.26%) and Environment organizations/Lobby groups (e.g., Worldwide Fund for Nature) by 3 (3.63%) respondents.

About the opportunity for climate change initiative, majority of them narrated as the formation of awareness club as a viable option with 33 (39.93%) respondents, participation in climate awareness rallies and using nonconventional sources of energy had equal number with 30 (36.30%), turning vegetarian 24 (29.04%) followed by others with a mere 4 (4.84%) respondents.

Their opinion on seriousness of the issue unlike other issues, 104 (79.43%) affirmed that such was not the case with climate change while 17 (20.57%) had the opposite view.

18.5 DISCUSSION

In the present questionnaire, 107 (88.42%) had heard about the IPCC (International Panel on Climate Change) and the Kyoto Protocol while 14 (11.57%) were ignorant of it. A similar observation by Gallup survey conducted in 2009 by face-to-face and telephone interviews with adults, aged 15 and older shows 32% of Indians say they know at least something about climate change, similar to awareness in previous years. Urban Indians, who tend to be better educated, are significantly more likely to report being aware of climate change. 41% of adults in urban India know at least something about climate change, compared with 28% in rural India where more than two thirds of the population lives.

According to the majority of respondents 70 (84.70%) respondents, the main responsibility of tackling climate change should rest with the individuals themselves. As per the study conducted by Read *et al.*, 75% opined that personal efforts should be made to reduce the climate change. Similar result was observed in the study conducted by Pandve et al. (2011) according to which 478 (65.21%) respondents; individual lifestyle changes would be most effective in tackling climate change and for preventing further changes in climate. In the study conducted by Read *et al.* (1994), the main reasons for climate changes were reduction in biomass (57%), automobiles (41%), and industries (32%).

18.5.1   CLIMATE CHANGE ISSUES PERTAINING TO THE AREA OF STUDY/RESEARCH/SUBJECT

The respondents narrated climate change issues pertaining to their area of study/research/subject. The issues highlighted were nuclear and electronic waste management, climate change and livestock management practices, greenhouse gas emission, global warming and its effect on altering the plant physiology and decreased yield of food grains, floods and uneven rainfall as a serious threat, yield reduction of agricultural crops, climate change and food security, land-use change, impact of climate variables on environment, erratic weather pattern, air pollution, nonjudicious use of ground water leading to lowering of water table among a few.

18.5.2   CLIMATE CHANGE ISSUES OF IMMEDIATE CONCERN TO FARMERS

The respondent's opinion on issues of climate change, which is a matter of concern to the farmers, was mixed and varied. The main focus lied on reduced productivity, droughts and higher infestation of insect, unpredictable weather conditions and swing of temperature, decreasing availability of water, overuse of inorganic chemicals, soil and water pollution, fall in the ground water level, degrading fertility of soil due to excessive use of inorganic pesticides, erratic climatic behavior such as unseasonal rains, winds, hail storms, etc.

18.5.3   INDIGENOUS/FARMERS KNOWLEDGE TO BE APPLIED TO MITIGATE CLIMATE CHANGE ISSUES

The indigenous/ farmers knowledge to be implemented in mitigating climate change threats were suggested by the respondents such as disposing off wastes smartly through recycling and plowing organic wastes back to the soil, use of gobar housing of animals, organic farming and conservation agriculture, composting at farmstead, biogas technique for successful management of wastes, use of local herbs in control of pest and diseases, folk lore, linking culture and tradition for predicting the climate awareness, short duration cropping, indigenous storage mechanisms of food grains, mulching of soil and water conservation measures, local food systems as per the climate suitability, increase ground water level by making check dams and planting of climate resilient crops and varieties, zero plowing, etc.

18.5.4 PRACTICAL SUGGESTIONS TO BE ADOPTED BY THE FARMERS TO MITIGATE CLIMATE CHANGE ISSUES

The respondents opined on suggestions, which could be practiced by the farmers to mitigate climate change issues. The responses were growing crops through organic means, multicropping, less use of inorganic and market based fertilizer and pesticide, use of vermin compost and organic manure, banning of plastic resources, conservation agriculture practices, plantation of tree on farm meadows, economic use of irrigation water by practicing sprinkler and drip methods of irrigation, integrated farming approaches, use of shelter belts and wind breaks, zero tillage method, use of aerobic rice and SRI technique in rice, good feeding habits of cattle's to avoid CH_4 emission, growing more pulses, changing cropping patterns based on weather information, proper water application, use of solar panels for power generation, chopping up of farm residues rather than burning on the field, growing trees inside each farm to provide for carbo-sequestration, etc.

18.6 CONCLUSIONS

The present study reflects that the Internet literate population is mostly aware about global climate change as well as role of human activities in climate change. The sources of information are usually e-based resources. The awareness regarding important agencies and groups and their commitment to climate change was found to be poor. Support for personal measures and individual efforts by majority of respondents for mitigating the effects of adverse climate change is a very promising sign. There is need to conduct large nationwide survey to generate information from various stakeholders for burning issues on climate change and providing effective measures to combat the changing climate scenario. Such surveys will also form the basis to establish a foundation for decision makers for climate change mitigation activities in the country. Further, the growing need for a specialized journal and literature reviews is increasingly felt for spearheading the developmental agenda. The common good demands solidarity with the poor who are often without the resources and face many problems, including the potential impacts of climate change. It is also recommended that awareness campaigns/programs regarding climate change and formation of groups/communities with a larger access through ICTs and media to be introduced for better preparedness.

KEYWORDS

- Action
- Climate Change
- ICTs
- Mitigation
- Perception
- Planning

REFERENCES

Hoggan, J. (2009). Climate Cover-up. Greystone Books, Vancouver, BC, Canada.

Jowit, J. (2010). Sharp Decline in Public's Belief in Climate Threat, British Poll Reveals Guardian (UK) http://www.guardian.co.uk/ environment/2010/feb/23/british-public-belief-climate-poll/print.

Kaufman, L. (2010). "Among Weathercasters, Doubt on Warming." New York Times. http://www.nytimes.com/2010/03/30/science/earth/30warming.html?scp=1&sq=met eorologists&st=cse (Accessed March 15, 2013).

Leiserowitz, A. (2007). Fighting Climate Change: Human Solidarity in a Divided World. Human Development Report Office, UNDP.

McCright, A., & Dunlap, R. (2000). Challenging Global Warming as a Social Problem, An analysis of the Conservative Movement's Counter-Claims, Social Problems 47(4), 499–522.

Pandve et al. (2011). Assessment of Awareness Regarding Climate Change in an Urban Community. http://www.ncbi.nlm.nih.gov/pmc/articles/PMC3299094/ (Accessed April 4, 2013).

Pew Center for the People and the Press (2008). "A Deeper Partisan Divide Over Global Warming." http://www.people-press.org/report/417/a-deeper-partisandivide-overglobal-warming (Accessed March 18, 2013).

Ray, J., & Pugliese, A. (Accessed February 23, 2013). Indians Largely Unaware of Climate Change. http://www.gallup.com/poll/125267/indians-largely unawareclimate-change.aspx.

Saad, L. (2009). Increased Number Think Global Warming Is "Exaggerated" Gallup Organization. http://www.gallup.com/poll/116590/Increased-Number-Think-Global-Warming-Exaggerated.aspx?version=print.

Satzman, D., Radziner, K., & Scott, G. (2010). "Political Change on Climate Change" http://www.kcrw.com/news/programs.

Scruggs, L., & Benegal, S. (2012). Declining Public Concern About Climate Change: Can We Blame the Great Recession? *Global Environ. Change*, doi: 10.1016/j.gloenvcha.

Weber, E. U., & Stern, P. C. (2011). Public Understanding of Climate Change in the United States. *American Psychologist*, 66(4), 315–328.

CHAPTER 19

CLIMATE CHANGE AND INDIAN AGRICULTURE

MD. HEDAYETULLAH[1], PARVEEN ZAMAN[2],
SHAILENDRA KUMAR YADAV[3], MD. ABU NAYYER[4], and
MD. WASIM SIDDIQUI[5]

[1]School of Agriculture Science, Centurion University, Odisha, India-761211; E-mail: hedaye.bckv@gmail.com

[2]Bidhan Chandra Krishi Viswavidyalaya, Mohanpur, Nadia, WB, India - 741252.

[3]Central Soil & Water Conservation Research & Training Institute (ICAR) Research Centre, Datia, MP, India-475661

[4]Department of Horticulture (Fruit and Fruit Technology), Bihar Agricultural University, Sabour, Bhagalpur, Bihar (813 210) India

[5]Department of Food Science and Technology, Bihar Agricultural University, Sabour, Bhagalpur, Bihar (813 210) India

CONTENTS

19.1   INTRODUCTION

Significant variation in either mean state of climate or in its variability, persisting for an extended period (typically decades or longer) is referred as climate change, which may be due to natural internal processes or external forcing or to persistent anthropogenic changes in the composition of the atmosphere or in land use (NRAA, 2014). Climate change is being perceived as greatest threat to mankind in twenty-first century however, its nature, extent and magnitude are variable in different regions and locations. Indian agriculture is very diverse due to wide range in edapho-climatic conditions. Over the past century, surface temperatures have risen, and associated impacts on physical and biological systems are increasingly being observed (PRC, 2007). Rise in temperature shall enhance respiration rate, reduce crop duration, alter physiology of flowering, fruiting and hasten ripening and maturity which may adversely affect the crop productivity and adaptability. Earlier studies conducted in India also generally confirm the trend of agricultural decline with climate change (Sinha, 1993). Climatic extremes like droughts, floods, tropical cyclones, heavy precipitation events, hot extremes and heat waves are mainly originated and triggered by climate change causes negative impact on agricultural production. It has been projected by the recent report of the IPCC and a few other global studies that unless we adapt, there is a probability of 10–40% loss in crop production in India by 2080–2100 due to global warming despite beneficial aspects of increased CO_2 (Parry et al., 2004). According to the Intergovernmental Panel on Climate Change's (IPCC) Fourth Assessment Report of 2007 (AR4), climate change can reduce agricultural yields as much as 50% under rainfed farming. In India rainfed agro-ecologies contribute 60% of the net sown area (Vision, 2050) hence climate change is expected to cause serious difficulties for Indian agriculture.

Arid and semiarid regions of India are more prone to climate change and experiencing erratic and poor rainfall and temperature induced climatic extremes. Increase in atmospheric temperature and change in precipitation pattern due to climate change has significantly raised the sea level and steadily shifted/moved the climatic zones of country. The climate change also affects the atmospheric temperature and relative humidity which are the major factors affecting infestation of pests and diseases. Climate change projections made up to 2100 for India indicate an overall increase in temperature by 2–4°C with no substantial change in precipitation quantity (Kavikumar, 2010). Since climatic factors serve as direct inputs to agriculture, any change in climatic factors is bound to have a significant impact on crop yields. Many studies in the past have shown that India is likely to witness one of the highest agricultural productivity losses in the world in accordance with the climate change pattern observed. The impact of climate change on Indian agriculture could result in problems with food security and livelihood for the second most populated country of the world.

19.2   CLIMATE CHANGE

Solar radiation is the main source of energy for life on earth, which arrives mainly in the form of visible light. About 30 percent of the solar radiation scattered back into space by the outer atmosphere, but the rest reaches to the earth's surface, which reflects it in the form of a calmer, more slow-moving type of energy called infrared radiation (this is the sort of heat thrown off by an electric grill before the bars begin to grow red). Infrared radiation is carried slowly aloft by air currents. Greenhouse gases (GHGs) make up only about 1 percent of the atmosphere, but they act like a blanket around the earth, or like the glass roof of a greenhouse which trap heat and keep the planet some 30°C warmer than it would be otherwise (Republics of Maldives Climate Change, 2014). (The GHGs in the atmosphere act as a thermostat for controlling the temperature on Earth's climate. The atmosphere of the earth is under very complex state of energy equilibrium. But rapid industrialization mainly during the second half of the twentieth century has significantly increased concentration of green house gases such as water vapor, carbon dioxide, ozone and methane in atmosphere through indiscriminate burning of coal, oil, and natural gas which eventually trapped the escape of infrared radiation into space and cause increase in atmospheric temperature. Farming activities like changes in land use, application of nitrogenous fertilizers and methane emissions from agricultural activities like ruminant animals and the cultivation of paddy has significantly added methane and nitrous oxide in atmosphere. The increase in atmospheric temperature of earth due to increase in concentration of green house gases is technically termed as "enhanced greenhouse effect." Since 1970 to 2013 several key GHG emissions, including carbon dioxide (CO_2), methane (CH_4), nitrous oxide (N_2O), hydrofluorocarbons (HFCs), per fluorocarbons (PFCs) and sulfurhexafluoride (SF_6), increased by 70 percent. In its fourth assessment report since 1990, the Intergovernmental Panel on Climate Change (IPCC) concluded that climate change is already happening and can be primarily attributed to human activity.

19.2.1   IMPACT OF CLIMATE CHANGE ON INDIAN DIVERSE AGRO ECOLOGICAL SYSTEM

Indian agro ecological system is drastically changing due to climate change. The impact of climate change is observed in the Semi arid regions like Gujarat plains, Malwa, Chhotonagpur plateau and Deccan plateau. Due to rise in atmospheric temperature, semiarid regions of western India are expected to receive higher rainfall, while central regions are likely to experience 10 to 20% decrease in winter rainfall by 2050. In North India, the average mean surface temperature is expected to rise between 3.5°C and 5°C by the end of this century. The unusual flood occurred in Barmer and other parts of western Rajasthan have already formed three large lakes covering about 7 to 8 square km in Kawas, Malwa, and Uttarlai, all in the Barmer district of Rajasthan. This is destructive to the agro ecosystem of the region and adversely affected livelihood of the populace. In Orissa, seawater incursion has reached 2.5 kilometers inland over the previous two decades. Drought prone districts such as Balangir, Kalahandi, Koraput, Bargarh and Jharsuguda have experienced frequent floods in the prior two decades. The negative effects are already visible in the Sunderbans, the wetlands at the mouth of the Ganga and Brahmaputra river systems. The Sunderbans wetlands harbor one of the most globally important wildlife habitats, the largest mangrove forest. The Sunderbans wetlands have long been highly susceptible to seasonal ocean currents, tides, waves, winds, and cyclonic storm surges that cause rapid soil erosion on the one hand and salt deposition on the other.

19.2.2   IMPACT OF CLIMATE CHANGE ON AGRICULTURE

Agriculture production system is affected due to changes in the weather parameters. Changing climatic parameters like temperature, moisture concentrations (rainfall and relative humidity) and CO_2 levels is likely to have diverse impacts on agro ecosystems and therefore on crops, livestock, pests and pathogens. The physiological response of crop plants to changing climate is expected to be varied on growing period, chiling requirements, flowering, fruiting, ripening, maturity and senescence. The nature of changes may be uncertain but what is certain is that changing environmental parameters are highly likely to affect agro ecosystems.

19.2.2.1 MONSOON

About 60% of net sown area of the country is under rainfed agro-ecology which is depending on southeast and north east monsoon hence any change and shift in the occurrence of monsoon pattern directly affect the crop success and productivity. Timely arrival of the monsoon is crucial for food production, economy and livelihood of the country which is mainly agriculture dependent. The impact of climate change has gradually shifted summer monsoon and delayed its onset. The delay in onset of monsoon has shortened growing period of rainfed crops which adversely affect their growth and development. Pre-monsoon rainfall disruptions and occurrence of dry spells during growing period imposed water stress to the rainfed crops. For example, the Chhattisgarh region in the past years has received half its usual amount of water during the months of May and June, seriously affecting rice production as well as others crop production like pulses, oilseeds, sugar crops and vegetable crops.

19.2.2.2 MELTING GLACIERS

Glaciers play an important role in regulation of climate and weather of the region which is very important for agricultural and its allied activities. Glaciers are important for maintaining integrity in ecosystem and feeding water in rivers, thus providing water for agriculture. India relies heavily on the Himalayan Rivers to meet irrigation requirements for the crops grown in their command area. The hydrological characteristics of Himalayan region have already undergone significant changes as a result of climate change and anthropogenic activities which has resulted in increased variability in rainfall and surface water runoff, landslides, frequent hydrological disasters and the pollution in reservoirs of lakes and watersheds. The increase in temperature causes high rainfall and glacial meltdown which reflects as gradual bug significant receding of Himalayan glaciers. These changes in climate will inevitably interact with changes in glaciers and glacial lakes. Results also show that the recession rate has increased with rising temperature. A forecast was made that up to a quarter of the global mountain glacier mass could disappear by 2050 and up to half could be lost by 2100 (Oerlemans, 1994; IPCC, 1996). This is very alarming forecast in respect of the existence of the Himalayan Rivers.

19.2.3   UNEVEN DISTRIBUTION OF RAINFALL

The uneven distribution of rainfall patterns negatively impacts Indian agriculture since agriculture systems have developed cropping patterns dependent on regional weather conditions. The different agro climatic zone of India receives different amount of rainfall and this precipitation pattern is changing with wet years becoming wetter and dry years become drier. The growth and development of crops is also affected by rainfall variability. This change could result in a greater number of heavy rainfall events a decrease in the overall number of rainy days, and longer gaps between rains, as well as increased rate of evapo-transpiration. This would disturb established cropping patterns. Heavy rainfall combined with a decrease in the total number of rainy days is apparent in major part of India. Significant variability in annual rainfall is also noticed in form of frequent occurrence of droughts and floods. These changes impose several biotic and abiotic stresses on fauna and flora of the ecosystem and changing their occurrence, frequency and distribution. They are also affecting the emergence and spread of pathogens which affect crop yields. Spatio-temporal variations in projected changes in temperature and rainfall are likely to lead to differential impacts in the different regions of India (Prince et al., 2006).

19.2.3.1   EXCESS RAINFALL

Changes in the rainfall pattern and frequent occurrence of intense weather events imposed occurrence of frequent floods. Flooding associated with heavy rain events can also damage crops (Wassmann, 2007). Extreme rainfall events will impact agriculture in tropical areas which is already vulnerable to floods and environmental hazards such as drought, cyclones, and storms. For example, in 2013 India, lost about half a million tons of rice due to occurrence of floods. The most of the rice varieties can withstand submergence for about 6 days, but after this period high mortality of approximately 50% is recorded. Submergence for 14 days or longer cause crop mortality up to 100%. Complete submergence at flowering stage completely devoid grain filling. However, changes in climate are already being observed in the last 60 years were the warmest in the last 1000 years and changes in precipitation patterns have brought greater incidence of floods or drought globally (Samra, 2002).

19.2.3.2   SCANTY RAINFALL

In certain vulnerable arid and semiarid regions, increased temperature has already resulted in diminished precipitation. Increased temperatures cause an intensification of the water cycle with more extreme variations in weather events and longer lasting droughts. Large areas in Rajasthan, Andhra Pradesh, Gujarat, and Maharashtra and some areas of Karnataka, Orissa, Madhya Pradesh, Tamil Nadu, Bihar, West Bengal, and Uttar Pradesh are already experiencing recurrent drought and currently experiencing water deficits. In dry land areas, marginal cropland could convert to range land, and some cultivated land and rangelands could no longer be suitable for food and fodder production. More frequent drought would necessitate greater multiyear reservoir storage capacity, in which India is currently deficient. Water conservation and management practices, as well as water storage will need urgent attention. Similarly, the drought of 2002 reduced area under rainfed agriculture more than 15 m ha and resulted in a loss of more than 10% in food production (Aggarwal, 2007).

19.2.4   SOIL AND MINERALS

The precipitation patterns are impacted by climate change and soil health is also heavily affected. This is because changes in temperature and precipitation influence water run-off and soil erosion reduced organic carbon, microbial biomass and nitrogen content in soil profile. Soil erosion and poor drainage also developed salinity in the soil. This impact also observed on the biodiversity of soil microorganisms. High air temperature and solar radiation increases soil temperature which may check growth of several agriculturally beneficial microbes. According to researchers, a small increase in temperature in low carbon soil results in higher carbon dioxide emissions as compared with medium and high carbon soil[12]. This phenomenon makes low carbon soil more vulnerable to warming. It has been shown that soil nitrogen availability is reduced in drier soil conditions. This is because dry soil adversely affects root growth and decomposition of organic matter that adversely affect the activity of biological nitrogen fixing bacteria. Reduced nitrogen fixation in turn reduces soil fertility. Supplementing higher dose of chemical fertilizers to compensate soil degradation induced fertility depletion, with intention to harvest sustained and higher crop productivity, is a common thought. Excessive use of chemical fertilizers damage living processes of the soil and makes it more vulnerable to edapho-climatic variability. At the same time, they contribute to emission of nitrous oxide, a potent greenhouse gas. Climate change accelerates soil erosion and land degradation processes which ultimately induce desertification of fertile soil and contributes in expansion of arid zones. The impact of climate change has already resulted as contraction of several million hectares of fertile cultivated land.

19.2.5 CROP DIVERSITY

The fourth IPCC Report (Lakshmanan et al., 2007) states that by the end of this century climate change will be the main cause of depletion in biodiversity. If there is an increase of the average global temperature by 1.5–2.5°C, then approximately 20–30 percent of known plants and animal species will be under threat of extinction. Climate change will increase the pressure of land degradation and habitat loss, as well as genetic erosion which is already intensifying by growing uniform agricultural systems across the India as well as the world. According to the Food and Agricultural Organization (FAO), three-quarters of the global crop diversity is already lost. This is particularly problematic as the loss of genetic diversity, both in natural ecosystems and cultivated crops, is facing problem of the impact of climatic change. Changes in the climate pattern also favor the diffusion of invasive alien species which are considered to be second most destructive threat to global biodiversity and agro ecosystems. Invasive alien species are able to conquer new territories when changed ecoclimatic zones become favorable for their breeding. Since these species constitute a large majority of the weeds in agriculture, they pose a growing threat to food production. Alien invasive plants like *Parthenium, Lantana*, etc. often replicate faster by wider spread of seeds and multiplication through vegetative means (roots, shoots and stolon's etc) and are usually more responsive to increases in atmospheric CO_2 concentrations (Hedayetullah et al., 2011).

19.2.6 CROP-WEED COMPETITION

Weed invasion is one of the major biotic stresses in agriculture causes yield loss up to 33% (Vision, 2050). Majority of the crops are C_3 plants whereas most of the weeds are C_4 plants which have better adoptability and growth under elevated CO_2 concentration due to climate change. It is observed that CO_2 enriched environment, weeds transfer significantly more carbon to roots and rhizomes than to shoots. This improves their root growth and increases their chances of survival. More vigorous roots lead to more viable plants and higher reproduction with more seeds. Weeds with stronger roots are more competitive and more difficult to control with traditional weed management techniques or herbicide which attack the foliage rather than the roots. Hence, climate change is expected to promote proliferation of new weed species in the composition of weed flora and causes invasion over agricultural crops. This is important since the dynamics of crop-weed competition ultimately decides crop yield. The higher genetic diversity in weeds gives them flexibility to adapt to new environments through quickly responding to the changes with higher rates of growth and reproduction. A larger number of weed species are found associated with respective major crops as well as minor crops. Often the worst weeds of crops are their wild relatives, for example, red rice is a weed in rice cultivation and Johnson grass a close relative of sorghum is a weed in sorghum fields.

19.2.7 CROP PRODUCTIVITY

The impact of climate change on agriculture especially in crop productivity is observable. According to the IPCC, the next few decades of climate change are likely to bring benefits to higher latitudes through longer growing seasons, but in lower latitudes, even small amounts of warming will tend to decrease yields. The regional inequality in food production resulting from climate change will have a very great implication for global food politics. In these regions even moderate warming of 1°C for wheat and maize and 2°C for rice will reduce yields significantly. It is known that many agricultural systems are seasonally dependent and thus sensitive to climate change. Crop and livestock production need a specific range of weather conditions at particular times, for optimal growth. The most vulnerable agricultural systems are the arid, semiarid, and dry subhumid regions of the developing world. In these regions high rainfall variability and recurrent drought, flood cycles disrupt crop development. Extreme high and low temperatures cause physical injuries like freezing and sunburn to the crop plants and damage the grain. Temperature rise in lower latitude regions accelerates the rate of respiration, transpiration and evapotranspiration excessively leading to sub optimal growth. The rice productivity is estimated to decrease under climate change due to its sensitivity to temperatures that cause damage to the plant, thus affecting yield (Chakraborty et al., 2008). Increased temperatures have multiple impacts on crop productivity depending on the biological characteristics of the specific crop and the time of the heat stress in relation to its growth and development. Higher daytime temperature accelerates plant maturity and results in poor grain filling due to pollination failure while higher night temperatures increase yield losses due to higher rate of respiration (Hedayetullah et al., 2014).

19.2.8 DISEASES AND PESTS

The diseases and pest infestation on crop plant is fully depend on climatic parameters. Most of the plant disease like blast, bacterial leaf blight and brown spot observed on paddy, wheat (*Puccinia and Septoria*) and horticultural (*Meloidogyne*) crops are affected by climate change. The trend indicates that severity of majority of diseases is found to be higher with elevated CO_2 levels, an off-shoot of climate change (Mock et al., 2006). The severity of diseases caused by fungi, bacteria, viruses, and insects are anticipated to increase with global warming and these pest and pathogens trying to adopt under diverse climatic condition. Elevated CO_2 may modify pathogen aggressiveness and or host susceptibility and affect the initial establishment of the pathogen, especially fungi, on the host (Heiser et al., 2005; Yates, et al., 2000). In most examples, host resistance has increased, possibly due to changes in host morphology, physiology and composition.

Increased fecundity and growth of some fungal pathogens under elevated CO_2 has also been reported (Singh, 2004). However, it has been reported that greater plant canopy size, especially in combination with humidity and increased host abundance, can increase pathogen load. Temperature Changes in environmental conditions are likely to result in the northward extension of certain diseases and pests, more generations of pathogens per season, and a better capacity to survive the winter, thus increasing their prevalence and range. The favorable temperature range for the infestation of the rice gundhi bug is 13°C–35°C. If the temperature exceeds the upper threshold of the favorable range, the development rate and the survival capacity of the bugs get reduced. Moisture also plays an important role in the activities of the pathogens as optimal breeding conditions are usually created when moisture is high. The increase in temperature and moisture creates especially favorable conditions and affects both hosts and pathogens in various ways.

19.2.9   FUNGAL INFECTION

Another aspect is that the occurrence of plant-based toxic contaminants can be influenced by changes in climate. For example, aflatoxin a metabolite of the fungal species Aspergillus is a dangerous contaminant that infects food grains and is harmful for human and animal health. Climatic changes including increased prevalence of drought and unseasonal rains, changes in relative humidity and shift in temperature may favor the population and spread of aflatoxin-producing fungi. Compounding this problem is the fact that changes in climate may allow the increased population of fungal pathogens to have increased opportunities to attack food grain crops. For example, dry conditions during grain filling and maturity enhance the probability of cracked grains, which can get more easily infected by this fungal pathogen. Heavy rains during or after the harvest may lead incomplete drying of grains before storage and causes proliferation of the fungus during the storage period.

19.2.10   POULTRY AND LIVESTOCK

Poultry, livestock and other farm animals will suffer due to climate changes because higher temperatures will increase the number of new diseases directly or indirectly affecting them. For instance exposure to drought and excessive humidity or heat renders cattle more vulnerable to infections. Also, alternating drought and heavy rainfall cycles provide a good environment for midge and mosquito vectors that are linked with outbreaks of vector-borne livestock diseases. Poultry is similarly affected by excessive heat or rainfall.

19.3   IMPACT AND VULNERABILITY OF INDIAN AGRICULTURE TO CLIMATE CHANGE

Indian agriculture today is faced with the challenge of having to adapt to the projected vagaries of climate change. It must develop mechanisms to reduce its vulnerability. The Indian Council of Agricultural Research (ICAR) has already begun research to assess the likely impact of climate change on various crops, fisheries, and livestock under NICRA (*National Initiative on Climate Resilient Agriculture*). A sector wise analysis is given below.

19.3.1   CEREAL CROPS

The Asia-Pacific region is likely to face the worst impacts on cereal crop yields. Loss in yields of wheat, rice and maize are estimated in the vicinity of 50%, 17%, and 6% respectively by 2050. This yield loss will threaten the food security of at least 1.6 billion people in South Asia. The projected rise in temperature of 0.5°C to 1.2°C will be the major cause of grain yield reduction in most areas of South Asia.

19.3.1.1   WHEAT

India is considered to be the second largest producer of wheat and the national productivity of wheat is about 2708 kg/ha. The Northern Indian states such as Uttar Pradesh, Punjab, Haryana, Uttaranchal and Himachal Pradesh are some of the major wheat producing states. Here the impact of climate change would be profound, and only a 1°C rise in temperature could reduce wheat yield in Uttar Pradesh, Punjab and Haryana. In Haryana, night temperatures during February and March in 2003–2004 were recorded 3°C above normal, and subsequently wheat production declined from 4106 kg/ha to 3937 kg/ha in this period. In March 2004, temperatures were higher in the Indo-Gangetic plains by 3–6°C, which is equivalent to almost 1°C per day over the whole crop season. As a result, wheat crop matured earlier by 10–20 days and wheat production dropped by more than 4 million tons in the country (Rani, 2007). An assessment of the impact of climate change on wheat production states that the country's annual wheat output could plunge by 6 million tons with every 1°C rise in temperature (Mathauda, 2000). However, utilizing adaptation strategies such as changing the planting dates and using different varieties, it is possible to moderate the losses. By adapting certain agronomic strategies it was estimated that at a 1°C rise, 3 million tons could be restored. The assessment also found that the impact of climate change on wheat production varies significantly by region. North India and other areas with higher potential productivity were less impacted by a rise in temperature than the low-productivity regions. If there is no mechanism or strategy to cope with rainfall variability, then rainfed crops will be more heavily impacted than irrigated ones.

19.3.1.2 PADDY

An increase of 2–4°C is predicted to result in a reduction in rice yields. Eastern regions are predicted to be most impacted by increased temperatures and decreased radiation, resulting in relatively fewer grains and shorter grain filling durations. By contrast, potential reductions in yields due to increased temperatures in Northern India are predicted to be offset by higher radiation, lessening the impacts of climate change. Although additional CO_2 can benefit crops, this effect was nullified by an increase of temperature. Mathauda et al. (2000) show the effect of temperature increase on the projection of climate change over rice crop. The increase of temperature will decrease the life span, grain yield, maximum leaf area index, biomass, and straw of the rice. Research conducted by Indian Agricultural Research Institute (IARI) has shown that the grain yield of rice is not impacted by a temperature increase less than 1°C. However from an increase of 1–4°C the grain yield reduced on average by 10% for each degree the temperature increased. Thus, higher temperatures accompanying climate change will impact world rice production creating the possibility of a shortfall. We have seen already that basmati varieties of rice are particularly vulnerable to temperature induced pollen sterility, and thus to lower grain formation. Many determinants of crop yield like rate of photosynthesis, biomass production, leaf area index, number and test weight of grains were reduced by elevated O_3 level up to 14%. According to projections, ozone pollution may cause rice yield losses up to 16% with no change in agricultural practices, which would put food security in Asia at substantial risk (Wissuwa, 2008; Vanja et al., 2009). Rainfall pattern is a very important limiting factor for rain-fed rice production. Higher variability in distribution and a likely decrease in precipitation will adversely impact rice production and complete crop failure if severe drought takes place during the reproductive stages. In upland fields, if the rice crop receives up to 200 mm of precipitation in 1 day and then receives no rainfall for the next 20 days, the moisture stress will severely damage final yields. Assessments predict a decrease in the rice production in tropical regions, but an increase of rice production outside tropical regions. This shift is of particular concern to India because lower rice production will immediately create a hunger situation on a large scale. Studies on the impact of night time temperature rise on rice yield indicate that the warmer nights have an extensive impact on the yield of rice; every 1°C increase in night time temperature led to a 10% reduction in yield. The eastern region of India has diverse physiographic and agro-climatic land which supports genetic resources. According to a study conducted by the Indian Agriculture Research Institute (IARI) that the impact of climate change with increased temperature and decreased radiation which will ultimately lead to decrease in productivity of rice in the North Eastern region.

19.3.1.3   MAIZE

Maize (*Zea mays* L) is the third most important cereal crop in India and has a major role to play in food security. Maize production in arid and semi arid tropical regions is particularly sensitive to weather conditions, especially rainfall. Therefore, variation in the rainfall as well as maximum and minimum temperature during the south-west and north-east monsoon period will negatively impact maize crops. The Upper Indo-Gangetic Plain characterized by low winter temperature, the maize yield can increase up to a 2.7°C rise in temperature. High temperatures plays a greater role in affecting maize yield as compared to rainfall, which may not have a major impact on winter yields as the crops in the Gangetic belt are well irrigated. Maize yield during monsoon could be reduced by up to 35% in most of the Southern Plateau regions and up to 55% in Mid Indo-Gangetic Plains, whereas the Upper Indo-Gangetic Plain is expected to be relatively unaffected.

19.3.2 VEGETABLES AND LEGUMES

Most of the vegetables crops are sensitive to extreme weather parameters, thus periodic high temperature and soil moisture stress conditions are likely to reduce yield. But research also shows that higher CO_2 concentration could offset the negative effect of higher temperature especially in the case of leafy vegetables that would benefit from increased rates of photosynthesis. Short duration vegetables and legumes could perform better under higher concentration of CO_2, especially crops such as chickpeas which have a well developed carbon sink capacity, due to their ability to use additional photo-assimilates more effectively. Chickpeas grown under elevated concentration of CO_2 (up to 550 ppm) showed better performance compared to plants grown under current CO_2 concentrations of 370 ppm. There was greater shoot elongation and leaf expansion under elevated CO_2 and an 18% increase in the number of seed in some varieties. Nevertheless, an increase in temperature is likely to reduce the beneficial effect of increasing CO_2. Elevated CO_2 (550 ppm) leads to higher number of leaves and to larger bulb size in onions. The pseudo stem length, number of leaves and leaf area are higher at bulb initiation and bulb development stages than at ambient CO_2 levels (370 ppm). Plant length, number of secondary branches and leaf area increase at elevated levels of CO_2, as compared to ambient levels. The yield increase is mainly due to increase in the number of fruits per plant with a mean of 74 fruits per plant, compared to 56 fruits per plant under current ambient conditions (Singh, 2010; Siddiqui et al., 2011).

KEYWORDS

- **Agriculture**
- **Agro Ecosystem**
- **Climate Change**
- **Crops**
- **Diversity**
- **Impact**

REFERENCES

Aggarwal, P. K. (2007). "Global Climate Change and Indian agriculture, Case Studies from the ICAR Network Project, Indian Council of Agricultural Research, New Delhi., 8.

Aggarwal, P. K., & Rani, D. N. S. (2007). 'Assessment of Climate Change Impacts on Wheat Production in India. In Global Climate Change and Indian agriculture, Case Studies from the ICAR Network Project. Indian Council of Agricultural Research, New Delhi.

Aggarwal, P. K., & Sinha, S. K. (1993). Effect of Probable Increase in Carbon Dioxide and Temperature on Productivity of Wheat in India *J. Agric. Meteorol*, 48, 811–814.

Ainsworth, A. E. (2008). Rice Production in a Changing Climate a Meta-Analysis of Responses to Elevated Carbon dioxide and Elevated Ozone Concentration, *Global Change Biolo*, 14, 1642–1650.

Chakraborty, S., Luck, J., Hollaway, G., Freeman, A., Norton, R., Garrett, K. A., Percy, K., Hopkins, A., Davis, C., & Karnosky, D. F. (2008). Impacts of Global Change on Diseases of Agricultural Crops and Forest Trees. CAB Reviews Perspectives in Agriculture, Veterinary Sciences, Nutrition and Natural Resources. 3, 1–15.

Chakraborty, S., Pangga, I. B., Lupton, J., Hart, L., Room, P. M., & Yates, D. (2000). Production and Dispersal of Colletotrichum gloeosporiodies Spores on Stylosanthes Scabra under Elevated CO_2. *Env. Pollution.*, 108, 381–387.

Frei, M., Tanaka, J. P., & Wissuwa, M. (2008). Genotypic Variation in Tolerance to Ozone in Rice, Dissection of Distinct Genetic Factors Linked to Tolerance Mechanisms, *J. of Expt. Botany*. 59, 3741–3752.

Geethalakshmi, V., Palanisamy, K., Aggarwal, P. K., & Lakshmanan (2007). A. 'Impacts of Climate on Rice, Maize, and Sorghum Productivity in Tamil Nadu.' In Global Climate Change and Indian agriculture, Case studies from the ICAR Network Project. Indian Council of Agricultural Research, New Delhi.

Goswami, B. N., Venugopal, V., Sengupta, D., Madhusoodanan, M. S., & Prince, K.X. (2006). Increasing Trend of extreme Rain Events over India in a Warming Environment, *Science*, 314, 1442–1445.

Hedayetullah, M., Kundu, C. K. & Islam, S. (2014). Climate Change Threatens Food Crops Across the World: An overview. In: *Climatic variability: Impacts on Agriculture and allied sector* (Eds. Datta, D. and Daschaudhuri, D.) New India Publishing House. (ISBN-13: 978-9-38145094-9).

Hedayetullah, M., Koley, S., Giri, U., & Khan, S. (December 24-25, 2011). Impact of climate change and technological challenges on farming system: An overview. International Seminar on *Global Environmental Issues: Challenges to Industry, Ecology and Society*, IRDF & SKAIL, Kalyani.

IPCC, (2007). A Report of Working Group One of the Intergovernmental Panel on Climate Change Summary for Policy Makers. Intergovernmental Panel on Climate Change.

IPCC. (1996). 'Climate Change, 1995. 'Impacts, Adaptation and Mitigtion of Climate Change Scientific and Technical Analyzes.' Contribution of Working Group II to the Second Assessment Report of the Intergovernmental panel on Climate Change. Byjesh, K. S., Naresh Kumar, S., & Aggarwal, P. K., Simulating Impacts, Potential Adaptation and Vulnerability of Maize to Climate Change in India. *Mitig. Adapt. Strat. Global Change*. 2010, 15, 413–431.

Kavikumar, K. S. (February 18–20, 2010). "Climate Sensitivity of Indian Agriculture: Role of Technological Development and Information Diffusion, in Lead Papers-National Symposium on Climate Change and Rainfed Agriculture, Indian Society of Dryland Agriculture," Central Research Institute for Dry land Agriculture, Hyderabad.

Mathauda, S. S., Mavi, H. S., Bhangoo, B. S., & Dhaliwal, B. K. (2000). Impact of Projected Climate Change on Rice Production in Punjab (India), *Trop Ecol.*, 41(1), 95–98.

Matros, A., Amme, S., Kettig, B., Buck-Sorlin, G. H., Sonnewald, U., & Mock, H. P. (2006). Growth at Elevated CO_2 Concentrations Leads to Modified Profiles of Secondary Metabolites in Tobacco cv. SamsunNN and to increased resistance against infection with potato virus Y. *Plant, Cell and Env.* 29, 126–137.

National Rainfed Area Authority (NRAA). Date: 05–01–2014. www. http://nraa.gov.in/. Republics of Maldives Climate change. http://www.clima.md/pageview.php?l= en&idc=152. Sited on 05–02–2014.

Oerlemans, J. (1994). Quantifying Global Warming from the Retreat of Glaciers *Science,* 264, 243–245.

Parry, M. L., Rosenzweig, C., Livermore, A. M., & Fischer, G. (2004). Effects of Climate Change on Global Food Production under SRES Emissions and Socio-Economic Scenarios *Global Environ. Change*, 14, 53–67.

Plessl, M., Heller, W., Payer, H. D., Elstner, E. F., Habermeyer, J., & Heiser, I. (2005), Growth Parameters and Resistance against *Drechslera teres* of Spring Barley (*Hordeum vulgare* L. cv. Scarlett) Grown at Elevated Ozone and Carbon Dioxide Concentrations. *Plant Biolo.* 7, 694–705.

PRC, (2007). China's National Assessment Report on Climate Change, Beijing. Science Publishing House.

Samra, J. S., & Singh, G. (2002). Drought Management Strategies, Indian Council of Agricultural Research, New Delhi, 68.

Samra, J. S., & Singh, G. (2004). Heat Wave of March 2004: Impact on Agriculture, Indian Council of Agricultural Research, New Delhi, 32.

Seo, N., & Mendelsohn, R. (2008). "A Ricardian Analysis of the Impact of Climate Change on South American Farms" *Chilean J. of Agric. Res.*, 68(1), 69–79.

Siddiqui, M. W., Hedayetullah, M., Kundu, R. & Dhua, R. S. (March 23-26, 2011). Climate change and its impact on fruits and vegetables. *International conference Tropical Islands Ecosystems*, Port Blair, Andaman & Nicobar.

Singh, H. P. (2010). "Impact of climate change on horticultural crops." Chapter 1, Page no.2. Challenges of Climate Change-Indian Horticulture, Westville Publishing House, New Delhi.

Vanja, M., Reddy, P. R. R., Maheswari, M., Lakshmi, N. J., Rao, M. S., & Rao, G. G. S. N. (2009). "Impact of Elevated Carbon Dioxide on Growth and Yield of Castor Bean." Chapter 7, Page no.32. In Global climate change and Indian agriculture, Case studies from the ICAR Network Project. Indian Council of Agricultural Research, New Delhi.

Vision 2050. Directorate of Weed Science (ICAR). http://www.nrcws.org/DWSR-Vision-2050. pdf Sited on 05–02–2014.

Wassmann, R., & Achim, D. (December 2007). Climate Change Adaptation through Rice Production in Regions with High Poverty Levels SAT eJournal ejournal.icrisat.org, 4.

CHAPTER 20

CLIMATE CHANGE: IMPACT ON POLLINATORS' BIODIVERSITY IN VEGETABLE CROPS

TAMOGHNA SAHA, NITHYA C., and S. N. RAY

Department of Entomology Bihar Agricultural University, Sabour, Bhagalpur-813210, Bihar; E-mail: tamoghnasaha1984@gmail.com

CONTENTS

20.1 INTRODUCTION

One of the most important ecosystem services for sustainable crop production is the mutualistic interaction between plants and animals pollination. The International community has acknowledged the importance of a diversity of insect pollinators to support the increased demand for food brought about by predicted population increases. Changing weather condition due to increased temperature, erratic rainfall, and enhanced incidence of diseases are all set to affect the production trend of various vegetable crops. According to Intergovernmental Panel on Climate Change, it is defined as "Change in climate over time, either due to natural variability or as a result of human activity." The most of the warming observed over the last 50 years is attributable to human activities. The global mean surface temperature is predicted to increase by 1.4 to 5.8°C from 1990 to 2100. If temperatures rise by about 2°C over the next 100 years, negative effects of global warming would begin to extend to most regions of the world (IPCC, 2001). Developing countries in the tropics will be particularly vulnerable. Latitudinal and altitudinal shifts in ecological and agro-economic zones, land degradation, extreme geophysical events, reduced water availability, and rise in sea level and salinization are postulated (FAO, 2004).

There are considerable uncertainties about agronomic implications of vegetable crops. Predicting impact of climate change on vegetable crops accurately on regional scale is a big problem. Current estimates of changes in climate indicate an increase in global mean annual temperatures of 10°C by 2025, and 30°C by the end of the next century. The date at which an equivalent doubling of CO will be attained is estimated to between 2025 and 2070, depending on the level of emission of greenhouse gasses (IPCC, 1990a, b).

20.2 CLIMATE CHANGE, CROP POLLINATION AND POLLINATORS

Pollination is a vital stage in the reproduction of most flowering plants, and pollinating animals are essential for transferring genes within and among populations of wild plant species (Kearns et al., 1998). Although the scientific literature has mainly focused on pollination limitations in wild plants, in recent years there has been an increasing recognition of the importance of animal pollination in food production. Klein et al. (2007) found that fruit, vegetable or seed production from 87 of the world's leading food crops depend upon animal pollination, representing 35% of global food production. Roubik (1995) provided a detailed list for 1330 tropical plant species, showing that for approximately 70% of tropical crops, at least one variety is improved by animal pollination. Losey and Vaughan (2006) also emphasized that flower-visiting insects provide an important ecosystem function to global crop production through their pollination services. Economic value of crop pollination in worldwide has been estimated at €153 billion annually (Gallai et al., 2009). The leading pollinator-dependent crops are vegetables and fruits, representing about €50 billion each, followed by edible oil crops, stimulants (coffee, cocoa, etc.), nuts and spices.

Climate change may be a further threat to pollination services (Hegland et al., 2009; Memmott et al., 2007; Schweiger et al., 2010). Indeed, several authors (Sutherst et al., 2007; Van der Putten et al., 2004) have argued that including species interactions when analyzing the ecological effects of climate change is of utmost importance. Empirical studies explicitly focusing on the effects of climate change on wild plant-pollinator interactions are scarce and those on crop pollination practically nonexistent. Our approach has therefore been to indirectly assess the potential effects of climate change on crop pollination through studies on related topics. We have focused on the effects of climate change on crop plants and their wild and managed pollinators, and studies on wild plant-pollinator systems that may have relevance.

Estimates from the IPCC indicate that average global surface temperatures will further increase by between 1.1°C (low emission scenario) and 6.4°C (high emission scenario) during the twenty-first century, and that the increases in temperature will be greatest at higher latitudes (IPCC, 2007). The biological impacts of rising temperatures depend upon the physiological sensitivity of organisms to temperature change. (Deutsch et al., 2008) point out that in contrast, insect species at higher latitudes, where the temperature increase is expected to be higher-have broader thermal tolerance and are living in cooler climates than their physiological optima. Warming may actually enhance the performance of insects living at these latitudes. It is therefore likely that tropical agro ecosystems will suffer from greater population decrease and extinction of native pollinators than agro ecosystems at higher latitudes.

Insect pollinators are valuable and limited resources (Delaplane and Mayer 2000). Currently, farmers manage only 11 of the 20 000 to 30 000 bee species worldwide (Parker et al., 1987), with the European honey-bee (*Apis mellifera*) being by far the most important species. Depending on only a few pollinator species belonging to the Apis genus has been shown to be risky. *Apis*-specific parasites and pathogens have lead to massive declines in honey bee numbers. Biotic stress accompanied with climate change may cause further population declines and lead farmers and researchers to look for alternative pollinators. Well-known pollinators to replace honey bees might include the alfalfa leaf-cutter bee (*Megachile rotundata*) and alkali bee (*Nomia melanderi*) in alfalfa pollination (Cane 2002), mason bees (*Osmia* spp.) for pollination of orchards (Bosch and Kemp, 2002; Maccagnani et al., 2003) and bumblebees (*Bombus* spp.) for pollination of crops requiring buzz pollination (Velthuis and van Doorn, 2006). Stingless bees are particularly important pollinators of tropical plants, visiting approximately 90 crop species (Heard, 1999). Some habits of stingless bees resemble those of honey-bees, including their preference for a wide range of crop species, making them attractive for commercial management. Pollinator limitation (lack of or reduced availability of pollinators) and pollen limitation (insufficient number or quality of conspecific pollen grains to fertilize all available ovules) both reduce seed and fruit production in plants. Some crop plants are more vulnerable to reductions in pollinator availability than others. (Ghazoul, 2005) defined vulnerable plant species as:
- having a self-incompatible breeding system, which makes them dependent on pollinator visitation for seed production;
- being pollinator-limited rather than resource-limited plants, as is the case for most intensively grown crop plants, which are fertilized; and
- being dependent on one or a few pollinator species, which makes them particularly sensitive to decreases in the abundance of these pollinators.

Semi-natural habitats provide important resources for wild pollinators such as alternative sources of nectar and pollen, and nesting and breeding sites. Especially in the United States, many of these intensively cultivated agricultural areas are completely dependent on imported colonies of managed honey-bees to sustain their pollination. The status of managed honey-bees is easier to monitor than that of wild pollinators. For example, bee numbers and diurnal activity patterns can be easily accessed by visually inspecting the hives. Although not commonly used by farmers, scale hives can yield important information on hive conditions and activity, the timing of nectar flow and the interaction between bees and the environment (http://honeybeenet.gsfc.nasa.gov).

In most developing countries, crops are produced mainly by small-scale farmers. Here, farmers rely more on unmanaged, wild insects for crop pollination (Kasina et al., 2009). To identify the most important pollinators for local agriculture, data on visitation rate alone does not necessarily suffice. Crop species may be visited by several species of insects, but several studies have shown that only a few visiting species may be efficient pollinators. An effective pollinator is good at collecting, transporting and delivering pollen within the same plant species.

In a recent review, Hegland et al. (2009) discussed the consequences of temperature induced changes in plant-pollinator interactions. They found that timing of both plant flowering and pollinator activity seems to be strongly affected by temperature. Insects and plants may react differently to changed temperatures, creating temporal (phenological) and spatial (distributional) mismatches-with severe demographic consequences for the species involved. Mismatches may affect plants by reduced insect visitation and pollen deposition, while pollinators experience reduced food availability. We have found three studies investigating how increased temperatures might create temporal mismatches between wild plants and their pollinators. Gordo and Sanz (2005) examined the nature of phenological responses of both plants and pollinators to increasing temperatures on the Iberian Peninsula, finding that variations in the slopes of the responses indicate a potential mismatch between the mutualistic partners. Both *Apis mellifera* and *Pieris rapae* advanced their activity period more than their preferred forage species, resulting in a temporal mismatch with some of their main plant resources (Hegland et al., 2009). However, Kudo et al. (2004) found that early flowering plants in Japan advanced their flowering during a warm spring whereas bumble-bee queen emergence appeared unaffected by spring temperatures. Thus, direct temperature responses and the occurrence of mismatches in pollination interactions may vary among species and regions (Hegland et al., 2009).

20.3   TEMPERATURE SENSITIVITY OF CROP POLLINATORS

Bees are the most important pollinators worldwide (Kearns et al., 1998) and like other insects, they are ectothermic, requiring elevated body temperatures for flying. The thermal properties of their environments determine the extent of their activity (Willmer and Stone, 2004). The high surface-to-volume ratio of small bees leads to rapid absorption of heat at high ambient temperatures and rapid cooling at low ambient temperatures. All bees above a body mass of between 35 and 50 mg are capable of endothermic heating, that is, internal heat generation (Bishop and Armbruster, 1999; Stone and Willmer, 1989; Stone, 1993). Examples of bee pollinators with a body weight above 35 mg are found in the genera *Apis, Bombus, Xylocopa* and *Megachile*. Examples of small bee pollinators are found in the family Halictidae, including the genus *Lasioglossum*. All of these groups are important in crop pollination.

In addition to endothermy, many bees are also able to control the temperatures in their flight muscles before, during and after flight by physiological and behavioral means (Willmer and Stone, 1997). Examples of behavioral strategies for thermal regulation include long periods of basking in the sun to warm up and shade seeking or nest returning to cool down (Willmer and Stone, 2004). With respect to the potential effects of future global warming, pollinator behavioral responses to avoid extreme temperatures have the potential to significantly reduce pollination services (Corbet et al., 1993). The natural distribution of the European dark bee (*Apis mellifera*) is found in a region where average July temperatures ranges from 15–20°C, which may represent their thermal tolerance. The Eastern honey-bee (*Apis cerana*) is native to parts of Asia. The giant honey-bee (*Apis dorsata*) lives only at tropical and adjacent latitudes in Asia and occurs less widely than the Eastern honey bee (*Apis cerana*), but can live at higher altitudes. The dwarf honey-bee (*Apis florea*) is more restricted than that of the larger *A. dorsata* and *A. cerana*. It is mainly found in Asia.

The effect of climate change on pollinators depends upon their thermal tolerance and plasticity to temperature changes. Our goal was to obtain thermal tolerance data for the most important pollinators worldwide. However, a literature review indicates that this information is missing for most species.

20.4   TREMENDOUS CLIMATE EVENTS

Extreme climate events might have detrimental effects on both crop plants and pollinator populations. High temperatures, long periods of heavy rain and late frost may affect pollinator activity either by reducing population sizes or by affecting insect activity patterns. The probability of extreme climate events may change in the future. Risk assessments should be conducted to better understand the changes in frequency of extreme climate events and minimize the effects (Kjohl et al., 2011).

20.4.1   SUPPLEMENTARY THREATS TO POLLINATION SERVICES

Pollination is under threat from several environmental pressures. Climate change is only one, and it cannot be seen in isolation, but should be addressed in relation to other pressures affecting plant-pollinator interactions. Here we list some of the most important pressures to be assessed in order to understand how crop pollination might be affected by climate change (Kjohl et al., 2011).

20.4.2   POLLINATORS ACTIVITY

In order to understand the nature of crop pollination, it is necessary to have precise information on the pollinator species involved. There are several ways of assessing the status of pollinator species and communities, and the structure of pollination networks (Committee on the Status of Pollinators in North America 2007). Two effective methods have been identified to estimate bee species richness (a useful proxy for measuring the diversity of pollinator communities in many areas): pan traps and transect walks (Westphal et al., 2008). Pan traps passively collect all insects attracted to them without assessing their floral associations or whether they pollinate crop species. They can, however, be an effective method for estimating relative population size and species richness as they collect a large number of individuals with little effort. The effectiveness of pan traps in collecting other types of pollinators such as butterflies and hoverflies has not been assessed to the same extent as for bees. While bees (especially honey bees) are the most frequent visitors to crop plants worldwide, the composition of pollinator communities may vary both locally and regionally. We recommend transect walks within agricultural fields to assess the status of pollinator communities of entomophilous crops. It is especially important to train field workers in sampling techniques and pollinator taxonomy since variations in skill have been shown to induce bias and reduce data quality. In addition to visitation frequencies, data on the quality of each visit is important for measuring the effectiveness of pollination.

20.5 IDENTIFICATION OF IMPORTANT POLLINATORS

Through intensive monitoring, the most frequent visitors to a particular crop species can be identified. However, pollinators vary in their effectiveness in initiating seed set. Fidelity to particular plant species, body size and morphology, and physical movement within and among flowers all affect pollination quality. The importance of each pollinator species is a product of the visitation frequency and the quality of each visit. Visitation quality of the most frequently observed pollinators should be investigated by presenting flowers to single visits of particular pollinator species (Kjohl et al., 2011).

20.5.1 POLLINATOR RESPONSES TO POTENTIAL CLIMATE CHANGE SCENARIOS

Pollinators may respond to climate change in different ways, depending on the system under study and climatic variable in focus. Pollinators may also respond in different ways depending on whether the scale is individuals vs. populations or local vs. landscape (Kjohl et al., 2011). Pollinators may change behavior in response to shifts in climate. Observations of pollinators in experimentally warmed greenhouses reveal behavioral responses to climate change that may be important for flower visitation. The time taken for thermoregulation at higher temperatures comes at the cost of foraging, with negative consequences for pollination. It is likely that pollinators will change their activity patterns as temperature increases, in turn changing the efficiency of pollen removal and deposition. For this reason, it is important to investigate taxonomic differences in pollinators' ability to regulate body temperature and avoid overheating.

Climate change may also impact activity patterns of pollinators. As temperatures increase, pollinators are at risk of overheating, particularly in regions where current ambient temperatures are high and climatic conditions are stable. In these regions, pollinators such as bees have a body temperature close to the ambient temperature and have a narrow thermal tolerance. Bees have different mechanisms for avoiding overheating, such as shade seeking and prolonged time spent in the nest. Bumblebees are particularly prone to overheating if temperatures increase because of their large size, dark color and hairy bodies.

20.5.2 IMPACT OF CLIMATE CHANGE IN VEGETABLE PRODUCTION

Climate change will have many impacts on horticulture and a few examples are given below (Thakur and Soni, 2012).

- A rise in a temperature of above 1°C will have shifted a major area of potential suitable zones.
- Production timing will change. Because of rise in temperature, crops will develop more rapidly and mature earlier.
- While temperature rises, photoperiods may not show much variation. Onions, a photosensitive crop, will mature faster leading to small bulb size.
- The winter regime and chilling duration will reduce in temperate regions affecting the temperate crops.
- The faster maturity and higher temperature induced ripening will make the produce a less storage period in trees/ plants. They will overripe.
- Pollination will be affected adversely because of higher temperature. Floral abortions will occur.
- Higher temperatures will reduce tuber initiation process in potato, reduced quality in tomatoes and poor pollination in many crops. In case of crucifers, it may lead to bolting; anthocyanin production may be affected in capsicum.

20.5.3 ADAPTING TO CHANGING CLIMATE

A diverse assemblage of pollinators, with different traits and responses to ambient conditions, is one of the best ways of minimizing risks due to climatic change. The "insurance" provided by a diversity of pollinators ensures that there are effective pollinators not just for current conditions, but for future conditions as well. Resilience can be in agro-ecosystems through biodiversity. Pollination management practices can also be undertaken to built respond to climate change.

TABLE 20.1   Important Non-*Apis* Bee Pollinators of Some Vegetable Crops in India

Crop/ Plant	Family	Bee species
Pea	Leguminosae	*Xylosandrus fenestrate, X. pubescens, Megachile lanata,* *Braunsapis* spp,
Sweet potato	Convolvulaceae	*X. fenestrate, B. albopleurali, Bombus asiaticus*
Egg plant	Solanaceae	*B. asiaticus, X. fenestrate, Nomia caliphora Pithitis* spp
Onion	Liliaceae	*X. fenestrate, Lasioglossum* spp, *Nomioides* spp,
Field mustard	Cruciferae	*Nomioides, Megachilids, Andrenids, Halictids*
Cabbage & Cauliflower	Cruciferae	*Andrena ilerda, Lassioglossum* spp
Raddish	Cruciferae	*Anthophora* spp, *Nomia* spp, *Lassioglossum* spp,
Pumpkin & squashes	Cucurbitaceae	*X. fenestrate, X. pubescens, Halictus* spp, *Nomioides* spp
Cucumbers	Cucurbitaceae	*Nomia* spp, *Lassioglossum* spp
Coriander	Umbelliferae	*Nomioides* spp, Halictidae, *X. fenestrate*
Carrot	Umbelliferae	*Lassioglossum* spp, *Pithitis smaragdula*

Source: Thakur and Soni (2012).

20.5.4   ECONOMIC VALUE OF CROP POLLINATION

Information on visitation frequency and subsequent seed set is valuable when categorising crops according to their degree of dependence on crop pollination (Delaplane and Mayer, 2000). However, the total value of pollinators' ecosystem services at both local and larger scales is little understood. A protocol for assessing pollination deficits in crops has been developed by FAO in collaboration with other institutions (Vaissière et al., 2011). Experiments carried out using such protocols will identify crop species under threat of pollination failure in different regions. Further research focused on vulnerable species can identify actions to minimize negative effects. A recent report published by FAO can be used as a tool for assessing the value of pollination services at a national or larger scale, and vulnerabilities to pollinator declines (Gallai and Vaissière, 2009).

20.6   CONCLUSION

Although concern has been raised about negative effects of climate change on the services provided by pollinating insects, there is still a paucity of scientific literature regarding how pollination interactions may be affected. In line with the recent review by Hegland et al. (2009), we found few studies on this topic with respect to crop pollination. Climate change may affect the phenology and distribution ranges of both crop plants and their most important pollinators, leading to temporal and spatial mismatches. It is therefore important to identify the temperature sensitivity of the most important pollinators and their crop plants, and the environmental cues controlling the phenology and distribution of the identified species. Long-term monitoring of agro ecosystems and experimental assessments of species' climate sensitivity may enhance our understanding of the impacts of climate change on crop pollination.

KEYWORDS

- **Climate Change**
- **Insect Pollinators**
- **Pollination**
- **Temperature Sensitivity**
- **Vegetable Crops**

REFERENCES

Bishop, J. A., & Armbruster, W. S. (1999). Thermoregulatory Abilities of Alaskan bees, Effects of Size, Phylogeny and Ecology, *Funct Ecol,* 13, 711–724.

Bosch, J. & Kemp, W. P. (2002). Developing and Establishing Bee Species as Crop Pollinators, The example of Osmia spp. (Hymenoptera Megachilidae) and Fruit Trees, *Bull Entomol Res,* 92, 3–16.

Cane, J. H. (2002). Pollinating Bees (Hymenoptera Apiformes) of US alfalfa Compared for Rates of Pod and Seed Set, *J. Econ. Entomol.,* 95, 22–27.

Committee on the Status of Pollinators in North America NRC. (2007). *Status of Pollinators in North America.* Washington, DC, National Academies Press.

Corbet, S. A., Fussell, M., Ake, R., Fraser, A., Gunson, C., Savage, A., & Smith, K. (1993). Temperature and the Pollinating Activity of Social Bees. *Ecol Entomol,* 18, 17–30.

Delaplane, K. S., & Mayer, D. F. (2000). *Crop pollination by bees,* New York, CABI.

Deutsch, C. A., Tewksbury, J. J., Huey, R. B., Sheldon, K. S., Ghalambor, C. K., Haak, D. C., & Martin, P. R. (2008). Impacts of Climate Warming on Terrestrial Ectotherms Across Latitude, *Proc Natl Acad Sci U. S. A.,* 105 6668–6672.

FAO, (2004). Impact of Climate Change on Agriculture in Asia and the Pacific, Twenty-seventh,FAO Regional Conference for Asia and the Pacific. Beijing, China, 17–21 May 2004.

Gallai, N., Salles, J. M., Settele, J., & Vaissiere, B. E. (2009). Economic Valuation of the Vulnerability of World Agriculture Confronted with Pollinator Decline, *Ecol Econom,* 68, 810–821.

Gallai, N., & Vaissière, B. E. (2009). *Guidelines for the Economic Valuation of Pollination Services at a National Scale,* Rome, FAO.

Ghazoul, J. (2005). Business as usual? questioning the global pollination crisis. *Trends Ecol Evolut,* 20, 367–373.

Gordo, O. & Sanz, J. J. (2005). Phenology and Climate Change a long-term study in a Mediterranean locality, *Oecologia,* 146, 484–495.

Heard, T. A. (1999). The Role of Stingless Bees in Crop Pollination, *Annu Rev Entomol,* 44, 183–206.

Hegland, S. J., Nielsen, A., Lázaro, A., Bjerknes, A. L., & Totland, O. (2009). How does Climate Warming Affect Plant-Pollinator Interactions? *Ecol Letters,* 12, 184–195. IPCC, (1990a). Climate Change, The IPCC Scientific Assessment. Intergovernmental Panel on Climate Change. Geneva and Nairobi, Kenya, World Meteorological Organization and UN Environment Program, 365p.

IPCC. (1990a). Climate change, The IPCC Scientific Assessment. Intergovernmental Panel on Climate Change. Geneva and Nairobi, Kenya: World Meteorological Organization and UN Environment Program, 365p.

IPCC. (1990b). The Potential Impacts of Climate Change on Agriculture and Forestry. Intergovernmental Panel on Climate Change. Geneva and Nairobi, Kenya: World Meteorological Organization and UN Environment Program, 55pp.

IPCC, (2001). Climate change 2001, The Scientific Basis, Contribution of Working Group I to the Third Assessment Report of the Intergovernmental Panel on Climate Change (IPCC). Cambridge University Press, Cambridge http://www.grida.no/climate/ipcc_tar/.

IPCC. (2007). Climate change 2007, Synthesis Report-Contribution of Working Groups 1, 2 and 3 to the Fourth Assessment Report of the Intergovernmental Panel on Climate Change. *In Change IPoC,* Geneva.

Kasina, J. M., Mburu, J., Kraemer, M. & Holm-Mueller, K. (2009). Economic Benefit of Crop Pollination by Bees, a case of kakamega small-holder farming in western Kenya. *J Econ Entomol*, 102, 467–473.

Kjohl, M., Anders, N., & Nils, C. S. (2011). Potential Effects of Climate Change on Crop Pollination. Food and Agriculture Organization (FAO) of the United Nations, Rome.

Klein, A. M., Vaissiere, B. E., Cane, J. H., Steffan-Dewenter, I., Cunningham, S. A., Kremen, C., & Tscharntke, T. (2007). Importance of Pollinators in Changing Landscapes for World Crops, *Proc R Soc Lond [Biol]*, 274, 303–313.

Kudo, G., Nishikawa, Y., Kasagi, T., & Kosuge, S. (2004). Does Seed Production of Spring Ephemerals Decrease When Spring come Early? *Ecol Res*, 19, 255–259.

Losey, J. E., & Vaughan, M. (2006). The Economic Value of Ecological Services Provided by Insects. *Bioscience*, 56, 311–323.

Maccagnani, B., Ladurner, E., Santi, F., & Burgio, G. (2003). Osmia Cornuta (Hymenoptera, Megachilidae) as a Pollinator of Pear (Pyrus communis) fruit and seed-set, *Apidologie*, 34, 207–216.

Memmott, J., Craze, P.G., Waser, N. M., & Price, M. V. (2007). Global Warming and the Disruption of Plant-Pollinator Interactions, *Ecol Letters*, 10, 710–717.

Parker, F. D., Batra, S. W. T., & Tepedino, V. J. (1987). New Pollinators for our Crops, *Agricult Zool Rev*, 2, 279–304.

Roubik, D. W. (1995). Pollination of Cultivated Plants in the Tropics. Rome, FAO.

Schweiger, O., Biesmeijer, J. C., Bommarco, R., Hickler, T., Hulme, P., Klotz, S., Kühn, I., Moora, M., Nielsen, A., Ohlemuller, R., Petanidou, T., Potts, S. G., Pyzek, P., Stout, J. C., Sykes, M., Tscheulin, T., Vilà, M., Wather, G. R., & Westphal, C. (2010). Multiple Stressors on Biotic Interactions, How Climate Change and Alien Species Interact to Affect Pollination. *Biol Rev*, 85, 777–795.

Stone, G. N. (19930. Endothermy in the Solitary Bee Anthophora-Plumipes-Independent Measures of Thermoregulatory Ability, Costs of Warm-Up and the Role of Body Size, *J Exp Biol*, 174, 299–320.

Stone, G. N., & Willmer, P. G. (1989). Endothermy and Temperature Regulation in Bees a Critique of Grab and Stab Measurement of Body-Temperature, *J. Exp Biol*, 143, 211–223.

Sutherst, R. W., Maywald, G. F., & Bourne, A. S. (2007). Including Species Interactions in Risk Assessments for Global Change, *Global Change Biol*, 13, 1843–1859.

Thakur, R. K. & Soni, J. (2012). Management of Pollinators of Vegetable Crops under Changing Climatic Scenario. Vegetable production under changing climate scenario (1st September to 21st September), Centre of advanced faculty training of horticulture (vegetable), YSP University of Horticulture and Forestry, Solan, Himachal Pradesh (India).

Vaissière, B. E., Frietas, B. M., & Gemmill-Herren, B. (2011). Protocol to Detect and Assess Pollination Deficits in Crops: a Handbook for its use, Rome, FAO.

Van der Putten, W. H., de Ruiter, P. C., Bezemer, T. M., Harvey, J. A., Wassen, M., & Wolters, V. (2004). Trophic Interactions in a Changing World. *Basic Appl Ecol* 5, 487–494.

Velthuis, H. H. W., & van Doorn, A. (2006). A Century of Advances in Bumblebee Domestication and the Economic and Environmental Aspects of its Commercialization for Pollination, *Apidologie* 37, 421–451.

Westphal, C., Bommarco, R., Carré, G., Lamborn, E., Morison, N., Petanidou, T., Potts, S. G., Roberts, S. P. M., Szentgyörgyi, H., Tscheulin, T., Vaissière, B. E., Woyciechowski, M., Biesmeijer, J. C., Kunin, W. E., Settele, J. & Steffan-Dewenter, I. (2008). Measuring Bee Biodiversity in Different European habitats and biogeographical regions. Ecol Monograph, 78, 653–671.

Willmer, P., & Stone, G. (1997). Temperature and Water Relations in Desert Bees. *J Thermal Biol*, 22, 453–465.

Willmer, P. G., & Stone, G. N. (2004). Behavioral, Ecological, and Physiological Determinants of the Activity Patterns of bees, In Advances in the Study of Behavior 34, San Diego, C. A., Elsevier Academic Press Inc. 347–466.

CHAPTER 21

CLIMATE CHANGE IMPACTS ON FIELD AND HORTICULTURAL CROPS WITH SPECIAL REFERENCE TO BIHAR, POSSIBLE ADAPTATION STRATEGIES AND MITIGATION OPTIONS

A. ABDUL HARIS, B. P. BHATT[1], and VANDNA CHHABRA

ICAR Research Complex for Eastern Region, Patna, India.
[1]E-mail: drbpbhatt.icar@yahoo.com

CONTENTS

21.1   INTRODUCTION

Day to day variation in the state of our atmosphere is called weather. Average state of the weather that is fairly stable and predictable is called climate. Climatic parameters include temperature, amount of precipitation, days of sunlight, humidity and wind, etc. Climate change refers to variation in earth's global climate or regional climate over different timescales. Climate change was happening from time immemorial, but the accelerated changes due to anthropogenic factors and consequent global warming is of serious concern. Global warming due to accumulation of different green house gases is not only indicated by changes in temperature but, represents a series of events, which ultimately can have more implications than can't be summarized in a few words. Major vulnerable areas as identified by IPCC, covering almost every aspect of our life, are global social systems, global biological systems and geophysical systems.

Anthropogenic activities have caused accumulation of Green House Gases (GHG) in the atmosphere, leading to the potential hazards of climate change looming over us. The atmospheric concentration of CO_2 has increased from the preindustrial levels of 280 ppm to 379 ppm in 2005 (IPCC, 2007). High carbon dioxide and other green house gases in the atmosphere tend to warm up the atmosphere, besides affecting other meteorological variables. Long-term weather trends indicate constant rise in temperature with increasing concentrations of GHG's mainly carbon dioxide, methane, oxides of nitrogen, etc. in the atmosphere. In India 0.68°C increase per century, increasing trends in annual mean temperature, warming more pronounced during post monsoon and winter, increase in extreme rains in north-west during summer monsoon in recent decades and lower number of rainy days along east coast was reported.

Agriculture sector, whether in developing or developed countries, depends on climate and climatic resources leading to the development of special consideration for this sector to study the impacts of climate change among researchers. Intergovernmental Panel on Climate Change (IPCC) and other researchers have stressed the need to study impacts on agricultural production at local, regional, national as well as on global scales to capture the local conditions. Projecting future crop yields has significant uncertainty attached due to changed fertilizer and water application strategies, occurrence of extreme climatic events, changes in pest and disease occurrence. Decision Support Systems (DSS) or Crop Models provide a way, where the relative effects of these variables on crop growth and yield can be studied in particular combinations on regional basis. Early simulation studies on impacts of climate change gave prime importance to the expected increase in carbon dioxide levels only, while off late researchers have suggested, from their studies, that agricultural production gets affected not only by CO_2 alone, but also by weather variables (Curtis and Wang, 1998).

Most crops grown under enriched carbon dioxide environment showed increased growth and yield (Allen et al., 1997; Parry et al., 2004). Enhanced CO_2 effects the growth and physiology of crops, enhancing photosynthesis as well as water use efficiency (De Costa et al., 2003; Ewert, 2004; Widodo et al., 2003). Elevated CO_2 besides affecting the crop affects the environment, which in turn may have either beneficial or damaging effect on agricultural production. Mall and Singh, 2000 reported that small changes in growing season temperature over the years appeared to be key aspect of weather affecting yearly wheat yield fluctuations. Pathak et al. (2003) attributed the decline in potential yield of wheat and rice to the negative trend in solar radiation and an increase in minimum temperature in the Indo-gangetic plains of India. FAO and IPCC have estimated a drop in cereal production for India as much as by 125 mt and an overall increase of 2°C in temperature may lead to almost 8% loss in farm level net revenue and around 5% in GDP (Gahukar, 2009). C4 plants are photosynthetically more efficient than C3, especially when the level of CO_2 is high (Ku et al., 1999). This chapter examines results of different studies, which has special significance for Bihar.

21.2 IMPACTS OF CLIMATE CHANGE ON AGRICULTURE

21.2.1 DIRECT EFFECTS

Some of the direct effects reported by various studies are higher water demand, increase in evaporation rate, reduction in crop growth period, crop damage/failure, inability to cultivate on land due to increased salinization /water logging (coastal areas), reduced net photosynthetic rate, extension/ shortening of growing season, lesser production and productivity in warmer regions, increased production in some rabi crops due to elevated CO_2 and temperature effects, etc.

21.2.2   INDIRECT EFFECTS

Soil erosion and increased leaching of nutrients, increased reproduction rate, rise in severity of infestation and extension of geographical range of insect pests and pathogens apart from outbreak of new pests, salinization of fresh water resources, ground water depletion, changes in social system due to urbanization, migration, land use change like deforestation, shifting of cropping zones, land degradation, etc.

21.3   IMPACTS ON INDIVIDUAL CROPS WITH SPECIAL REFERENCE TO BIHAR

21.3.1   RICE

In rice, decline in rice productivity with increasing temperature under current management is predicted. In Bihar, present productivity can be maintained with proper agronomic interventions up to middle of twenty-first century as indicated by simulation studies. Increase in crop duration by increasing temperatures during reproductive phase (>30°C) is a possibility in rice as indicated in studies. Long duration varieties are most affected (Haris et al., 2010). Simulation studies show that 22% decline possible in later half of the century. Spikelet sterility and lesser grain filling are primary reasons.

21.3.2   WHEAT

Significant decline in productivity with increasing temperature is predicted under current management in India. Decline in duration (reduced biomass accumulation), reduction in Kernel number (4% per degree centigrade rise in the range of 14°C–22°C) and reduced kernel weight (2–5% decline per degree centigrade increase) is a possibility at higher temperatures. Increased respiration and loss of chlorophyll after anthesis is observed under warmer conditions. High night temperatures reduce grain number. Simulation studies points to 4 to 28% decline in yield at different time periods.

21.3.3 MAIZE

Decline in productivity is observed with increasing temperature under current management in kharif, however, rabi season maize may benefit from warmer winters. Decline in duration of rabi maize with higher temperatures is observed with no reduction in biomass accumulation as growth is not limited by chilling phase. Radiation Use Efficiency may possibly increase with temperature increase from 16–21°C. CO_2 fertilization is evident (Ku et al., 1999) as higher CO_2 showed benefits. Simulation studies showed a decline in kharif maize by 2 to 14% and in rabi maize, 12 to 41% increase in yield. Relative growth rate and net assimilation rate increased with temperature from 16–28°C.

21.3.4 GRAM

In gram impacts of climate change vary spatially. Simulation studies have shown that North Bihar can be more productive under scenarios.

21.3.5 POTATO

Photosynthesis in potato is suppressed by high temperature. Tuberization and partitioning of photosynthates to tuber is highly sensitive to temperature so, reduces tuber number and size leading to lower yield. Simulation studies showed a decline in production with increasing temperature under current management. However, regions prone to frost may show an enhancement under warmer climate. The effect of climatic change on the global potato production was assessed by Hijmans (2003). Between 1961-90 and 2040-69, global potential potato yield decreased by 18 to 32% (without adaptation) and by 9 to 18% (with adaptation) considering a predicted temperature increase between 1 and 1.4°C.

21.3.6 VEGETABLES

Higher temperatures can alter the cropping calendar, increase water demand and pest problems. Early flower initiation promoted by high temperature in winter crops. Vegetable crops needing a cold climate for bulking may be affected. Crop cycle may shorten, leading to reduction in yields. Weather extremes and stagnation of water can be damaging. Many vegetable crops namely tomato, water melon, potato, soybeans, peas, carrot, beet, turnip, etc are more susceptible to air pollution damage. Yield of vegetable can be reduced by 5–15% when daily ozone concentrations reach to greater than 50 ppb (Raj Narayan, 2009).

21.3.7   FRUITS

High wind velocity and hails can be damaging to fruit crops. Higher temperatures may affect fruit setting. Increase in pest and disease problems reducing the yield and quality is a possibility. Banana cultivation may suffer from high temperature, soil moisture stress or flooding/water logging. High temperature and moisture stress also increase sunburn and cracking in apples, apricot and cherries and increase in temperature at maturity will lead to fruit cracking and burning in litchi (Kumar and Kumar, 2007)

21.4   ADAPTATION MEASURES

Adaptation is the adjustment of natural or human systems to a new or changing environment which moderates harm or exploit beneficial opportunities.

Diversity of agriculture itself is a manifestation of adaptation to climate over time. Measures for adaptation include, conservation of genetic variability and biodiversity, strengthening of research for enhancing adaptive capacity, systematic collection and recording of weather information, soil and crop data, etc.

Focused action is required for development of crop varieties suited to different stresses like heat, drought, flooding, salinity, etc. Integrated nutrient management, pest and disease management and delineation of niche areas (most ideal climate) for different crops and intensification of cropping in such areas (technology, input delivery, market, etc.) are required.

Adaptation with different approaches,

1. Agronomic measures: Early planting of field crops, mixed cropping, intercropping, crop diversification, irrigation and fertilizer management, integrated management and systems approach.

2. Socio-economic measures: Creating awareness among farming communities, exploring traditional wisdom of farmers, sustainable practices at village level and crop planning, etc.
3. Technology: Weather surveillance, warning system for farmers and development of machinery, etc.
4. Policy level interventions: Incentives for resource conservation technologies, credit availability, crop insurance schemes, marketing, dynamic Export Import Policy, promotion of Renewable Resources use, education and Training.

21.5   MITIGATION APPROACHES

Mitigation is a human intervention to reduce the human impact on the climate system. It includes strategies to reduce greenhouse gas sources and emissions and enhancing greenhouse gas sinks.

1. Activities bringing about reduction in GHG emission.
    a. Fossil fuel substitution: Using biofuels produced in the agricultural sector instead of fossil fuels can help lower *GHG* concentrations.
    b. Emissions reductions: Agricultural CH_4 and N_2O emissions can be reduced through effective manure and feed management and efficient fertilizer application. Carbon dioxide (CO_2) emissions can be reduced by adopting more fuel-efficient technologies and practices.
2. Activities bringing about Removal of GHG from atmosphere.
    a. Sequestration: CO_2 removed from the atmosphere can be stored in soils, biomass, and harvested products, and protected or preserved to avoid CO_2 release back to the atmosphere. These become carbon stores or carbon sinks.
    b. Agro-forestry: Potential for sequestration is 39–49 MT/year.
    c. Organic farming and practices, which leave more organic carbon in the soil.
    d. Farming system approach: Promotes synergy and sustainability.
    e. Biodiesel: The production of bio diesel from algae could reduce greenhouse gas emissions; help to address future fuel shortages, bio fuels.

21.6 CONCLUSIONS AND RECOMMENDATIONS

From the different studies made in this region, it can be concluded that there is likely reduction in yield of most crops in long-term with current varieties and under present practices. With proper interventions, present agriculture can be adapted to increased temperature and changing rainfall pattern for minimizing losses. Cost of adaptation could be high, but can be appreciated only if cost of inaction and its future impacts be considered. Mitigation approach would require system approach and coordinated efforts among Government institutions and other stakeholders.

Some points for future consideration are:
1. Generating policy and financial support at different levels.
2. Partnership in adaptation and mitigation efforts.
3. Educating farmers and other stakeholders.
4. Strengthening adaptation and mitigation research activities.
5. Developing network for climate monitoring and providing advisory services.
6. Strengthening extension system for promoting informed policy making at grass root level.
7. Creation of center of excellence at state level for coordinated efforts in climate change research.

KEYWORDS

- **Adaptation**
- **Cereals**
- **Climate Change**
- **Fruits**
- **Mitigation**
- **Vegetables**

REFERENCES

Allen Jr., L. H., Kirkham, M. B., Olszyk, D. M., & Whitman, C. E. (eds.) (1997). Advances in Carbon Dioxide Research ASA Special Publication 61, Madison, WI, 228.

Curtis, P. S., & Wang, X. (1998). A meta-analysis of elevated CO_2 effects on woody plant mass, form, and physiology. Oecologia, 113, 299–313.

De Costa, W. A. J. M., Weerahoon, W. M. W., Herath, H. M. L. K., & Abeywardena, M. I. (2003). Response of Growth and Yield of Rice to Elevated Atmospheric Carbon dioxide in the Sub Humid Zone of Sri Lanka, J. Agron. Crop Sci., 189, 83–95.

Ewert, F. (2004). Modeling Plant Responses to Elevated CO_2, How important is Leaf Area Index, Ann. Bot., 93, 619–627.

Gahukar, R. T. (2009). Food security: The challenges of Climate change and bioenergy, Curr. Sc. 96(1), 26–28.

Haris, A. A., Biswas, S., & Chhabra, V. (2010). Climate Change Impacts on Productivity of Rice (Oryza sativa) in Bihar, Indian J. Agron, 55(4), 295–298.

Hijmans, R. J. (2003). The effect of climate change on global potato production. American Journal of Potato Research. 80(4), 271–279

IPCC, (2007). Summary for Policymakers. In Climate Change (2007), The Physical Science Basis. Contribution of Working Group I to the Fourth Assessment Report of the Intergovernmental Panel on Climate Change [Solomon, S., D., Qin, M., Manning, Z., Chen, M., Marquis, K. B., Averyt, Tignor, M., & Miller, H. L. (eds.). Cambridge University Press, Cambridge, United Kingdom and New York, NY, USA.

Ku, M. B. S., Agarie, S., Nomura, M., Fukayama, H., Tsuchida, H., Ono, K., Horose, S., Toki, S., Miyao, M. & Matsuoka, M. (1999). High-level Expression of Maize Phosphoenolpyruvate Carboxylase in Transgenic Rice Plants, Nature Biotech, 17, 76–80.

Kumar, R. & Kumar, K. K. (2007) Managing physiological disorders in litchi. Indian Horticulture, 52 (1), 22–24.

Mall, R. K., & Singh, K. K. (2000). Climate Variability and Wheat Yield Progress in Punjab using the CERES Wheat and WTGROWS models, *Vayumandal*, 30, 35–41.

Pathak, H., Ladha, J. K., Aggarwal, P. K., Peng, S., Das, S., Yadvinder Singh, Bijay Singh, Kamra, S. K., Mishra, B., Sastri, A. S. R. A. S., Aggarwal, H. P., Das, D. K., & Gupta, R. K. (2003). Climatic Potential and on Farm Yield Trends of Rice and Wheat in the Indo-Gangatic Plains. Field Crops Res. 80(3), 223–234.

Raj Narayan (2009). Air pollution–A threat in vegetable production. In: Sulladmath, U.V. and Swamy, K.R.M. International Conference on Horticulture (ICH-2009). Horticulture for Livelihood Security and Economic Growth, 158-159.

Widodo, W., Vu, J. C. V., Boote, K. J., Baker, J. T., & Allen, L. H. (2003). Elevated Growth CO_2 Delays Drought Stress and Accelerates Recovery of Rice Leaf Photosynthesis, Environ. Exp. Bot., 49, 259–272.

INDEX